Handbook of Pollution Control and Waste Minimization

T0132567

Civil and Environmental Engineering

A Series of Reference Books and Textbooks

Editor

Michael D. Meyer

Department of Civil and Environmental Engineering
Georgia Institute of Technology
Atlanta, Georgia

Additional Volumes in Production

Handbook of Pollution Control and Waste Minimization

edited by
Abbas Ghassemi
New Mexico State University
Las Cruces, New Mexico

CRC Press
Taylor & Francis Group
Boca Raton London New York

CRC Press is an imprint of the
Taylor & Francis Group, an **informa** business

CRC Press
Taylor & Francis Group
6000 Broken Sound Parkway NW, Suite 300
Boca Raton, FL 33487-2742

First issued in paperback 2019

© 2002 by Taylor & Francis Group, LLC
CRC Press is an imprint of Taylor & Francis Group, an Informa business

No claim to original U.S. Government works

ISBN-13: 978-0-8247-0581-7 (hbk)
ISBN-13: 978-0-367-39699-2 (pbk)

Visit the Taylor & Francis Web site at
http://www.taylorandfrancis.com

and the CRC Press Web site at
http://www.crcpress.com

Foreword

Erestor, in Tolkien's *The Lord of the Rings*,* declares that only two possibilities exist for dealing with the menace of the Ring: "to hide it forever, or to unmake it. But both are beyond our power." This analogy was used by Amory Lovins in an unpublished review of proposed U.S. policy on nuclear waste management. The dilemma described by Erestor is analogous not only to the world's nuclear waste issues but to many other concerns related to long-life hazardous waste being generated worldwide at an alarming rate. Once nuclear or hazardous waste is produced, we cannot "unmake" it. Congress has responded to these concerns through a number of legislative initiatives that attempt to minimize the amount of waste containing hazardous constituents and also place restrictions on its disposal in underground repositories. In the latter case, the most notable legislation is the Resource Conservation and Recovery Act (RCRA), requiring that the best available technology be used to remove the chemical constituents in both hazardous and mixed waste before it can be permanently disposed of in underground repositories. Compliance with RCRA and other environmental regulations

---

*The council of Elrond from *The Fellowship of the Ring: Lord of the Rings—Part One* by J. R. R. Tolkien, New York: Ballantine Books edn., pp. 349–50, 1986.

has come at a high cost in terms of dollars and, sometimes, potential risks to worker health and safety.

It is with these issues in mind that the editor has compiled a comprehensive textbook that covers the broad spectrum of pollution prevention including process design, life cycle analysis, risk and decision analysis. The information presented will increase awareness of the need to "do it right" the first time. The amount of waste generated in any process results in a net reduction in profits. In every process, the ultimate goal is to produce goods and materials that can be sold or bartered for a profit. Looking into the future, it is ideal to proceed with industrial development that maximizes the productions of goods and materials while minimizing or eliminating the waste produced. This includes looking into raw materials, and production efficiencies as well as process modification and enhancement that would result in the ultimate goal. This text encourages future generations to develop the policies and priorities necessary to effectively deal with scientific and political issues associated with hazardous and radioactive waste management.

Jim Bickel
Former Assistant Manager
Projects and Energy Program
U.S. Department of Energy

Ron Bhada
Emeritus Director
WERC
Las Cruces, New Mexico

Preface

The most significant issues facing mankind today are related to the quality of our environment. Past decisions did not always consider environmental factors as critical elements. However, current decisions made daily should reflect the importance of the environment. All environment-related issues are multidisciplinary, ranging from science and engineering to social, economic, and regulatory issues. Further, these issues are not related to any one region or country, but are global in nature, requiring multidisciplinary, multiorganizational, and multinational educational efforts.

This book provides an introduction and current information to the academic community as well as to any professional who must deal with these issues on a day-to-day basis. My aim is to have environmental issues become a major factor in process design consideration. Our contributors present the fundamentals of pollution prevention: life-cycle analysis, designs for the environment, and pollution prevention in process design. Selected case studies are provided as well.

All of the contributors to this volume are, in one way or another, associated with WERC, a Consortium for Environmental Education and Technology Development program.

The first part of the book deals with elements required for legal, organizational, and hierarchal components of pollution prevention. Parts II–IV deal with the basics of pollution prevention, leading with fundamentals of pollution prevention, followed by methodology and life cycle cost analysis. Part V deals with risk and decision fundamentals, as well as, utilization of pollution prevention in process design. Part VI presents selected case studies in various fields.

As the editor, I realize that I have just begun to scratch the surface with some of the recent advances. I would like to take this opportunity to thank the chapter authors for their contributions to this volume. Further gratitude goes to Dr. Franc Szidarovszky, University of Arizona, and Kay Perkins, WERC, for their editing assistance.

Abbas Ghassemi

Contents

Contributors

Paul K. Andersen, Ph.D. Associate Professor, Department of Chemical Engineering, New Mexico State University, Las Cruces, New Mexico

Tarunjit Singh Butalia, Ph.D. Research Scientist, Department of Civil and Environmental Engineering and Geodetic Science, The Ohio State University, Columbus, Ohio

W. David Constant, Ph.D., P.E. Professor of Civil and Environmental Engineering and Assistant Director of U.S. Environmental Protection Agency Hazardous Substance Research Center/South and Southwest, Department of Civil and Environmental Engineering, Louisiana State University and A&M College, Baton Rouge, Louisiana

Mary Ann Curran, M.S. Research Chemical Engineer, Office of Research & Development, U.S. Environmental Protection Agency, Cincinnati, Ohio

Teresa J. Cutright, Ph.D. Assistant Professor, Department of Civil Engineering, The University of Akron, Akron, Ohio

Warren A. Dick, Ph.D. Professor of Soil Science, School of Natural Resources, The Ohio State University, Wooster, Ohio

Abdollah Eskandari, Ph.D.   Postdoctoral Researcher, Department of Systems and Industrial Engineering, University of Arizona, Tucson, Arizona

C. Evenson   Graduate Degree Candidate, BioEnvironmental Engineering and Science Laboratory, University of Oklahoma, Norman, Oklahoma

Steven P. Frysinger, Ph.D.   Professor, Department of Integrated Science and Technology, James Madison University, Harrisonburg, Virginia

Alan Fuchs, Ph.D.   Assistant Professor, Department of Chemical Engineering, University of Nevada, Reno, Reno, Nevada

Herold J. Gerbrandt, P.E., Ph.D.   Professor, Department of General Engineering, Montana Tech of the University of Montana, Butte, Montana

Abbas Ghassemi, Ph.D.   Executive Director, WERC, New Mexico State Univesity, Las Cruces, New Mexico

Gilbert J. Gonzales, Ph.D.   Ecologist, Ecology Group, Los Alamos National Laboratory, Los Alamos, and Adjunct Professor, Department of Fishery and Wildlife Sciences, New Mexico State University, Las Cruces, New Mexico

Sarah W. Harcum, Ph.D.   Associate Professor, Department of Chemical Engineering, New Mexico State University, Las Cruces, New Mexico

René Reyes Mazzoco, Ph.D.   Professor, Department of Chemical Food and Engineering, Universidad de las Américas–Puebla, Cholula, Puebla, Mexico

Terrence J. McManus, P.E., D.E.E.   Intel Fellow and Director, EHS Technologies, Corporate Environmental Health and Safety, Intel Corporation, Chandler, Arizona

Marc Obladen, Ph.D.   Scientific Co-Worker, Institute of Technology of Energy Supply Systems and Energy Conversion Plants, and Professor, University of Essen, Essen, Germany

Shuo Peng, Ph.D.   Research Assistant, Department of Chemical Engineering, University of Nevada, Reno, Reno, Nevada

Toni K. Ristau, M.S., J.D.   Director of Environmental Services, Department of Power Production and Energy Services, Public Service Company of New Mexico, Albuquerque, New Mexico

Ingo F. W. Romey, Ph.D.   Head of Chair, Institute of Technology of Energy Supply Systems and Energy Conversion Plants, University of Essen, Essen, Germany

Rita C. Schenck, Ph.D.   Executive Director, Institute for Environmental Research and Education, Vashon, Washington

James H. Scott, Ph.D. President, Abaxial Technologies, Los Alamos, New Mexico

Harish Chandra Sharma, M.S., M.B.A, REM. Site Liaison and Project Manager, U.S. Department of Energy, Albuquerque, New Mexico

Bart Sims, B.S. Manager, Hazardous Waste and Waste Minimization, Oil and Gas Division, Environmental Services Section, Railroad Commission of Texas, Austin, Texas

Thomas P. Starke, Ph.D. Program Manager, Environmental Stewardship Office, Los Alamos National Laboratory, Los Alamos, New Mexico

K. A. Strevett, Ph.D. Professor of Environmental Engineering, BioEnvironmental Engineering and Science Laboratory, University of Oklahoma, Norman, Oklahoma

Ferenc Szidarovszky, Ph.D. Professor, Department of Systems and Industrial Engineering, University of Arizona, Tucson, Arizona

Panuwat Taerakul Department of Civil and Environmental Engineering and Geodetic Science, The Ohio State University, Columbus, Ohio

Victor R. Vasquez, Ph.D. Assistant Professor, Department of Chemical Engineering, University of Nevada, Reno, Reno, Nevada

Harold W. Walker, Ph.D. Assistant Professor, Department of Civil and Environmental Engineering and Geodetic Science, The Ohio State University, Columbus, Ohio

Joseph Wang, Ph.D. Professor, Department of Chemistry and Biochemistry, New Mexico State University, Las Cruces, New Mexico

Jeff Weinrach, Ph.D. Vice President and Director of Quality Standards, JCS/Novation, Inc., Albuquerque, New Mexico

L. Wolf Graduate Degree Candidate, BioEnvironmental Engineering and Science Laboratory, University of Oklahoma, Norman, Oklahoma

William E. Wolfe, Ph.D. Professor, Department of Civil and Environmental Engineering and Geodetic Science, The Ohio State University, Columbus, Ohio

Acronyms

ACAA	American Coal Ash Association	**E&P**	Exploration and production
AMD	Acid mine drainage	**E2**	Energy efficiency
ASTM	American Society for Testing and Materials	**EA**	Environmental assessment
		EDDS	Environmental Decision Support System(s)
C&D	Construction and demolition debris	**EMIS**	Environmental Management Information System(s)
CCA	Clean Air Act	**EMS**	Environmental Management System(s)
CCB	Coal combustion by-product		
CDF	Code of Federal Regulation	**EPCRA**	Emergency Planning and Community Right-to-Know Act
CERCLA	Comprehensive Environmental Response, Compensation, and Liability Act	**ER**	Environmental restoration
		ERP	Enterprise resource planning
CLSM	Controlled low-strength material	**ESP**	Electrostatic precipitation
D&D	Decontamination and decommissioning	**FBC**	Fluidized bed combustion
		FGD	Flue gas desulfurization
DfE	Design for environment	**FHWA**	Federal Highway Administration
DQO	Data quality objective		

GIS	Geographic Information System(s)	**PPE**	Personal protective equipment
IADC	International Association of Drilling Contractors	**PPOA**	Pollution Prevention Opportunity Assessment
IEIS	Integrated Environmental Information System(s)	**Psi**	Pounds per square inch
ISO 14001	International Standards Organization's standard for environmental management systems	**PSM**	Pozzolanic-stabilized Mixture
		RCRA	Resource Conservation and Recovery Act
Lb/mm BTU	Pound per million British thermal unit	**RI/FS**	Remedial investigation/ feasibility study
LFA	Lime-fly ash-aggregate	**ROI**	Return on investment
LIMB	Lime injection multistage burner	**RRC**	Railroad Commission of Texas
LOI	Loss on ignition	**SI**	Site investigation
LQG	Large quantity generator	**SO2**	Sulfur dioxide
MSW	Municipal solid waste	**SPE**	Society of Petroleum Engineers
NAAQS	Natural Ambient Air Quality Standards	**SR**	State route
NOx	Nitrogen oxide	**TCLP**	Toxicity of characteristic leaching procedure
NOV	Notice of violation	**SWMU**	Solid waste management unit
NPL	National Priorities List		
NSPS	New Source Performance Standards	**TRI**	Toxic Release Inventory
OCDO	Ohio Coal Development Office	**US DOE**	U.S. Department of Energy
		US EPA	U.S. Environmental Protection Agency
P2	Pollution prevention		
PA	Preliminary assessment	**WM**	Waste management
PC	Pulverized coal	**Wmin**	Waste minimization
POTW	Publicly-owned treatment works		

Glossary

Anode   The positive element of any electrical device from which electricity flows.

Cathodic protection   The negative element that draws current away from pipe and other metal equipment to protect the pipe or equipment from corrosion.

Centrifugal filter   A device that spins a fluid at high speed to separate and remove materials from the fluid.

Closed-Loop drilling fluid system   A system of tanks that contains the drilling fluid used in drilling an oil or gas well so that drilling fluid is not placed in conventional pits.

Coalescor panel   A device used in fluid separation equipment that stabilizes fluid flow through the device (e.g., reduced turbulence).

Construction and demolition debris (C&D)   Waste building materials, packaging, and rubble resulting from construction, remodeling, repair, and demolition operations on pavement, houses, commercial buildings, plants, and other structures.

Data quality objective (DQO) Qualitative and quantitative statements derived from the DQO process that clarify study objectives, define the appropriate type of data, and specify the tolerable levels of potential decision errors that will be used as the basis for establishing the quality and quantity of data needed to support decisions. It provides a systematic procedure for defining the criteria that a data collection design should satisfy, including when and where to collect samples, the tolerable level of decision errors for the study, and how many samples to collect.

Decontamination and decommissioning (D&D) The process of reducing or eliminating and removing from operation of the process harmful substances, such as infectious agents, so as to reduce the likelihood of disease transmission from those substances. After the D&D operation, the process is no longer usable.

Demolition The wrecking or taking out of any load supporting structural member and any related razing, removing, or stripping of a structure. Also called Deconstruction.

Design for environment (DfE) It is the systematic consideration of pollution prevention/waste minimization options during the design consideration of any process associated with environmental safety and health over the product life cycle.

Drilling fluid The circulating fluid used in drilling oil and gas wells. Drilling fluid lubricates the drill bit, carries rock cuttings from the wellbore to the surface, and controls subsurface formation pressures.

Drilling rig The collection of equipment, such as a derrick, used to drill oil and gas wells.

Enhanced oil recovery Methods applied to oil and gas reservoirs depleted by primary production to make them productive once again.

Environmental assessment (EA) Document that briefly provides sufficient evidence and analysis for determining whether to prepare an environmental impact statement or a finding of no significant impact. Includes a brief discussion of the need for the proposal, alternatives as required by EPA regulations, the environmental impacts of the proposed action and alternatives, and a listing of agencies and persons consulted.

Flowline The surface pipe through which oil travels from a well to processing equipment or storage.

Functional unit The measure of a life-cycle system used to base reference flows in order to calculate inputs and outputs of the system.

Hazardous waste Solid waste that is hazardous as defined in Title 40 of the Code of Federal Regulations (CFR). Hazardous waste is either specifically listed as such or exhibits a characteristic of hazardous waste as specified in 40 CFR Part 261, Subparts D and C.

Heat-Medium oil Oil that is used to transfer heat from one medium to another medium.

ISO International Standards Organization (or International Organization of Standardization).

ISO 14000 International Standardization of Environmental Management System Standard which is "that part of the overall management system which includes organizational structure, planning activities, responsibilities, practices, procedures, processes and resources for developing, implementing, achieving, reviewing and maintaining the environmental policy."

Large quantity generator (LQG) A hazardous waste generation site classification. In general, LQG sites generate more than 2204 pounds of hazardous waste each month.

Life (a) Economic: period of time after which a product, machine, or facility should be discarded because of its excessive costs due to costs or reduced profitability; (b) Physical: that period of time after which a product, machine, or facility can no longer be repaired in order to perform its designed function properly.

Life cycle cost Evaluation of the environmental effects associated with any given activity from the initial gathering of raw materials from the earth to the point at which all materials are returned to the earth; this evaluation includes all releases to the air, water, and soil.

Life cycle impact assessment A scientifically based process or model that characterizes projected environmental and human health impacts based on the results of the life cycle inventory.

Life cycle inventory An objective, data-based process of quantifying energy and raw material requirements, air emissions, waterborne effluents, solid waste, and other environmental releases throughout the life cycle of a project, process, or activity.

Municipal solid waste (MSW) Residential and commercial solid wastes generated within a community.

Natural gas processing plant A facility containing equipment and vessels necessary to purify natural gas and to recover natural gas liquids, such as butane and propane.

Paraffin A heavy, wax-like hydrocarbon commonly found in produced crude oil. Paraffin often accumulates within wells and the associated surface equipment.

pH A unit of measure of the acid or alkaline condition of a substance. On a logarithmic scale between 1 and 14, a neutral solution has a pH of 7, acid solutions are less than 7, and alkaline solutions are greater than 7.

Pipeline compressor station An equipment station on a natural gas pipeline that uses a device to raise the pressure of the gas in order to move it along the pipeline.

Pollution prevention The use of materials, processes, or practices that reduce or eliminate the creation of pollutants or wastes at the source.

Pollution Prevention Opportunity Assessment (PPOA) The systematic process of identifying areas, processes, and activities that generate excessive waste streams or waste by-products for the purpose of substitution, alteration, or elimination of the waste.

Publicly owned treatment works Any device or system used to treat (including recycling and reclamation) municipal sewage or industrial wastes of a liquid nature and is owned by a State, municipality, intermunicipality, or interstate agency (defined by Section 502(4) of the Clean Water Act).

Pump-jack A surface unit that imparts a reciprocating motion to a string of rods that operate a pump in an oil well.

Reagent A substance that, because of the chemical reactions it causes, is used in analysis and synthesis.

Recycling of materials The use or reuse of a waste as an effective substitute for a commercial product, as an ingredient, or as feedstock in an industrial or energy producing process; the reclamation of useful constituent fractions within a waste material; or removal of contaminants from a waste to allow it to be reused. This includes recovery for recycling, including composting.

Reserve pit The pit in which a supply of drilling fluid is stored for use in drilling an oil or gas well. A reserve pit is typically an excavated, earthen-walled pit, which may be lined to prevent contamination of soil and water.

Return on Investment (ROI) The calculation of time within which the process would save the initial investment amount if the suggested changes were incorporated into it. In this calculation, depreciation, project cost, and the useful life are taken into account.

Rod-pump   The pump in an oil well that lifts oil to the surface as a result of the reciprocating action of a rod-string (*see also Pump-Jack*).

Sand-blasting media   Abrasive material used to remove paint and other coating material from metal surfaces. Sand-blasting media is forced onto the surface using high air pressure.

Screen-Type filter   A filter unit, typically constructed of steel, from which the screen portion can be removed, cleaned of filtrate, and reused.

Separator   A cylindrical or spherical vessel used to isolate the components in streams of mixed fluids.

Soda ash   Sodium carbonate, typically used for pH control.

Sour gas   Natural gas that contains hydrogen sulfide or another sulfur compound.

Specific gravity   The ratio of the weight of a given volume of a substance at a given temperature to the weight of an equal volume of a standard substance at the same temperature.

Strata   Distinct beds of rock, which are usually parallel. An individual bed of rock is a stratum in which the subsurface oil and natural gas are contained within certain strata.

Sulfur dioxide scrubber   A device designed to remove sulfur dioxide from the exhaust gases of engines.

Thermal destruction   Destroying of waste (generally hazardous) in a device that uses elevated temperatures as the primary means to change the chemical, physical, or biological character or composition of the waste. Examples of the processes could be incineration, calcinations, oxidation, and microwave discharge. Commonly used for medical waste.

Toxic release inventory (TRI)   Required by EPCRA, it contains information on approximately 600 listed toxic chemicals that facilities release directly to air, water, or land or transport off-site.

Turn-key contract   For the purpose of drilling an oil or gas well, a contract that calls for the payment of a stipulated amount to the drilling contractor on completion of the well. A turn-key contract may be based on a set cost per foot of well drilled.

Vitrification   The process of immobilizing waste that produces a glass-like solid that permanently captures the radioactive materials.

Waste combustion Combustion of waste through elevated temperature and disposal of the residue so generated in the process. It also may include recovery of heat for use.

Waste management (WM) Activities associated with the disposition of waste products after they have been generated, as well as actions to minimize the production of wastes. This may include storage, treatment, and disposal.

Handbook of
Pollution Control and
Waste Minimization

1

Pollution Prevention and Waste Minimization—Back to Basics

Jeff Weinrach
JCS/Novation, Inc., Albuquerque, New Mexico

1 TERMINOLOGY

Prevention, *n.* (prevent, *v.*—to keep from occurring; avert; hinder)

Minimization, *n.* (minimize, *v.*—to reduce to the smallest possible amount or degree)

Source, *n.* any thing or place from which something comes, arises, or is obtained; origin

Reduction, *n.* (reduce, *v.*—to bring down to a smaller extent, size, amount, number, etc.)

Recycle, *n.* to treat or process (used or waste materials) so as to make suitable for reuse

:
:

Control, *n.* (control, *v.*—to exercise restraint or direction over; dominate; command: to hold in check; curb; to eliminate or prevent the flourishing or spread of)

Management, *n.* (manage, *v.*—to take charge or care of: to handle, direct, govern, or control in action or use)

Treatment, *n.* (treat, *v.*—to subject to some agent or action in order to bring about a particular result)

:

Waste, *n.* (waste, *v.*—to consume, spend, or employ uselessly or without adequate return; use to no avail or profit; squander: useless consumption or expenditure; use without adequate return; an act or instance of wasting: anything unused, unproductive, or not properly utilized: anything left over or superfluous, as excess material or by-products, not of use for the work in hand)

:

Pollution, *n.* the introduction of harmful substances or products into the environment

Pollute, *v.* to make foul or unclean, especially with harmful chemical or waste products

:

Process, *n.* A systematic series of actions directed to some end

:

System, *v.* An assemblage or combination of things or parts forming a complex or unitary whole

Source: *The Random House Dictionary of the English Language,* 2nd edition, Unabridged. New York: Random House, 1987.

2 BACKGROUND

We are a society rooted in language: both the words and their meanings. We choose our words carefully in an attempt to convey the particular meaning that we have in mind. Often, over time, words will develop a "life" of their own, and their meanings (as well as their usage) may become murky and indistinct. In the environmental arena, few words or groups of words have gone through the constant deluge of interpretation and meaning as "pollution prevention." Over the last decade, pollution prevention has gone through an evolutionary process from specific activities such as solvent substitution (CFCs versus replacement solvents, as an example) to more systematic approaches that often provide the core element to cost-effective environmental management systems (EMS).

Throughout this evolutionary process, due primarily to the regulatory oversight that has placed administrative boundaries on what pollution prevention is and what it is not, the practical issue has not always been the focus of attention:

What are the most cost-effective methods to reduce or eliminate environmental impacts without sacrificing health, safety, or other related concerns?

With the advent of EMS and environmental standards such as the ISO 14000 series, pollution prevention is now more often than not viewed as part of a systematic approach to environmental improvement that includes planning, information management, and process management. The last two entries in the glossary above, "process" and "system," reflect the conditions under which pollution prevention is now most practically used.

Many organizations have embraced pollution prevention and have already addressed or are in the midst of addressing the "low-hanging fruit," the relatively easy pollution-prevention activities that often do not require a thorough understanding of the processes that are generating the pollutants. A number of articles in the literature describe success using nonhazardous solvents, for example, for particular applications such as paint stripping or cleaning parts. As the number of case studies grows and the nonhazardous solvents are shown to be successful for particular processes or types of processes, the trust in using these solvents in similar applications naturally grows as well. But with organizations now looking for more innovative and cost-effective solutions to reducing waste and inefficiency, the application of environmental management systems where processes and systems are key to identifying and implementing opportunities to reduce or eliminate the waste and the inefficiency is becoming more commonplace.

Nevertheless, it is still helpful to have an understanding of what pollution prevention is and how it can be applied to reduce or eliminate environmental impacts. This book will provide many examples of pollution prevention as it relates to particular processes or industries. The remainder of this chapter will provide an overview of the various aspects to pollution prevention, with an emphasis on getting beyond the environmental vernacular and focusing on the practical.

In a simplified way, we can think of pollution prevention in two different contexts: we can prevent pollutants from being generated in the first place, or we can prevent the pollutants from being introduced into the environment. In the first case, if the pollutants are considered waste (which is most often the case), then preventing these materials from being generated would result in a reduction, minimization, or elimination of the waste (waste minimization). This usually provides direct economic benefit, since reducing waste usually coincides with increased efficiency, productivity, and profitability. In the second case, the pollutants are still being generated but are not being released to the environment. This should provide some environmental benefits, since the environment is not being negatively impacted by these materials. However, the economic benefit may not be as great, since the wastes need to be stored, treated, or disposed, which is an added cost. Also, the management of these wastes is likely to be a temporary measure and ultimately there will be some release to the environment unless the wastes can be reused or recycled. This second case is often referred to as pollution control. This book will focus on the pollution prevention practices that primarily involve preventing pollutants and wastes from being generated in the first place.

Pollution prevention is often viewed as part of an overall environmental management hierarchical framework (in order of preference): source reduction, recycling, treatment, disposal. This hierarchy typically reflects the degree of economic benefit and environmental protection that can be realized through these efforts. However, the hierarchy is not always practical in terms of prioritizing opportunities. For a particular process, recycling may be a more economically and environmentally viable option to source reduction given the existing techniques and approaches. By developing an effective environmental management system, identifying the most cost-effective options to reducing waste and preventing pollution will be much more likely. Even so, the hierarchy has been shown to be quite effective in prioritizing pollution prevention and waste management projects and is still useful as a first attempt to improve environmental performance.

3 SOURCE REDUCTION

Source reduction involves the use of processes, practices, or products to reduce or eliminate the generation and/or the toxicity of pollutants and wastes. Source reduction includes, but is not limited to, material substitution, process substitution, and process elimination. Examples of some source reduction applications are described below.

3.1 Material Substitution

Materials that will result in less toxic wastes can be substituted for materials that are currently being used. Examples include the following:

> Shifting from solvent-based paints to water-based paints reduces the toxicity of paint wastes.
> Shifting from chlorofluorocarbons (CFCs) to nonhazardous solvents will reduce waste management costs and health risks.

3.2 Process Substitution or Elimination

Process that result in less waste and increased efficiency can be substituted for processes that are currently being used. Also, entire processes can be eliminated if pollution prevention is implemented effectively. Examples of some process substitution or elimination opportunities are described below:

Replacing traditional parts-cleaning processes using solvents with processes that use supercritical fluids

Using Dry Ice pellets or other blasting techniques to remove paint, in lieu of solvents.

3.3 Good Housekeeping and Equipment Maintenance

Good housekeeping and equipment maintenance are two environmental management practices that are often low-cost/high-benefit approaches to pollution prevention. A common example of good housekeeping practices involves the use of drip pans to catch leaks or drips from equipment. Equipment maintenance is important for two distinctly different reasons: (a) routine maintenance will reduce the occurrence of leaks and drips, and (b) routine maintenance will extend the lifetime of the equipment. When thinking about pollution prevention and waste minimization, it is important to consider that when equipment comes to the end of its useful life, it also becomes a waste!

3.4 Water and Energy (Resource) Conservation

Water conservation is critically important in all of our industrial and personal activities. Non-point-source pollution (caused by water moving over and through the ground picking up man-made and natural contaminants) can be significantly reduced by limiting water usage. Also, the cost of treating wastewater is often related to the volume of water that requires treatment.

Energy conservation and pollution prevention are often thought of as two sides of the same coin. Waste management is typically an energy-intensive step that, naturally, provides an additional incentive to pollution prevention and waste minimization. When we use life-cycle analysis (see below), we identify additional costs and wasteful steps associated with the energy consumed as part of managing wastes. Also, energy production usually coincides with particular waste streams and pollutants entering the environment. This cyclic interdependency between energy efficiency and pollution prevention signifies the importance of using systematic approaches to achieve significant environmental improvement. Resource conservation, in general, is a fitting complement to pollution prevention. As critical components to an integrated environmental management system, best practices such as pollution prevention, water conservation, and energy conservation can be more effective in tandem than as separate activities.

3.5 Pollution Prevention in Design and Planning

Designing or planning for a new process or operation is the best time to address pollution prevention considerations. With an existing process, implementing pollution prevention can require some possible downtime due to either equipment reengineering or technician training. This will add greatly to the cost and, therefore, reduce the economic benefit of the particular pollution prevention approach. Also, in the design phase, all environmental improvement options are open for evaluation. There are no practical reasons to dismiss any particular option due to inability to transition from the current process to the improved process. In the design and planning phase, there is no status quo and, therefore, no downtime and associated costs.

3.6 Training and Awareness

Training and awareness programs are critical to ensuring that pollution prevention is realized to its fullest potential. The best ideas will come from people who work with machines, use materials, and generate waste. These people must be aware that often there are alternatives and that they constantly need to be thinking about ways to improve operations, efficiency, etc. It is always more effective to provide pollution prevention training to people with process knowledge (often, the implementers and stakeholders) than to provide "pollution prevention experts" with process knowledge to develop a pollution prevention plan.

3.7 Life-Cycle Analysis

Pollution prevention often utilizes a principle known as "life-cycle analysis" to address all associated costs and possible solutions associated with a particular pollutant or waste. Life-cycle analysis, sometimes referred to as "cradle-to-grave" analysis, is often used to track a particular material from its inception to its ultimate demise. This tracking usually requires documentation from other companies (both vendors and customers) in the material chain. In material substitution, for example, a possible material alternative that would drastically reduce a particular waste stream may require a process change by the vendor first. Also, a positive pollution prevention approach implemented by a particular company could have negative impacts to its customers or contractors. For these reasons, it is helpful to include vendors, customers, and contractors as part of the pollution prevention team.

3.8 Inventory Control

Inventory control addresses the effective use of data and information to track the procurement, use, and management of materials throughout the operation. Inventory control practices include the following:

> "Just-in-time" procurement—purchase only what is needed, in the amounts needed. This is extremely important for chemicals or materials that have relatively short shelf-lives and have to be disposed if not used in a timely manner.

Affirmative procurement—purchase only materials that have been or can be recycled. Purchase nonhazardous chemicals and materials whenever possible.

Barcoding—use barcodes to track material usage throughout the facility. This is extremely helpful in limiting the amount of material purchased if it is known how much of that material may already be stored at the facility. Through a chemical or material exchange program, chemicals and materials can be obtained from operations within the facility instead of having to purchase the material.

4 RECYCLING

For the purpose of this book, recycling is addressed in two different fashions whenever possible: (a) in-process recycling (recycling materials), and (b) end-of-pipe recycling (recycling wastes).

In-process recycling implies that a material is recycled before it becomes a waste. If the material is not being treated as a waste, then waste management regulatory requirements are not applicable to these processes (no treatment permit required, for example) because the recycling is in-process. The development of these recycling activities requires knowledge of the process itself.

End-of-pipe recycling implies that the material being recycled has already become a waste. In many cases, waste management regulatory requirements are applicable to these recycling processes. Because the recycling is end-of-pipe, knowledge of the process that generated the waste is normally not necessary. End-of-pipe recycling as a pollution prevention alternative does not, therefore, depend on the processes that generated the waste.

5 TREATMENT (INCLUDING WASTE SEGREGATION)

Waste treatment is usually the third option after source reduction and recycling opportunities have been exhausted. Treatment includes techniques such as precipitation, neutralization, stabilization, and incineration. Waste segregation is also considered as a treatment alternative. In many cases, waste treatment is performed off-site by a contracting organization. The waste-generating organization must maintain very careful records regarding the contents of the waste so the proper waste management procedures can be carried out. In many cases, information regarding the process that generated the waste is maintained with the waste information. This information is helpful in demonstrating an understanding of how (and why) the waste was generated, and it lessens the risk to the contracting organization that may be treating wastes it would otherwise not be permitted to treat.

Waste segregation is an environmental best management practice designed to reduce costs through storing incompatible waste separately, including separating hazardous from nonhazardous wastes, or regulated from nonregulated wastes. In many circumstances, mixing regulated with nonregulated wastes renders the entire waste contents regulated and unnecessarily increases waste management costs.

6 DISPOSAL

If there are no other practical options, disposal needs to be carried out in an environmentally responsible manner. In the majority of cases, waste disposal will be provided by a contractor. It is critically important that proper documentation and records are maintained regarding waste disposal, both by the parent company and by the contractor. In many regulatory environments, for example, liability for the disposal of waste is not totally eliminated after the waste is removed from the site.

7 CONCLUSION

Waste minimization and pollution prevention are two components of a broader, effective system of process improvements that often have both environmental and economic benefit. As technologies continue to be developed and as new, innovative approaches to improving efficiency and productivity are implemented, these and other environmental best practices will likely be drawn in to the overall operational improvement and excellence that we strive for. If we can effect significant improvement through these types of approaches, our companies will be more productive, more profitable, and more competitive in the global marketplace.

2

Role of Pollution Prevention in Waste Management/Environmental Restoration

Harish Chandra Sharma

U.S. Department of Energy, Albuquerque, New Mexico

1 POLLUTION PREVENTION IN PROCESS MANAGEMENT

The single most important challenge facing today's environmental engineer of any major industry is how to keep environmentalist organizations, regulators, and stakeholders on one side and business managers and shareholders on the other side satisfied simultaneously. Complying with existing regulations is a simple matter of incorporating a few additional process changes. Unfortunately, changes of this kind are costly. In the process, waste changes from one form to another. The only way out of this complex situation is through incorporation of waste minimization/pollution prevention (WMin/P2) measures, every step of the way, wherever economically feasible. As per a National Association of Manufacturers' survey, companies that take advantage of the U.S. Environmental Protection Agency (EPA) 33–50 emissions reduction program for toxic chemicals, over half the companies saved money (1). In spite of this encouraging fact, resistance to incorporate waste minimization or pollution prevention at any level is very real. Many industrial houses today have placed a vice president in charge of environmental affairs. These companies have a vision and mission statement to go with the position, but when it comes to changing a process or incorporating a significant change, it just won't happen. This is true for an existing process in which waste is managed as an end-of-pipe fact under the waste management (WM) program and also in the case of site cleanup that is conducted under the environmental restoration (ER) process or under the decontamination/decommissioning (D&D) process for an abandoned facility.

1.1 Waste Management (WM)

Waste management and minimization program is "compliant, cost effective management of the minimum waste" (2). Waste does not contribute added value to the products in the marketplace but rather is an enduring liability to the company. It is a liability when it is produced, due to the cost of transportation and disposal, and in the future, as the landfill where it has been disposed may leak, thus harming the environment and the community. As a result, cost effectiveness is the primary focus of commercial waste management programs.

Waste management covers newly generated waste or waste from an on-going process. When steps to reduce or even eliminate waste are to be considered, it is imperative that considerations should include total programmatic oversight, technical, and management services of the total process. From raw material to the final product, this includes technical project management expertise (for cost and technical effectiveness), technical project review and pollution prevention technical support and advocacy. Waste management also includes handling of waste, including treatment, storage, and disposal.

In WM, the waste, both the quantity and composition, are known. Given these facts, disposal can be handled in a planned manner. In most cases, other than the waste from routine process, waste that is generated from handling of known waste can be controlled and virtually eliminated.

Minimizing of waste in WM programs is primarily due to the opportunity it provides not only to reduce production costs but also to reduce liability at the same time. Liability exists at the point of generation and at the off-site disposition of that waste. As a result, "waste minimization is not a stand alone program but an element in a corporate liability reduction initiative" (2).

1.2 Environmental Restoration (ER)

In the case of environmental restoration (ER) projects, the quantity and composition of the waste itself is an estimate, as the waste was usually "dumped" at the site some time ago. In most cases, these sites were "discovered" during the 1980s, when the EPA required them to be identified. Under the Resource Conservation and Recovery Act (RCRA), they are termed as "solid waste management units" (SWMUs). These sites, if highly contaminated, under the Comprehensive Environmental Response, Compensation, and Liability Act (CERCLA), may be listed under the National Priorities List (NPL). These are also called Superfund sites. Most of the waste to be processed by ER activities, i.e., contaminated soil, water, building material, is the result of past production activities or due to a spill which had been covered up at that time. This waste is called primary waste.

While performing restoration or cleanup operations, new waste, i.e., drilling cuttings, personal protective equipment, gloves, or cleaning equipment that are generated is called secondary waste. In cases where the ER site is very large or heavily disturbed or very old, the site may require extensive study to assess the level and extent of contamination. In those cases, waste will be generated during the preliminary assessment (PA)/site investigation (SI) phase and the characterization phase. This phase of investigation is sometimes referred to as the remedial investigation (RI)/feasibility study (FS) phase. During the RI/FS or PA, generation of waste, which is mostly secondary waste, can be controlled. The number of samples and the extent of sampling must be carefully determined to ensure that the waste generated during this phase is minimized. However, this is possible through meaningful negotiations with regulators, affected parties, and stakeholders.

Waste generated during interim action or cleanup action that is conducted quickly or with limited analysis to reduce or eliminate imminent threats to the environment, the public, or workers may be difficult. However, these situations arise only in cases of emergency or time-critical action, generally due to either an accident or a spill of a hazardous substance.

Superfund sites require additional studies. Studies include prioritizing all the hazardous substances found on site, preparing a toxicological profile of each substance, and making a complete health assessment of all substances on site. Potential migration of these substances off-site and its impact on the surrounding population at risk must be assessed. These studies require extensive sampling and analysis of the site. Data quality objective (DQO) should be considered during on-site sampling.

During the ER process, there are four phases for the incorporation of Wmin/P2 in the process:

1. *Negotiation and planning:* This is the stage when negotiations with the regulators will be taking place.
2. *Assessment:* Typically during sampling, drilling, treatability, and test-run phases of the process.
3. *Evaluation and selection of process:* This is the decision-making phase, when the cleanup process is decided upon.
4. *Implementation:* Actual cleanup is done in this phase. Some fine-tuning of the Wmin/P2 already documented may be essential to further decrease generation of waste.

1.3 Decontamination and Decommissioning (D&D) Activity

Waste generated from D&D activity is generally similar to ER operations but more predictable, both in quantity and composition aspect. In this case, secondary waste and waste during characterization can easily be controlled. However, before a D&D action is considered for cleanup, the history of the site must be fully investigated. A thorough investigation will significantly reduce the amount of waste ultimately generated from site cleanup. Collecting historical data may take a considerable amount of time and should be completed prior to site cleanup. If the site is heavily contaminated, it may qualify as an NPL site and require the extensive study called for by the CERCLA.

Prior to decommissioning of any radioactive contaminated site, criteria and procedures for decontamination and cleanup specified by the Nuclear Regulatory Commission (NRC) must be determined. The NRC suggests that proposed D&D work incorporate simple and inexpensive methods prior to NRC approval to begin work.

Requirements for D&D may include incorporation of long-term monitoring and surveillance of the site. Consideration of DQO may result in cost savings as well as the number of samples to be taken during the surveillance phase.

Construction and demolition (C&D) is a bit more complex than D&D activity. The difference is that in C&D activity, decontamination may not be necessary, as the site is presumed to be clean. C&D generally refer to demolition of buildings or structures where no hazardous or toxic substance would have ever been handled or processed. To increase pollution prevention/waste minimization in the process, demolition is sometimes referred to as deconstruction.

2 REGULATORY REQUIREMENTS

Through the passage of numerous regulations, state and federal regulatory agencies such as the EPA have encouraged and sometimes mandated that WMin/P2 be incorporated into the process. Some of the regulatory requirements include the passage of the Pollution Prevention Act of 1990. Disposal of hazardous waste requires the existence of a pollution prevention plan before a manifest can be signed and the waste taken over by a transporter. Those plans exist in each and every plant that produces waste. In most cases, those plans remain on paper.

In 1993, the EPA formally endorsed pollution prevention as a guiding principle for all EPA programs, to encourage sustainable development while continuing the agency's mission to protect human health and the environment. It stated that the "mainstream activities at EPA, such as regulatory development, permitting, inspections, and enforcement, must reflect our commitment to reduce pollution at the source, and minimize the cross-media transfer of waste" (3).

As shown in Figure 1, the EPA recommends source reduction, reuse, and recycle as the order while handling waste. Reduction includes not only lower use of raw material in any process but also the substitution for hazardous or toxic materials by either lesser toxic material or a nontoxic substance. Sometimes this is referred to as chemical or material substitution. Disposal is the final option. The reason for this is that in 1997, the nation's largest 21,490 industrial users of toxic chemicals released 2.58 billion pounds of toxic chemicals into the environment (U.S. EPA, 1997 Toxic Release Inventory) (4) and spent billions of dollars managing pollution control technology systems to prevent that quantity from being even higher. If one were to add in the purchase price of the raw materials that eventually escaped as 2.58 billion pounds of chemical waste (instead of product), the price tag grows even higher. If the cost of effects on human health were added, the price tag would be truly phenomenal.

Consider municipal solid waste (MSW), which consists mostly of nonhazardous material: a total of 208 million tons was generated in 1995. This corresponds to 4.3 pounds per person per day. Of this, 56.2 million tons or 27% was recycled, and 33.5 million tons or 16.1% was incinerated and the energy reused. The other 118.3 million tons or 56.9% was landfilled (5). This corresponds to a reduction of about a million tons from the previous year, possibly due to recycling and reuse of material.

To encourage WMin/P2 through source reduction, reuse, and recycling of waste, regulatory agencies offer many incentives. Under the permitting, inspection, and enforcement process, the regulatory agencies have the power to encourage the process. Many state and federal agencies encourage this through their actions. For example, a primary goal of the Clean Air Act (CAA) is "to encourage or otherwise promote reasonable federal, state, and local government actions, consistent with the provisions of this chapter, for pollution prevention." The Clean Water Act (CWA) seeks "to eliminate the discharges of toxic pollutants." Stormwater regulations specifically require a pollution prevention plan. Strategies of the EPA and state regulators to encourage WMin/P2 are accomplished by issuing flexible permits, on occasion including explicit pollution prevention requirements as a condition of the permit. Many states, such as New Mexico and Texas, have a separate division in their environmental regulatory agency office which conducts pollution prevention walk-throughs at facilities. These agencies guarantee that any violation observed will not be used to issue a Notice of Violation (NOV). Some states encourage P2 in their multimedia permitting process. For example, the Massachusetts Department of Environmental Protection (DEP) uses a voluntary P2 worksheet in its Title V (for air) permit application process. The Ohio EPA issues an air permit and includes with the permit a cover letter urging the permittee to investigate P2 and energy conservation alternatives. The Michigan Department of Environmental Quality (DEQ) has amended its

permit application to include a state that simply reads: "Pollution Prevention is the Best Solution."

3 BARRIERS TO WMIN/P2

There is intense pressure from stakeholders in the industry to incorporate and integrate WMin/P2 measures as an integral element into each and every process. They want the industries in their neighborhood to be responsible for environmental protection and stewardship. In spite of this pressure from the public and the incentives from regulatory agencies, there are many barriers to implementing P2 in an actual situation, whether it is waste management, environmental restoration, or a D&D activity. Some of the barriers include (6):

1. Organizational barriers
2. Communication barriers
3. Economic barriers
4. Waste Generation barriers
5. Regulatory barriers

3.1 Organizational Barriers

Organizational barriers may be considered the biggest impediment for implementation of WMin/P2 in an industry. It is true that large industrial firms and almost all federal agencies have a separate environmental branch chief at the level of a vice president in an industrial setting or at the Assistant Secretary level in the federal agency, but they are so far removed from the actual work or plant setting that, unless they are very significant, ideas may never reach them. In addition, the person at the top may wholeheartedly support the WMin/P2 concept, but the mid-management level may have different views on the subject. For management, running the operation—keeping the line of production moving—is more important than thinking about changing the process line, however significant that may be. Policy directives coming from the top can be easily discarded or passed over on some pretext or the other. In most cases this is done under the guise of low priority, and WMin/P2 is the lowest priority when operation of the plant is the consideration. In most plant staff meetings, this point is hardly ever raised.

In many organizations, pollution prevention is an additional responsibility, combined with other programs. As a result, P2 may have a poor infrastructure. In addition, the lack of status and visibility makes it difficult for P2 coordinators to develop effective programs. In many instances, the P2 responsibility has been combined with environmental compliance. Typically, compliance with environmental regulations takes all the time, as the first priority of the person is not to receive a Notice of Violation (NOV) for the company. In these cases, if P2 roles and responsibility are not clearly defined and incorporated into a job description, they will most likely be overlooked entirely.

3.2 Communication Barriers

Searching for ways to reduce or eliminate generation of waste requires knowledge of the process itself. In the case of regular plant operation, the P2 coordinator may not know the various intricacies or all the steps that are followed in any production line. If the P2 coordinator suggests any new procedure, it could be turned down very easily. One example of this would be the case of an environmental restoration project, when the P2 coordinator may not know about the cleanup until a very late stage, when incorporation of pollution prevention methods or procedures becomes increasingly difficult to implement.

3.3 Economic Barriers

Funding for P2 is hard to come by. This is especially true when the operation is proceeding smoothly. Generally, the benefits of P2 projects are realized over the long term. After meeting basic regulatory requirements, it becomes difficult to push P2 beyond projects required to conduct normal plant operations. In almost all cases, there is no clear-cut budget line for P2 projects, thus making it difficult for such projects to compete for funding. P2 projects have to compete with compliance projects for funds that are in short supply. In addition, funding is based on return on investment (ROI) of the project. When there is intense competition for scarce funds, P2 projects are set aside because the return on investment may not be high enough. In reality, life-cycle costs must be considered before economics are worked out, but for P2 projects these hidden costs are many times not examined, making the P2 project look unattractive and therefore not funded. It is much easier to pick up waste and truck it off-site, rather than modify an entire process, even though the new process would be of a permanent basis and savings would be perpetual. Whenever a PPOA is conducted in a project, a full ROI consideration should be a part of that study.

3.4 Waste Generation Barriers

For many industries, waste generation may be highly episodic, thus making implementation of a WMin/P2 project uneconomical. In the case of environmental restoration projects, quantities of waste are at best an estimate. Waste will usually consist of many different types and implementation of P2 in each case may seem to be an uneconomical consideration and therefore not implemented. Operational personnel do not want to add the number of steps required in handling the waste, unless savings will be very significant. In most cases, reduction in waste generation is not high on the list.

3.5 Regulatory Barriers

Differences in state laws raise barriers to pollution prevention innovations. Practices that are permitted under one state's laws may be prohibited in another state. Even in a single state, interpretation of the law may be different significantly between one inspector and another, thus complicating things further. It is true that many states, such as California, Massachusetts, New York, and Texas, among others, facilitate P2 through legislation and their multimedia permitting process, but this process has its difficulties. Difficulties include initial time investment by companies and agencies while preparing the overall permit package, which in itself is very complicated and time consuming. Additionally, there are risks associated for companies if they are unable to undertake steps as stated in their permit package, as this would result in regulatory penalties.

4 TOOLS FOR WMIN/P2 IN WM, ER, AND D&D ACTIVITIES

With proper steps, barriers to implementation of WMin/P2 in waste management, environmental restoration, and decommissioning projects can be overcome. Steps toward implementation are not significantly different in WM, ER, or D&D projects.

With the enactment of many new regulations, minimizing environmental risk through WMin/P2 has become imperative. The goal of legislation is to reduce toxic emissions. For example, since the RCRA, pollution prevention is required because of the requirement for toxic use reduction plans. The Right to Know Act and other criminal penalties for pollution in the law make it necessary for industries to incorporate P2 in their plans. Waste management includes source reduction and environmentally sound recycling, thus eliminating the problem of handling waste that should never have been generated. Pollution prevention, on the other hand, deals with the handling of waste that comes out of a process or from the "end of the pipe." In waste minimization, cost of handling, treatment, and disposal does not exist. As the quantity of handling hazardous materials as a raw material is reduced, worker risk and accidents are substantially reduced.

4.1 Role of Stakeholders

To achieve success in WMin/P2 programs, an effective environmental leadership must be in place. The main drivers are stakeholders. In any major entity, the primary stakeholders are investors, the community, employees, customers, and regulators.

> *Investors:* Every stockholder wants the company to have a good image and the company to be competitive. Also, the enterprise should consistently handle its affairs in a responsible way.

Communities: The enterprise must be a responsible and respected member of the community in which it operates its plants and offices. The community must have a positive impression of the corporation, whether it is a private concern or a governmental agency.

Employees: Employees must enjoy working for the company, both through the physical atmosphere and through positive support by management. Employees must feel completely safe in their work area.

Customers: When customers receive services from a company, they should not only feel that they have received their money's worth of services but also that the company they are dealing with is environmentally conscious and its products and services are handled in a responsible way. The packaging used in the product should contain as much recycled product as possible, and the contents should be as environmentally friendly as possible.

The Regulators: The company must be compliant and take voluntary actions whenever necessary.

Each stakeholder has his or her own distinct viewpoint on enforcement, effect, and operation of any facility. For example, the company has cost effectiveness and profits as motivators, while the regulators may have only the laws and regulations as their primary interest. The citizens, who suffer the consequences of pollution emanating from a facility, are not concerned about the profits of the company. They simply want no pollution, whatever the cost may be. The customer, on the other hand, is looking for a product at the best price possible.

5 PROFILE OF A COMPANY

Companies that have good environmental management and play a leadership role achieve WMin/P2 through proactive steps. Table 1 gives the difference between a reactive and a proactive company.

The difference between a reactive and a proactive company is that while a reactive company looks for profit today, the proactive company looks at the long-term cycle and considers the whole product line for increased profits. The proactive company looks for continuous improvement through business management and the decision-making process, taking the view of personnel on the shop floor.

To become a proactive company, one has to develop and implement strategic environmental management tools. For this, the strategy of the management must integrate environmental planning with the business planning of the enterprise. It has to consider environmental factors as a multimedia concept. The company has to focus on pending and anticipated environment concerns. Potential problems could be foreseen through a multimedia audit as well as past regulatory

TABLE 1 Reactive and Proactive Management Strategies

Reactive	Proactive
Forced to comply with law	Eliminates compliance cost through audit programs
Responds to law enforcement	Evaluates risks of actions
Lacks resources	Utilizes resources sensibly
Follows trends	Trendsetters
Minimum cost approach	Takes life-cycle cost approach
Non-existing environmental aware-ness among workers	Employees fully aware of environ-mental program

inspection results. As these cannot be thrust upon management, an educational process to highlight these must be undertaken to be "accepted" by management. Educational processes could highlight the benefits, particularly company image, reductions in costs, and increased safety and health of the workers. It should be pointed out that WMin/P2 is the most critical segment of environmental management, as this can assist when considering the entire operation. Benefits to be highlighted include:

> Greater efficiency means fewer materials used and less time and energy wasted.
> Improved worker safety and health means fewer or no accidents and reduced or no lost time.
> Total compliance with all regulations.
> Improved company image among stakeholders resulting in better public relations.
> Reduction in treatment, handling, and disposal cost of waste.

Senior management can demonstrate its commitment by setting quantitative, process, or specific section reduction and recycling goals; institute performance measures to measure the goals; support cost-saving P2 projects; and consider design of P2 into new projects, processes, and facilities (7). Management should provide a comprehensive policy direction and put in a viable infrastructure so that a meaningful P2 program can be implemented and goals realized.

6 INCORPORATION WMIN/P2 INTO THE PROCESS

Successful implementation is possible when it is handled as a team. The members of the team should come from all segments of the company—management, engineering, environmental regulatory compliance, shop workers, health and safety personnel, and marketing. Suggested steps are as follows (8):

1. Introduce and market WMin/P2 practices as best management practices and not as a separate philosophy or practice. This encourages wider acceptance. One reason for this is that present acceptance may be low.
2. Assign a WMin/P2 coordinator to the overall environmental program and thus eliminate a separate requirement on project managers.
3. Initiate mandatory, management-supported, WMin/P2 training, geared specifically toward personnel who may have to deal with substances of a hazardous nature and those involved with the production of waste, especially in large quantities. The training should be specific in nature and geared toward those attending the session.
4. Establish source reduction and recycling goals. These may be set up and endorsed by the highest management personnel.
5. Include WMin/P2 as an element of the budget, planning, and all the key decision-making documents.
6. Include WMin/P2 criteria and performance measures in negotiated agreements with regulatory agencies.
7. Focus on implementation of the projects.
8. Include cost drivers and other incentives to reduce waste. Total cost of disposal, which should include the cost from waste generation through disposal and also landfill replacement, should be considered and highlighted to personnel who produce the waste. This is a life-cycle cost of waste disposal and should be compared with the cost of recycling the waste.
9. Also, the cost of disposal must be the responsibility of the generating program.
10. Highlight the successes so achieved by these programs and reward the personnel involved in those successes. Top management must be made aware of the successful implementation of this program so that WMin/P2 can receive wider acceptance.

6.1 Priorities for the Process

If a company has a very poor environmental record, it may be difficult to embark upon all the steps outlined above. For this reason, management may feel it necessary to prioritize these steps. The company may still place immediate profits and lack of prior results above everything else, but highlighting the benefits may move management more quickly toward the implementation of the process.

For the team, establishing goals may be difficult at the beginning. Market standards should be used in this regard. However, goals must be in place at the beginning of the project. The earlier the goals are set, the better the chances of attaining these goals. To achieve these goals, each and every segment of the project should have a separate performance measure that can be related to concrete pollution prevention goals established by company management. The focus of the group should be on the implementation of the process within the time frame, with goals of Wmin/P2 for added savings. The team should highlight this focus to the entire company; without their active involvement and support, achieving the goals may become difficult. Incorporation of WMin/P2 in the design concept of the process is essential. While working toward these goals, care should be taken that all safety as well as environmental regulations are followed.

7 KEY ACTIVITIES TO BE CONSIDERED FOR WASTE MINIMIZATION/POLLUTION PREVENTION

For successful implementation of pollution prevention in a project, early negotiations with regulatory agencies, at the beginning of the project cleanup phase or design phase for a manufacturing plant, are necessary. Most states incorporate pollution prevention into their regulations, in both permitting as well as inspection process. Concessions may be received from the regulatory agency if it is involved early in the project. Discussions with regulators for permitting of the facility should be conducted as early as possible. Table 2 lists some issues that the companies may raise with the regulator (4).

In the case of environmental restoration and decommissioning projects, preliminary assessment and inspection must be viewed with WMin/P2 as a significant element of the process. In a major ER or D&D project, the characterization phase is significant, as substantial quantities of waste may be generated during the sampling and analysis phase.

In the case of large ER and D&D cleanup, project evaluation of cleanup activities should be seriously considered. All preferred alternatives should be

TABLE 2 Preliminary Discussion Topics with Regulators

Pre-permit facility analysis
Preapplication meetings and permit scoping sessions
Permit application
Draft permit writing
Public comment
Final permit issuance
Permit renewal
Permit modification for process expansion or increased capacity

---

documented and evaluated independently. This should include the amount of waste that may be generated, both primary and secondary waste. In the case of ER or D&D project, cleanup action should be designed with a view of reuse of equipment to be used for the operation. In the case of WM, the design considerations should include maximum utilization of raw materials so that waste generated is minimized.

Some of the places where waste minimization should be considered include fugitive emissions and process modifications (see Table 3).

7.1   Measures to Consider WMIN/P2

After the design is optimized, the option that generates the least amount of waste and at minimum cost to operate the process should be considered for implementation. Even the handling of waste so produced from either the cleanup process or from manufacturing plant needs to be designed. In the case of ER and D&D operations, if they are properly designed, generation of secondary waste can be significantly reduced.

Some of the Wmin/P2 applications that need to be considered include the following.

1. *Segregation of waste:* This includes collecting hazardous and non-hazardous waste separately. Depending on specific regulations concerning disposal of waste, even the concentration of hazardous substance may mean savings in disposal costs, and this must also be looked into.
2. *Use of a contamination zone:* A clearly marked contamination zone may result in not generating waste unnecessarily. This will keep separate the area of operation into which the public may not enter.

TABLE 3   Candidate Items for Waste Minimization

Fugitive emissions reduction
Process modifications
Material and product substitution
Preventive and corrective maintenance
Routine equipment inspections
Proper material handling and storage procedures
Loss-prevention practices
Employee training programs
Material tracking and inventory control
Improved documentation
Environmental audits

3. *Reduction of equipment used for cleanup:* A specific example is the use of personal protective equipment (PPE). Recyclable PPE should be used wherever possible. Large equipment used in cleanup areas should be optimized. In some cases, once equipment has been used, it may be contaminated and may eventually become a waste. This must be minimized as much as possible.

4. *In-situ waste treatment for ER cleanup:* Contamination such as petroleum products could be cleaned up in situ, thus also eliminating generation of waste. Bioremediation should be considered as much as feasible.

5. *Minimization uncontaminated material removal:* All the material marked for cleanup in a site may not have to be removed, as many parts of the same may still be untouched. Care should be taken to remove only the contaminated portion of the site.

6. *Implementation of new technologies:* This may become a viable option where the cost of disposal of any waste or cleanup of a site is very high. It is generally considered if the waste is radioactive in nature.

7. *Unpacking outside contaminated areas:* Packaging material should be removed in clean areas to reduce the amount of waste generation.

8. *Avoidance ground contamination through ground cover:* If this is done, then the cost of cleanup is significantly reduced in case of a spill.

9. *Strict housekeeping:* Cleanliness can result in significant reduction of generation of waste.

When several ER and D&D projects are being considered simultaneously, plans to use equipment from one project on another can be incorporated if considered at the planning stages. This is critical in some instances: the equipment becomes contaminated and has to be disposed of in a landfill unless it can be used on similar projects.

Some of the tools that are available for total WMin/P2 considerations include the "Design for the Environment" (DfE), also called "P2 by Design." Another is the "Pollution Prevention Opportunity Assessment (PPOA)" concept. For effective environmental protection of a contaminated site, sufficient quantity and quality of data needs to be collected to support defensible decision making and at the same time meet regulatory requirements. When conducting a PPAO, in some cases the ROI ranking system may be considered.

To minimize expenditures related to data collection by eliminating unnecessary, duplicative, or overly precise data, the EPA has developed the Data Quality Objective (DQO). Use of this process depends on the complexity of the study, but it can be used for both small as well as large sites. However, the DQO process should be used during the planning stage of any study that requires data collection, before the data are collected.

7.2 Design for Environment (DfE)

The environmental responsibility of the corporation and the EPA's source reduction as the first option to reduce waste makes consideration of DfE necessary. This is a testing ground for new approaches to risk reduction through pollution prevention. As per the EPA, a DfE program provides guidance and tools to help companies achieve continuous environmental improvement (9). A DfE-based environmental management system provides a company with opportunities to go "beyond compliance" and save money. It helps companies promote the evaluation of cleaner production alternatives. By implementing these alternatives, a company can continuously improve its environmental performance. DfE also strongly supports pollution prevention principles, as that is its goal. This goal is accomplished through front-end innovations through redesign rather than relying on end-of-pipe controls to reducing potential risks to human health and the environment. This approach uses cleaner technology substitute assessments and life-cycle tools to evaluate the performance, cost, and environmental impacts of competing technologies. This serves as a guide to decision makers in order to arrive at an optimum decision. For this to be considered, the project team should recruit partners from among the various stakeholders. This team will select the process that will generate the least amount of waste.

During process design consideration, not only will the flow of material be a consideration, but also the waste that will be disposed of in the atmosphere. For this reason, a regulatory profile identifying applicable federal and state regulations will be undertaken. For process consideration, the project team will look into the availability of raw materials and the location of availability, the effects of the use of the product by the consumer, the effect of disposal on the environment, and potential workplace safety hazards.

The state of New Mexico is implementing the Green Zia Environmental Excellence program. This award program has three steps that lead an organization from the beginning of a prevention-based environmental management system, termed commitment level, to an established system that is demonstrating results, termed achievement level, to a fully deployed system with significant results. This is the Excellence Award level under the Green Zia Program.

7.3 Pollution Prevention Opportunity Assessment (PPOA) (10)

A Pollution Prevention Opportunity Assessment (PPOA) is one of the first steps in preparing a pollution prevention plan. For each independent process, PPOA is a tool for the company to identify the nature and amount of wastes and energy usage, stimulate the generation of pollution prevention and energy conservation opportunities, and evaluate those opportunities for implementation. It is a systematic framework that can be used by a facility's own employees to identify waste minimization opportunities. As a structured program, it provides intermediate milestones and a step-by-step procedure to:

1. Understand the facility's processes and wastes
2. Identify options for reducing waste
3. Determine if the options are technically and economically feasible to justify implementation

These procedures consist of four major steps:

1. *Planning and organization*—organization and goal setting
2. *Assessment*—careful review of a facility's operations and waste streams and the identification and screening of potential options to minimize waste
3. *Feasibility analysis*—evaluation of the technical and economic feasibility of the options selected and subsequent ranking of options
4. *Implementation*—procurement, installation, implementation, and evaluation.

The initial goal of a PPOA is to encourage participation in the assessment process by management and staff at the facility. The steps, shown in Figure 2, include corporate commitment to a waste minimization initiative and selection of a task force or similar group to carry out the assessment. Top management must endorse the study and issue a policy statement regarding waste minimization, establishing tentative waste reduction goals to be achieved by the program. This will aid in identifying waste-generating sites and processes, conducting a detailed site inspection, developing a list of options which may lead to the waste reduction goal, formally analyzing the feasibility of the various options, measuring the effectiveness of the options, and continuing the assessment.

The EPA has developed very useful worksheets to conduct a PPOA for any process (11). Suggested steps include the following.

1. *Assessment overview:* This summarizes the overall program. This will include the establishment of the P2 program at the site, preliminary assessment, establishment of a program plan, defining P2 options, feasibility analysis, a written assessment report, implementation of the plan, a measure of progress, and maintenenance of the P2 program.
2. *Assessment phase:* This includes site description, collection of process information, input of materials inventory, products summary, waste stream summary, options generation, and descriptions of the options.
3. *Feasibility analysis phase:* This is the phase which looks at the profitability of the various options under considerations. It should facilitate the ability of management to select the optimum option that a PPOA should bring about.

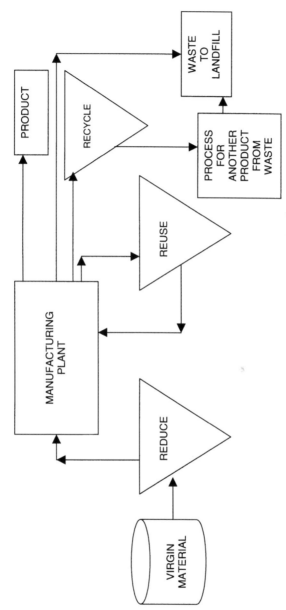

FIGURE 1 Hierarchy for pollution prevention/waste minimization.

The need to minimize waste

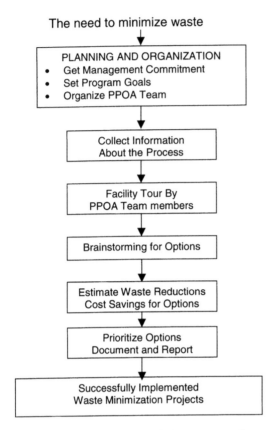

FIGURE 2 Planning a pollution prevention opportunity assessment for a process.

For municipal solid waste (MSW), the EPA's integrated waste management includes the following components (12).

1. *Source reduction:* This includes changes in packaging design, backyard composting, grass recycling, changes in purchasing habits, increased reuse, changes in industrial practices, and changes in use patterns. This is considered the generation of waste management.

2. *Recycling of materials:* This includes recovery for recycling, including composting.

3. *Waste combustion:* Combustion and disposal. Combustion means recovery of heat for use.

Using the above means, landfilling of waste is only 56.9% of the waste generated.

PPOAs have been successfully used in manufacturing, sampling and analysis, ER cleanups, D&D processes, as well as in the design and construction of new facilities.

7.4 Return on Investment (ROI)

Though similar to the study carried out under a PPOA, ROI is a tool which assists management in deciding on the options for WMin/P2 that are put forward. The ability to show cost savings on WMin/P2 projects is a major reason for funding these projects, in addition to the obvious benefit of reducing waste. Projects with high ROI will show the value of implementing the project in an atmosphere of diminishing resources.

For ROI consideration, the project should have a defined scope. The scope may include the following:

1. Percentage of return-on-investment (ROI)
2. Type of project
3. Type of waste that will be reduced or eliminated
4. Volume of waste that will be reduced
5. Technical approach to accomplish the reduction

Each of the above categories will have different levels. It is suggested that there be four levels for each of the five suggested scopes. Each could be given a specific number based on the importance attached to that category. For example, an ROI value of greater than 100% would be rated as 4, while an ROI value of less than 25% as 1. Similarly, if the project type results in source reduction, it would be rated as 4; while if the waste would be sent to a landfill, it would get a 1. Depending on the type of waste generated and the cost of disposal, ratings would be assigned. Similar is the case with the quantity of waste that will be either reduced or eliminated. The technical approach is a consideration. If the process under consideration is readily available, then the rating would be high, as results would certainly be achievable. On the other hand, for a research-related project the return is not a certainty and management may not be very inclined to fund it.

One way of calculating the percent ROI is through the following equation (13):

$$\text{ROI} = \frac{(B - A) - [(C + E + D)/L]}{C + E + D} \times 100$$

where A = annual costs after implementation of P2 project
B = annual costs before implementation of P2 project
C = capital investment for the P2 project
D = estimated project termination/disassembly cost
E = installation operating expenses
L = number of useful years of the project

Annual costs may include cost of the equipment; raw materials and supplies; utilities and labor costs; routine maintenance costs for the processes; PPE costs; waste management costs such as waste containers, treatment, storage, and disposal; inspection and compliance costs; recycling costs such as material collection/separation/preparation; materials and supplies; operation and maintenance labor costs; and administrative and other costs. For projects lasting less than five years, D could be used as is, after more than five years, D could be zero.

Care should be taken when the ROI is calculated: it is suggested that though implementation of the project would mean a permanent reduction in the generation of waste, savings over a three-year period should be considered for ROI calculations.

8   DATA QUALITY OBJECTIVE (DQO)

The DQO process is a strategic planning approach based on the scientific method that is used to prepare a data collection activity. It provides a systematic procedure for defining the criteria that a data collection design should satisfy, including when to collect samples, where to collect samples, the tolerable level of decision errors for the study, and how many samples to collect (14). DQO data collection specifies the final configuration of the environmental monitoring or measurement effort required to satisfy the specific study. It designates the types and quantities of samples or monitoring information to be collected; where, when, and under what conditions they should be collected; and what variables are to be measured.

The DQO process assures that the quality of environmental data used in decision making will be appropriate for the intended application. This process, as shown in Figure 3, can improve the effectiveness, efficiency, and defensibility of decisions in a resource-effective manner. The DQO process leads to the development of a quantitative and qualitative framework for a study, thereby facilitating an optimum selection of a cleanup for the site.

The benefits of DQO are that the data collection is most appropriate for the site, and condition of the site is considered during the planning stage of data collection. All the tolerable limits on decision errors, which will be used as the basis for establishing the quantity and quality of data needed to support the decision, are documented before sampling is undertaken. The process assures that optimum samples are collected so that decisions reached from the data are acceptable to all stakeholders. This will not only optimize resources for the project but also eliminate generation of waste.

DfE or P2 by Design is mostly useful in the planning of new processes, while PPOA can be used in any program, including ER and D&D projects. DQO can be used in all the cases where extensive sampling is one of the requirements. It should be noted here that results achieved by implementation of the WMin/P2

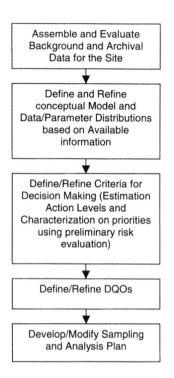

FIGURE 3   Data quality objective process.

concept in projects have resulted in cost savings of up to 50% in ER projects, and reuse of all waste in D&D projects.

9   MANAGEMENT EXAMPLES OF WMIN/P2

As indicated earlier, source reduction should be the first priority to reduce the generation of waste. Material substitution, consideration of new technologies, and use of best-management practices should be considered. This includes replacement of the use of hazardous substances with nonhazardous substances, and optimizing sample collection and analysis. The second option is recycling, and that includes reuse of material and reclaiming the waste produced from one process to make that a raw material in another. The third step is treatment of the waste. Here, options include stabilization, volume reduction through compaction, incineration, filtration, precipitation, thermal destruction of organic material, and vitrification. Disposal should always be considered a last option.

In most cases, regulatory agencies have indicated a willingness to adopt pollution prevention in any cleanup activity. By making pollution prevention a part of the planning process, it will be ratified by the regulatory agency and so could have a part in scheduling the pollution prevention consideration itself. This would give the agencies more time to devote to PPOA or DfE studies. Regulatory interpretation would now favor pollution prevention. If the project is an ER or D&D cleanup, future use of the land, cleanup levels, and release criteria could be negotiated.

WMin/P2 is a continuous improvement cycle (15). It starts with planning, implementation, evaluaticn, and review. Pollution prevention assessment has to be reviewed again with a multi-disciplinary group. If done early enough to review optimization of the process and energy use, it will generally result in further reduction in waste production and thus reduction of disposal costs.

10 CONCLUSION

There will be a constant struggle regarding reduction or even elimination of pollution from any source at any site. International Organization for Standardization (ISO) 14000 certification, a customer-driven worldwide movement, sees WMin/P2 as a very significant activity. Individual stakeholders have unique viewpoints regarding pollution emanating from a process. Differences in implementation of waste minimization/pollution prevention programs in waste management, environmental restoration, and decommissioning projects include the concept of waste in each case. Environmental restoration and decommissioning project managers may feel that as the waste exists, it cannot be reduced. However, once the concept of WMin/P2 is introduced and success is very visible, a change in attitude should take place. As a first step, secondary waste can easily be reduced, and this in itself should result in substantial savings for the project. Additionally, the concept of reuse of material in the cleanup process is another instance where savings are easily attainable.

It is true that the earlier the WMin/P2 concept is considered in the program management phase of any project, the greater the savings that will result. The concept will result in a better understanding of not only the problem being faced but also more unanimity in the solutions that will be presented by the parties involved. Beginning with management sponsorship of P2 activities as an important element, a successful implementation of Wmin/P2 requires continuous augmentation and support from all those involved.

ACKNOWLEDGEMENT

The author would like to thank Gaeton Falance for all his help and assistance in putting this chapter together.

REFERENCES

1. *Chemical and Engineering News,* vol. 77, no. 17, p. 10, April 26, 1999.
2. Independent Technical Review of Three Waste Minimization and Management Programs, p. 3-2. Albuquerque, NM: U.S. Department of Energy, Albuquerque and Oakland Office, August 1995.
3. EPA Pollution Prevention Policy Statement: New Directions for Environmental Protection, June 15, 1993.
4. EPA Pollution Prevention Solutions During Permitting, Inspections and Enforcement. EPA/745-F-99-001, p. 29, June 1999.
5. Characterization of Municipal Solid Waste in the United States: 1996 Update, U.S. EPA, Office of Solid Waste, EPA530-R-97-015, p. 10. Prepared by Franklin Associates, Prarie Village, KS, June 1997.
6. EPA Federal Facility Pollution Prevention: Tools for Compliance, EPA/600/R-94/154, p. 54, September 1994.
7. U.S. Department of Energy, Pollution Prevention Program Plan. DOE/S-01/8, p. 4. Washington, DC, 1996.
8. Los Alamos National Laboratory, Applicability of Waste Minimization to Environmental Restoration, LA-UR-96-17-21, Los Alamos, NM, pp. 9–15, June 1996.
9. EPA Environmental Management Systems Bulletin 1, EPA 744-F-98-004, July 1998.
10. U.S. EPA Waste Minimization EPA Assessment Manual, PEA/625/7-88/003, pp. 6–10. Cincinnati, OH: Hazardous Waste Engineering Research Lab, July 1988.
11. U.S. EPA Facility Pollution Prevention Guide, EPA/600/R-92/088, Washington, DC, May 1992.
12. Characterization of Municipal Solid Waste in the United States: 1996 Update, U.S. EPA, Office of Solid Waste, EPA530-R-97-015, p. 89. Prepared by Franklin Associates, Prarie Village, KS, June 1997.
13. Guidance for ROI Submissions. Albuquerque, NM: U.S. Department of Energy, 1996.
14. Environmental Protection Agency, U.S. Office of Research and Development, Guidance for the Data Quality Objectives Process, EPA/600/R-96/055, Washington, DC, September 1994.
15. U.S. EPA Facility Pollution Prevention Guide, EPA/600/R-92/088, Washington, DC, May 1992.

ABBREVIATIONS

A	annual costs after implementation of P2 project
B	annual costs before implementation of P2 project
C	capital investment for the P2 project
CAA	Clean Air Act
CERCLA	Comprehensive Environmental Response, Compensation, and Liability Act
C&D	construction and demolition debris
D	estimated project termination/disassembly cost

D&D	decontamination and decommissioning
DfE	design for environment
DQO	Data quality objective
E	installation operating expenses
EMS	environmental management system
EPCRA	Emergency Planning and Community Right-to-Know Act
ER	environmental restoration
ISO 14000	International Organization for Standardization 14000
L	number of useful years of a project
MSW	municipal solid waste
NOV	Notice of Violation
NPL	National Priorities List
PA	preliminary assessment
PPE	personal protective equipment
PPOA	Pollution Prevention Opportunity Assessment
RCRA	Resource Conservation and Recovery Act
RI/FS	remedial investigation/feasibility study
ROI	return on investment
SI	site investigation
SWMU	solid waste management unit
TRI	toxic release inventory
WM	waste management
WMin/P2	waste minimization/pollution prevention

GLOSSARY

Construction and demolition debris (C&D)   The waste building materials, packaging, and rubble resulting from construction, remodeling, repair, and demolition operations on pavement, houses, commercial buildings, plants, and other structures.

Data quality objective (DQO)   Qualitative and quantitative statements derived from the DQO process that clarify study objectives, define the appropriate type of data, and specify the tolerable levels of potential decision errors that will be used as the basis for establishing the quality and quantity of data needed to support decisions. It provides a systematic procedure for defining the criteria that a data collection design should satisfy, including when to collect samples, where to collect samples, the tolerable level of decision errors for the study, and how many samples to collect.

Decontamination and decommissioning (D&D)   The process of reducing or eliminating and removing from operation of the process harmful substances, such as infectious agents, so as to reduce the likelihood of disease transmission from those substances. After the D&D operation, the process is no longer usable.

Demolition   The wrecking or taking out of any load supporting structural member and any related razing, removing, or stripping of a structure. Also called *deconstruction.*

Design for environment (DfE)   The systematic consideration of pollution prevention/waste minimization options during the design consideration of any process associated with environmental safety and health over the product life cycle.

Environmental assessment (EA)   A document that briefly provides sufficient evidence and analysis for determining whether to prepare an environmental impact statement or a finding of no significant impact. This document will include a brief discussion of the need for the proposal, of alternatives as required by EPA regulations, of the environmental impacts of the proposed action and alternatives, and a listing of agencies and persons consulted.

Environmental management system (EMS)   A systematic approach to ensuring that environmental activities are well managed in any organization. It is very similar to ISO 14000.

Environmental restoration (ER)   Cleaning up and restoration of sites contaminated with hazardous substances during past production or disposal activities.

ISO 14000   International Standardization of Environmental Management System Standard which is "that part of the overall management system which includes organizational structure, planning activities, responsibilities, practices, procedures, processes and resources for developing, implementing, achieving, reviewing and maintaining the environmental policy."

Municipal solid waste (MSW)   Residential and commercial solid wastes generated within a community.

Pollution prevention opportunity assessment (PPOA)   A tool for a company to identify the nature and amount of wastes and energy usage, stimulate the generation of pollution prevention and energy conservation opportunities, and evaluate those opportunities for implementation.

Recycling of materials   The use or reuse of a waste as an effective substitute for a commercial product, as an ingredient, or as feedstock in an industrial or energy-producing process; the reclamation of useful constituent fractions in a waste material; or removal of contaminants from a waste to allow it to be reused. This includes recovery for recycling, including composting.

Return on investment (ROI)   The calculation of time within which the process would save the initial investment amount if the suggested changes were incorporated into it. In this calculation, depreciation, project cost, as well as useful life are taken into account.

Source reduction Any practice which: (a) reduces the amount of any hazardous substance, pollutant, or contaminant entering any waste stream or otherwise released into the environment prior to recycling, treatment, or disposal; and (b) reduces the hazards to public health and the environment associated with the release of such substances, pollutants, or contaminants.

Thermal destruction Destroying of waste (generally hazardous) in a device which uses elevated temperatures as the primary means to change the chemical, physical, or biological character or composition of the waste. Examples include incineration, calcination, oxidation, and microwave discharge. Commonly used for medical waste.

Toxic release inventory (TRI) Required by the EPCRA, a TRI contains information on approximately 600 listed toxic chemicals that the facilities release directly to air, water, or land or transportation of waste off-site.

Vitrification A process of immobilizing waste that produces a glasslike solid that permanently captures radioactive materials.

Waste combustion Combustion of waste through elevated temperature and disposal of the residue so generated in the process. It also may include recovery of heat for use.

Waste management (WM) Activities associated with the disposition of waste products after they have been generated, as well as actions to minimize the production of wastes. This may include storage, treatment, and disposal.

3

The Waste Management Hierarchy

W. David Constant
Louisiana State University and A&M College, Baton Rouge, Louisiana

1 INTRODUCTION

The management of waste can be approached from several venues, including regulations, history, technical methods, and interpretations of past management practices and our current methods to manage waste in what is considered the proper approach today. This chapter will explore the above approaches to waste management, present the Natural Laws (1) for the reader's consideration, and then describe a simple hierarchy for waste management based on these laws. The impact of the "implementation" of natural attenuation in many remediation schemes of today is also discussed. The objective is to raise awareness of both the capabilities and limitations that are placed on society in the management of waste.

2 HISTORICAL PERSPECTIVE

While we have recently increased our awareness of environmental problems and waste management, these issues have been in effect to some degree since society began to reach beyond simple existence. Humankind for centuries has developed and exploited available resources in useful and necessary ways, along with wasteful approaches. However, significant problems arose once communities, towns and cities developed into urban centers wherein contamination of water supplies from waste and animals caused significant deaths to occur. Further industrialization and heavy dependence on fossil fuels has in the past century greatly increased pressure on the environment to cope with the anthropogenic materials and methods of humankind's development. The development of regulations in the United States, described below, best illustrates the interactions for such a heavily industrialized nation.

In earlier history the best examples of industrial pollution are found in England (2), where factories contaminated nearby rivers and raised awareness about the limitations of drinking water sources. Air pollution resulted from use of coal for fuel, but it was only after many years, in the mid-1800s and later in the 1900s, that regulations and cause-and-effect mechanisms led to control of pollutant levels. Most unfortunate was the episode occurring in London during December 1952 due to stagnant conditions over the city, wherein pollutant concentrations resulted in death of about 4000 people from particulates and SO_2 buildup. This event was followed by the passage of the Clean Air Act by the government of England, which laid the basis for pollution control in that country.

In the United States, the historical perspective can be best represented through actions and activities in the United States and resulting regulations, to tie two perspectives together. Initial efforts were focused on water pollution by the River and Harbor Act of 1899, the Public Health Service Act of 1912, and the Oil Pollution Act of 1924, all being fairly localized in action. Only after World War II did the U.S. government take significant action to control pollution problems with the Water Pollution Control Act of 1948 and the following Federal Water Pollution Control Act (FWPCA) of 1956, which set funds for research and assisted in state pollution control with construction of wastewater treatment facilities. In 1965, the Water Quality Act provided national policy for control of water pollution. Focusing on drinking water, the Safe Drinking Water Act (SDWA) of 1974 directed the U.S. Environmental Protection Agency (EPA) to establish drinking water standards, which occurred in 1975. In 1980, Congress placed controls on underground injection of waste, requiring permits for the method. Finally, the SDWA amendments of 1986 led to interim and permanent drinking water standards.

It was not until the 1972 amendments were made to the FWPCA that the nation implemented major restrictions on effluents to restore and maintain water bodies in the United States. The Clean Water Act of 1977 added to this focus with consideration of toxins being 65 substances or classes as a basis to reduce and control water pollution. This action led to the initial priority pollutants list, which included benzene, chlorinated compounds, pesticides, metals, etc. In combination, then, the FWPCA and CWA provided the National Pollution Discharge Elimination System (NPDES) permit system in place today.

These regulatory activities, while focused on water media and abatement of problems in rivers and other water bodies, did not directly address the other media in our ecosystem—soil (land) and air. As industry responded to the water regulations, unengineered disposal of waste on land (unengineered pits) became an acceptable and legal method for waste management in many industrial streams, including petroleum wastes, petrochemical wastes and off-spec products, and solid waste disposal (old garbage dumps). These activities led to numerous acts to control and mitigate pollution from dumping, etc. Initial efforts involved control of the transportation of solid food wastes for swine, for control of trichinosis. Modern regulations began with the Solid Waste Disposal Act (SWDA) of 1965 and the National Environmental Policy Act of 1969, which required environmental impact statements. The Resource Recovery Act of 1970 amended the SWDA about the time that the Environmental Protection Agency was formed. True regulation for solid waste management did not come into effect until the Resource Conservation and Recovery Act (RCRA) of 1976, with guidelines for solid waste management and a legal basis for implementation of treatment, storage, and disposal regulations. Also, hazardous wastes and solid wastes were defined by the RCRA. With numerous amendments, the RCRA was followed by the Comprehensive Environmental Response, Compensation and Liability Act (CERCLA) in 1980 to deal with abandoned sites and provide the funds and regulations to perform cleanups. CERCLA, or Superfund, has been through numerous revisions, and its effectiveness has come under question due to the great deal of litigation involving cleanup of old sites.

Air quality needs became apparent in the 1950s due to the Donora, Pennsylvania, accident, and the linkage shown between automobile emissions and photochemical smog, but it was not until the Clean Air Act of 1963, and amendments in the 1960s, 1970s, and 1990s that true national programs were established for pollution control in the air medium. These regulations were focused on motor vehicle emissions, and on emissions from industrial sources. Thus, the United States has "chased" waste management and pollution in all media, and while regulations are now complex, they do provide for control, management, and abatement of pollution from recognized sources to water, land and air.

Two points develop from this brief historical–regulatory review. First, waste is tied directly to population, and population is growing at a rapid rate, so these growth centers must manage and direct waste properly to avoid release and contamination problems. Second, while many countries have significant controls in place as in the United States, many Third World countries and underdeveloped regions are "behind the curve" in regulatory and technical development to manage waste. Many are still dealing with "end-of-pipe" technologies while the United States and others are dealing with remediation, mitigation, and pollution prevention. Still others lack the fundamentals of basic treatment technologies and have significant population growth. Thus our history, in the United States and England, has the potential to continue to repeat itself, unless proper technology is brought to these developing population areas. While the United States and England had time to deal with waste issues, our continued use and development of agricultural land has diminished our resources, and places high stress on those agricultural lands to provide food for the expanding of society. Hopefully, balance will be achieved on a global scale in time to meet the population demand with managed resources and sufficient waste management to protect all media and humankind.

3 TECHNICAL APPROACH

In order to manage waste properly, we must explore the geography of a process so that appropriate engineering (and the constraints of different areas of geography) can be applied to solve a waste management issue or problem. Let us focus now on a chemical manufacturing process, wherein raw materials are taken to manufacture products, such as petroleum to petrochemicals for containers. There are three distinct areas—the process itself, the facility boundary (fence line), and "nature" outside the fence line. Historical sites such as those covered in Superfund regulations also include a boundary and "nature." Nature is defined here as everything except humankind or society. In order to properly apply a sound technical approach to the waste management of such a manufacturing facility, each of these three areas must be considered from an engineering perspective. First, in the process itself, classical chemical engineering is applied, including reactor design, thermodynamics, unit operations, mass transfer, etc., which are well established methods in the chemical process industry (CPI). The focus here is on the process, products, and profit. The second area, the boundary of the facility, is where the bulk of waste management is located, including recycle, reuse, treatment, source control, etc. Lines of these two areas are blurred today with optimization of processes, recycle, and substitution of chemicals to minimize pollution. However, both of these geographic areas are engineered and controlled in terms of materials handling, processing, and safety, as would be found in any chemical process. The third geographic area brings us to nature—the area around the facility or waste site, where the fate and transport of contaminants released from the first two regions now takes control. In the realm of environmental chemodynamics (3), the controlling factors are the transport of chemicals in the environment, governed by the physical-chemical relationship to reaction, transport, etc. Waste management in this region now involves sorption, sediment oxygen demand, groundwater modeling, biodegradation, partition coefficients, and other multimedia processes. The shift in understanding in this region is significant. We no longer have a reactor vessel, a temperature controller, or a homogeneous catalyst bed. The systems are heterogeneous, are difficult to scale, and may not provide consistent or reproducible results when management methods or technologies are applied to a waste problem. In addition to our lack of control over these systems, problems faced are usually dealing with low levels of contamination, which are difficult to model, predict, or treat. However, as risk assessment and exposure assessment methods improve in accuracy and realism, these problems are being tackled with growing frequency. It is important to recognize in the natural environment that our efforts are usually secondary to existing natural forces. An excellent basis to approach management of waste, both in the CPI model and beyond, in nature, is found in the Natural Laws, as illustrated below. Also, a significant contrast develops when we look at the Natural Laws, especially if one compares them to the five elements in the federal approach to management of hazardous wastes, as listed below:

1. Classification of hazardous waste
2. Cradle-to-grave manifest system
3. Federal standards for treatment, storage, and disposal (TSD) facilities
4. Enforcement with permits
5. Authorization of state programs

4 THE NATURAL LAWS

Dealing with waste falls under the Natural Laws (1,4) and it is from these laws that the waste management hierarchy is formed:

1. I am, therefore I pollute.
2. Complete waste recycling is impossible.
3. Proper disposal entails conversion of offensive substances into environmentally compatible earthenlike materials.
4. Small waste leaks are unavoidable and acceptable.
5. Nature sets the standards for what is compatible and for what are small leaks.

Briefly, these laws state the rules we must follow to properly manage waste in the future. Since we exist, we generate waste, and thereby pollute. This is due to the second law, which makes complete recycling impossible, as in thermodynamics, wherein no real process is completely reversible—some loss occurs. With some waste therefore being generated, the third law requires that the material be returned to the environment (nature) in a compatible format—that is, earthenlike—in either a solid, liquid, or gaseous state. When returned, small leaks will occur, as with minor auto emissions, and these are unavoidable and acceptable, provided we observe nature's standards as to what is compatible and how small (or large) the leaks can be. A logical flow of management choices follows from these laws.

5 WASTE MANAGEMENT CHOICES

The following list incorporates all options available and is similar to lists developed by the EPA and others (5). The management list also supports the relationship presented by Reible (2) in that environmental impact is proportional to population times per-capita resource usage divided by environmental efficiency. In words, then, the environmental impact is minimized for a given standard of living when the environmental efficiency is high or improved. Reible's relationship supports the third law, to minimize impact via high environmental efficiency, returning material (and energy) in compatible forms. It is important to note here that much of the waste discussion focuses on material, and that energy pollution should not be neglected, due to problems found in changing river temperatures due to discharge, global warming, etc. To answer the old question, "How clean is clean?," a material is clean when it is returned in a form, amount, and concentration which is acceptable to that found in nature. In other words, a material is "clean" when its concentration does not exceed the natural limits of that material in the space established by the balances (material) that assimilate it (6).

Clearly, then, minimization is the first choice and the optimal one from an environmental standpoint. However, society demands a certain standard of living, so for those wastes remaining from minimization, destruction becomes the best alternative. Why destruction, as such a choice would support technologies such as incineration? Because it is the molecular structure, among other things, that provides the toxicity of the compound, and if it can be broken down (hopefully not yielding a more toxic compound), toxicity can be reduced or eliminated in efficient and correct incineration processes. However, not all wastes causing toxicity problems can be destroyed, such as heavy metals passing through an incinerator. Thus, these materials must be properly treated prior to release, changing their chemical states or bonding for a less toxic or hazardous form. Finally, one notes that in all processes such as those above and others, some residuals always remain, and lead to the final option, disposal. Disposal requires compliance with the Natural Laws—earthenlike materials acceptable to nature's standards for assimilation.

Thus, the hierarchy for waste management is simply:

1. Minimization
2. Destruction
3. Treatment
4. Disposal

While technologies may overlap these steps, all are contained within, which brings us to an important concept: how does natural attenuation fit into the waste management scheme above? Natural attenuation, or monitored natural attenuation

(MNA), is at the front of waste management schemes for remediation of sites, coming into favor in the 1990s as a method to employ risk assessment with source, pathway, and receptor models to decrease active remediation techniques (and associated costs) and increase passive technologies. Clearly, budgets of governments and industry cannot support active remediation technologies in order to return contaminated systems to pristine conditions, and this has been realized through the use of MNA. In reality, MNA is nothing more than our understanding of the fifth Natural Law, and the standards set by nature. What we are observing, understanding, and utilizing in MNA, coupled with active remedies, is simply our quantification of nature's limits as to what it can assimilate. Our regulations tie in here with acceptable drinking water or use standards, along with artificial boundaries placed on problems, such as fence lines and our use needs. In any case, MNA provides treatment or destruction (reduction in toxicity) within the four choices for waste management.

Overall, choices for waste management within the hierarchy of minimization, destruction, treatment, or disposal are best made on a risk-based approach, such as that expressed by Watts (7). For a site, or a waste management program at a facility or other problem, the key elements can be broken down into three categories—sources, pathways, and receptors. In this manner, a risk-based approach may be taken by clearly identifying the sources and receptors, and then testing the pathways for effect, which falls under the realm of chemodynamics, as discussed earlier. We find then that while government and industry are driven by regulation and enforcement of waste management options, as with significant active remediation in the 1980s, the trend is turning strongly now to a risk-based approach, within the Natural Laws, and by understanding the sources, pathways, and receptors, and the fate and transport of low-level contaminants in the biota.

REFERENCES

1. W. D. Constant and L. J. Thibodeaux, Integrated Waste Management via the Natural Laws. *The Environmentalist,* vol. 13, no. 4, pp. 245–253, 1993.
2. D. D. Reible, *Fundamentals of Environmental Engineering,* pp. 10–12. Boca Raton, FL: Lewis Publishers, 1999.
3. L. J. Thibodeaux, *Chemodynamics: Environmental Movement of Chemicals in Air, Water and Soil,* pp. 1–5. New York: Wiley, 1979.
4. L. J. Thibodeaux, Hazardous Material Management in the Future. *Environ. Sci. Technol.,* vol. 24, pp. 456–459, 1990.
5. C. A. Wentz, *Hazardous Waste Management.* New York: McGraw-Hill, 1989.
6. W. D. Constant, L. J. Thibodeaux, and A. R. Machen, Environmental Chemical Engineering: Part I—Fluxion; Part II—Pathways. *Trends Chem. Eng.,* vol. 2, pp. 525–542, 1994.
7. R. J. Watts. *Hazardous Wastes: Sources, Pathways, Receptors,* pp. 38–40. New York: Wiley, 1998.

4

Legislative and Regulatory Issues

Toni K. Ristau

Public Service Company of New Mexico, Albuquerque,
New Mexico

1 OVERVIEW

In many respects, pollution prevention and waste minimization are less creatures of legislative fiat than are many other areas related to waste management. In part, this is due to the way that the legislative and regulatory waste management framework developed in the United States (1). In the United States, the major regulatory strategy for addressing wastes, particularly hazardous or toxic wastes, is a "command-and-control" system imposed upon the regulated community from the top down. By contrast, many pollution prevention initiatives are voluntary efforts initiated by companies that seek to improve the "bottom line," rather than requirements imposed by a regulatory agency.

To understand the current emphasis on pollution prevention, one must have an understanding of the history of the regulation of hazardous and toxic wastes. Now that environmental management is maturing as a discipline, there is an increasing recognition that pollution prevention and appropriate waste management (including minimizing waste streams wherever possible) during a facility's operational life can greatly reduce the potential for costly remediation and cleanup after operations are discontinued.

2   HISTORY

Much of the early legislative effort related to waste disposal or releases of toxic substances was engendered by incidents such as Love Canal in New York State, or the release of toxic gas from a factory in Bhopal, India. These incidents, which were widely reported by the media, outraged the public and caused a demand for Congressional action (2).

2.1   Love Canal and the Enactment of the Comprehensive Environmental Response, Compensation and Liability Act ("Superfund")

The Love Canal hazardous waste disposal site became the center of attention of the media, as well as the regulatory agencies, in the late 1970s and early 1980s, and inspired the passage of the Comprehensive Environmental Response, Compensation and Liability Act (CERCLA), also known as "Superfund."

The Love Canal site was not originally constructed to be a waste disposal facility. Originally, the Love Canal was to be the centerpiece of a "model city," and the use of the canal for waste disposal was not contemplated. The canal was originally constructed by William T. Love at the easternmost edge of the town of Niagara Falls, New York, in 1893. The canal was to be used to supply water to generate a cheap and essentially unlimited supply of hydroelectric power for this model community. The discovery and adoption of the use of alternating current in the mid-1890s, which allowed electricity to be generated at some distance from the point where the electricity was to be used, rendered Love's plans for the canal uneconomic, and Love's dream for the canal was never realized. The abandoned canal filled with rainwater and was used as a swimming hole and for winter ice skating by the local community. In the 1940s, Hooker Chemical Company obtained rights to the canal, and began to use the old canal as a dump for chemical wastes from Hooker's chemical manufacturing operations. Hooker Chemical drained the old canal, lined it with clay, and used the old canal as a waste dump. Between 1942 and 1953, an estimated 22,000 tons of chemical wastes, as well as municipal wastes, were dumped at the site. When Hooker Chemical discontinued active use of the site, the company capped the old canal with a thick layer of clay, and covered the entire site with sod (3).

Hooker Chemical sold the dump site to the local board of education in 1953 for a nominal sum (one dollar), on the condition that the company would not be liable for any problems related to the wastes that were disposed at the site. Though the board of education was aware that the site had been used for the disposal of hazardous and toxic wastes, a school was constructed at the site, as were numerous houses. Though Hooker Chemical tried to stop the development on the contaminated land, local governmental authorities ignored the warnings, and allowed the construction at and adjacent to the old disposal site.

High groundwater levels in the Love Canal area, resulting from unusually heavy rains and snowfalls during the 1970s, caused an increasingly serious situation at the old disposal site. Drums and other containers began to surface as the area over the old disposal facilities subsided, and ponded areas and other surface waters near the site exhibited high levels of contamination. Residents of some of the nearby houses noted that the basements were oozing an oily residue, and there were numerous complaints of noxious chemical odors. An engineering firm was hired to perform a study of the problems noted by the residents in the Love Canal area, and to formulate recommendations on how to address the problems. The engineering company recommended that the canal be covered with clay, that sump pumps used by nearby residents to prevent flooding of basement areas be sealed off, and that a tile drainage system be installed to control the migration of wastes. As these measures would be costly, the city elected not to implement the engineering recommendations. However, in some homes where the levels of chemical residues and problems related to noxious odors were found to be very high, the city had window fans installed.

Despite these minimal efforts by the city to address residents' complaints, evidence was mounting that the contamination from the old Love Canal disposal site was causing more than just inconvenience for the residents of the area. In March 1978, the New York State Department of Health initiated the collection of air and soil samples from the homes and other facilities located at and near the site. The department also conducted a health study of the 239 families who lived nearest to the old canal. Alarmed at the preliminary results from the study, in August 1978 the department issued a health order calling for the evacuation of pregnant women and children under the age of two, recommending that residents minimize the amount of time spent in the basements of their homes, and also recommending that residents not eat vegetables and fruits grown in their home gardens. Residents found themselves in the difficult situation of being unable to continue occupying their homes, but, because of the increasing publicity regarding the contamination, they were also unable to sell or rent their homes. Shortly after the issuance of the health order, the State of New York agreed to purchase the 239 homes closest to the old canal.

In 1979, subsequent to the evacuation of the 239 families living closest to the old disposal areas, the Love Canal Homeowners Association commissioned another study (the 239 families who had already been evacuated were not included in this study). This study indicated that there were increases in miscarriages, still births, crib deaths, birth defects, hyperactivity, nervous breakdowns, epilepsy, and urinary tract disorders in families living in the area. When the Homeowners Association presented its study findings to state health authorities, the significance of the findings were downplayed due to potential flaws in the study methodology. However, the resultant public outcry ultimately caused action to be taken at the national level. In October 1980, President Jimmy Carter ordered a total evacuation of the community. The Love Canal residents had the option of selling their houses to the government at fair market value, and moving to a new location.

The public outrage related to the situation in which the hapless residents of the Love Canal neighborhood found themselves resounded through the halls of Congress, and Congress responded in 1980 by passing the Comprehensive Environmental Response, Compensation and Liability Act (42 U.S.C. 9601 et seq.), also known as "Superfund." Superfund, or CERCLA, provides a mechanism for investigating threatened or actual releases of hazardous substances into the environment, identifying potentially responsible parties, and funding the requisite technical and engineering studies to address the problems caused by the hazardous substance release. In addition, the Superfund provides a means of funding cleanup activities through the imposition of a tax upon petrochemical industries.

And what was the final outcome for the Love Canal neighborhood? In 1982, the U.S. Environmental Protection Agency (EPA) completed studies of the contamination residing in the soils, water, and air near Love Canal, and initiated appropriate remedial action. The 239 homes that were located nearest the old canal were demolished. The remaining homes that were purchased by the government, and the neighborhood school, were renamed Black Creek Village, and the sale of the decontaminated homes to new families commenced in 1990. There are still approximately 22,000 tons of waste buried in the center of the community; periodic testing of the air, water, and soils in the community assure the safety of the new residents. Today, the Love Canal neighborhood has been "recycled," and Black Creek Village is again a vital, living neighborhood.

2.2 The Incident at Bhopal and the Emergency Planning and Community Right-to-Know Act

In December 1984, about four years after the enactment of CERCLA, an incident involving the release of the volatile and highly toxic gas, methyl isocyanate, occurred in Bhopal, India. Methyl isocyanate gas was produced and used at the Union Carbide plant there as an intermediate product in the manufacture of pesticides. The pesticides manufactured at the facility were important in several ways to the local and national economy; they are and were used to aid nations such as India to increase crop yields and improve conditions for their populace by providing a means to control insect pests (4).

When the Bhopal incident occurred in 1984, the gas, which burst from a tank at the Union Carbide plant, spread over a large, densely populated area near the plant. Many people died in their beds, and others died trying to escape the foglike cloud of poison gas. There were thousands of dead and injured in the poor and crowded neighborhoods near the plant; though local officials were unable to immediately determine the total number of deaths, the official death toll was ultimately computed to be almost 10,000.

Though this incident occurred in India, it garnered much attention within the United States. An analysis of the factors that caused the inordinate number of deaths and injuries indicated that much, if not all, of the suffering of the local population could have been prevented had there been appropriate emergency procedures and evacuation plans in place. Further, it appeared that many of the local populace may have been able to protect themselves had they but known what kinds of substances were being produced at the plant, and what measures they themselves might be able to take in case of a leak or a release at the plant.

As was the case when the Love Canal situation was brought to light by the media, there was a huge public outcry, and demands for legislation to assure that an incident like the tragedy at Bhopal would never happen in the United States. Congress was considering amendments to CERCLA at this time, and Congress responded by adding a new Title III to CERCLA as a part of the Superfund Amendments and Reauthorization Act (SARA) (42 U.S.C. 11001 et seq.) This new Title III was also separately titled as the "Emergency Planning and Community Right-to-Know Act," or EPCRA.

EPCRA, unlike the other portions of CERCLA, which are largely oriented toward cleanup of abandoned hazardous or toxic waste sites, focuses on community preparedness and reporting by industrial facilities to assure that national, state local response authorities, as well as local communities, are aware of the substances that are being utilized at industrial facilities within the community and are prepared to respond if there is a release, spill, or leak from such facilities.

2.3 Solid and Hazardous Waste Management and the Resource Conservation and Recovery Act

In contrast to legislation enacted as a reaction to environmental crises or catastrophes, the regulation of facilities and activities related to solid wastes (and of hazardous wastes as a subset of solid wastes), is conducted under the provisions of the Resource Conservation and Recovery Act (RCRA) (42 U.S.C. 6901 et seq.). The RCRA is designed in substantive part to regulate the day-to-day operation of solid and hazardous waste management facilities and activities through a permitting and standards system. The RCRA also contains provisions related to response from releases from active or inactive waste management units.

Though the RCRA contains provisions related to both solid waste and hazardous wastes, much of the regulatory attention in recent years has been on the hazardous waste component of solid waste streams. A particularly important set of provisions in the RCRA gave the EPA the authority to control hazardous waste from "cradle to grave." The EPA thus has regulatory authority and control over the generation, transportation, treatment, storage, and disposal of hazardous waste.

In 1984 and 1986, Congress passed major amendments to the RCRA. The 1984 amendments were known as the Hazardous and Solid Waste Amendments (HSWA). The HSWA required phasing out land disposal of untreated hazardous wastes. The HSWA also added increased enforcement authority for the EPA, provided for more stringent hazardous waste management standards, and provided for a comprehensive underground storage tank program. The HSWA also provided for corrective action for releases from solid and hazardous waste management units (both active and inactive) at operational solid and hazardous waste management facilities.

2.4 The Pollution Prevention Act of 1990 and a New Way of Managing Hazardous/Toxic Waste Streams

In response to many commentators, who noted that the existing RCRA and CERCLA regulatory frameworks in many cases provided disincentives to recycling and other waste minimization activities, Congress passed the Pollution Prevention Act (42 U.S.C. 13101 et seq.) in 1990. Opportunities for source reduction as a method of minimizing pollution are often not realized because the industries responsible for compliance with RCRA necessarily focus on treatment and disposal of the hazardous wastes generated by their processes, rather than on reducing the overall use of hazardous or toxic chemicals in their processes.

Unlike the RCRA and CERCLA, which provide at best indirect liability-driven disincentives to the use and production of toxic and hazardous substances and wastes, the Pollution Prevention Act attempts to focus public, governmental, and industry attention on reducing the amount of pollution produced, by encouraging cost-effective changes in production, operation, and raw materials use (known as "source reduction"). Source reduction requires solid and hazardous waste generators to concentrate on fundamental process changes to prevent waste (particularly hazardous waste) from being generated in the first place, rather than regarding hazardous waste streams as a necessary concomitant to industrial production and focusing on the treatment and disposal of that waste. Pollution prevention, as opposed to hazardous waste treatment and disposal, emphasizes the use of production practices that increase efficiency in the use of energy, water, or other natural resources. Pollution prevention practices include recycling and internal reuse of waste streams, source reduction through the minimization or elimination of hazardous or toxic substances as industrial inputs, and revision of industrial processes to minimize thermal and other energy losses.

3 APPROACHING POLLUTION PREVENTION AS A "SYSTEMS MANAGEMENT" PROBLEM

There has been growing recognition that treating pollution prevention and energy efficiency as fundamental process inputs, rather than as "add-on" or "end-of-pipe" systems, is a much more effective way of minimizing the environmental impact of industrial operations. This approach requires a refocusing of environmental management efforts from a reactive, compliance-based mode to a proactive, preventative approach. This model, which recognizes and works within productive industrial processes, rather than working against fundamental industrial process and adding operational complexity as well as costs, is beginning to find acceptance not only within industry, but within the EPA and equivalent state and local regulatory agencies in the United States.

Alternative approaches (such as the systems management approach) have been increasingly embraced by the international economic community as a more rational method of assuring that pollution effects are minimized but that needed industrial growth and development is not hindered. These alternative approaches are increasingly seen as a way of assuring that industry is not only economically efficient, but is "environmentally efficient" as well. The new emphasis on resource conservation and waste minimization accomplishes both goals—it minimizes the use of raw materials and energy required to maximize production (and thus lowers production costs), and it minimizes the environmental impacts from the extraction and production of energy and other raw materials as well as minimizing the impacts from waste disposal (and thus lowers the overall environmental impact of industrial development and growth).

3.1 The Environmental Management System Approach

The development and use of an environmental management system within a company, a facility, or an activity is one method of treating environmental management as a systems management/systems optimization approach. One such method that is gaining increasing acceptance in the United States as well as internationally is the environmental management system development and certification process embraced by the International Standards Organization (ISO). The environmental management system standards are codified in the ISO 14000 standards (5).

The standards are developed by internationally based technical committees. Each nation is free to adapt the standards as appropriate to fit each country's unique political and resource considerations. Within the United States, the standards, once adopted, are modified and implemented through U.S.-based organizations such as

American National Standards Institute (ANSI)

Underwriters Laboratories (UL)

American Standard Testing Methods (ASTM)

These U.S.-based standard-setting organizations conform the international standards with U.S. regulatory requirements and assure that the overall objectives of the standards can be met within the constraints of the U.S. political and regulatory system.

The environmental management systems approach is beginning to make headway in the United States. However, the "drivers" for adoption of ISO performance standards and/or certification are not as well developed as in other countries, where ISO certification may be a prerequisite for doing business in that country.

The basic principles of environmental management systems are not unique to any one type of business or industrial activity, but have applications in all activities where the environment may be affected. The definition of an "environmental management system" (EMS) is: " . . . that part of the overall management system which includes organizational structure, planning activities, responsibilities, practices, procedures, processes and resources for developing, implementing, achieving, reviewing, and maintaining the environmental policy." The basic principles of environmental management systems include:

Integration of environmental issues with other business issues

Looking at the environmental conundrum as an interactive system, rather than as "add-ons" of discrete activities

The ISO 14000 Environmental Management Standards include several standards that can be applied in the management of any company's environmental aspects of "doing business." The ISO 14000 substantive standards for environmental management include the following:

ISO 14001, Environmental Management Systems: Specification with Guidance for Use

ISO 14004, Environmental Management Systems: General Guidelines on Principles, Systems, and Supporting Techniques

ISO 14010, Guidelines for Environmental Auditing: General Principles

ISO 14011, Guidelines for Environmental Auditing: Audit Procedures for Auditing Environmental Management Systems

ISO 14012, Guidelines for Environmental Auditing: Qualification Criteria for Environmental Auditors

ISO 14024, Criteria for Certification Programs: Criteria for Self-Certification and Third-Party Certification

Each company or facility is free to develop its own environmental management system. The company selects and develops its own environmental performance objectives. In the United States, regulatory agency participation is not required, but is encouraged. If the regulatory agency chooses to participate, the agency will provide advice to the company in the setting of overall environmental performance objectives. Ideally, the company then selects how to meet those objectives, rather than being subjected to prescriptive control technology requirements (however, in the United States, this interactive systematic approach is not currently allowed for within the existing regulatory framework).

Though the company is free to identify the components of its own environmental management system, certain components must always be present if the company wishes to seek certification of its environmental management system under ISO 14000. These required environmental management systems components are as follows.

Environmental policy: Senior management must define the corporation's environmental policy and ensure that the policy includes, among other matters, a commitment to continual improvement, to pollution prevention, and to compliance with relevant regulatory requirements.

Planning: The company must establish and maintain a procedure to identify environmental impacts of its activities, as well as the legal and other requirements. The corporation must establish and maintain documented environmental targets and objectives, as well as environmental management programs for achieving its objectives.

Implementation: Roles, responsibilities, and authorities must be defined, documented, and communicated. The corporation must provide appropriate training and must establish and maintain procedures for proper communication. The company must have proper documentation of procedures, document control, operational control, and emergency preparedness and response (contingency planning).

Corrective action: The company must establish and maintain documented procedures to monitor and measure operations and activities that impact on the environment, and must have documented procedures for investigating nonconformances and implementing appropriate corrective action. Procedures must be in place for identifying, maintaining, and disposing of environmental records, and for periodic audit of the EMS.

Management review: Senior corporate management must review the EMS on a periodic basis (and document its review) to ensure that the EMS is suitable, adequate, and effective in meeting the company's environmental performance goals.

3.2 Tools Available for Employing Alternative Approaches

The EPA has, in recent years, become more interested in encouraging the voluntary development of systems approaches to environmental management, including pollution prevention versus "end-of-pipe" controls. The EPA now provides Web-based information and tools for companies or entities that wish to pursue pollution prevention or environmental management systems initiatives (6).

The EPA's Office of Enforcement and Compliance Assistance has produced "Sector Notebooks" for the following industry sectors:

Agricultural Chemical, Pesticide and Fertilizer Industry (1999)
Agricultural Crop Production Industry (1999)
Agricultural Livestock Production Industry (1999)
Aerospace Industry (1998)
Air Transportation Industry (1997)
Dry Cleaning Industry (1995)
Electronics and Computer Industry (1995)
Fossil Fuel Electric Power Generation Industry (1997)
Ground Transportation Industry (1997)
Inorganic Chemical Industry (1995)
Iron and Steel Industry (1995)
Lumber and Wood Products Industry (1995)
Metal Casting Industry (1997)
Metal Fabrication Industry (1995)
Metal Mining Industry (1995)
Motor Vehicle Assembly Industry (1995)
Nonferrous Metals Industry (1995)
Non-Fuel, Non-Metal Mining Industry (1995)
Oil and Gas Extraction Industry (1999)
Organic Chemical Industry (1995)
Petroleum Refining Industry (1995)
Pharmaceutical Industry (1997)
Plastic Resins and Man-made Fibers Industry (1997)
Printing Industry (1995)
Pulp and Paper Industry (1995)
Rubber and Plastic Industry (1995)
Shipbuilding and Repair Industry (1997)
Stone, Clay, Glass and Concrete Industry (1995)
Textiles Industry (1997)
Transportation Equipment Cleaning Industry (1995)
Water Transportation Industry (1997)
Wood Furniture and Fixtures Industry (1995)

The notebook contents, which can be viewed online and can also be downloaded and printed out, include the following types of information for each industry sector (7):

A comprehensive environmental profile
Industrial process information
Pollution prevention techniques
Pollutant release data
Regulatory requirements
Compliance/enforcement history
Government and industry partnerships
Innovative programs
Contact names
Bibliographic references
Description of research methodology

These sector notebooks focus on key indicators, including air, water, and land pollutant releases typical of the industry sector, and are a good source of information for managers who wish to implement pollution prevention or environmental management programs at their facilities.

3.3 "Project XL" and EPA Regulatory Reinvention Efforts

The EPA has long recognized that many of the media-specific environmental regulations (i.e., regulations applicable to the pollution of air, water, or land) can be counterproductive in that imposing more stringent standards or increasing regulatory scrutiny for one medium (such as air) may minimize the amount of pollution within that medium, but can actually cause increased releases to other media. Several years ago, the EPA initiated a regulatory reform effort that allowed regulated companies to propose multimedia projects meeting certain criteria to allow trade-offs between media and to enable better environmental performance overall within a facility. This regulatory reform effort, denominated "Project XL," allowed companies and other entities to propose projects that substitute performance based standards for the prescriptive, one-size-fits-all, standards that are often imposed through the EPA's media-specific regulatory programs.

The EPA required that pilot projects proposed by facilities, sectors, and government agency projects meet the following criteria (8).

Environmental results. Projects that are proposed should he able to achieve environmental performance that is superior to what would be achieved through compliance with current and reasonably anticipated future regulation. "Cleaner results" can be achieved directly through the environmental performance of the project or through the reinvestment of the cost savings from the project in activities that produce greater environmental results. Explicit definitions and measures of "cleaner results" should be included in the project agreement negotiated among stakeholders.
Cost savings and paperwork reduction. Projects that are proposed should

produce cost savings or economic opportunity, and/or result in a decrease in paperwork burden.

Stakeholder support. The extent to which project proponents have sought and achieved the support of parties that have a stake in the environmental impacts of the project is an important factor in the selection of projects. "Stakeholders" may include communities near the project, local or state governments, businesses, environmental and other public interest groups, or other similar entities.

Innovation/Multimedia pollution prevention. The EPA is looking for projects that test innovative strategies for achieving environmental results. These strategies may include processes, technologies, or management practices. Projects should embody a systematic approach to environmental protection that tests alternatives to several regulatory requirements and/or affects more than one environmental medium. The EPA has a preference for protecting the environment by preventing the generation of pollution rather than by controlling pollution once it has been created.

Transferability. The proposed pilot projects are intended to test new approaches that could be incorporated into other agency programs or in other industries, or other facilities in the same industry. The EPA is therefore most interested in pilot projects that test new approaches that could one day be applied more broadly.

Feasibility. The proposed project should be technically and administratively feasible and the project proponents must have the financial capability to carry it out.

Monitoring, reporting, and evaluation. The project proponents should identify how to make information about the project, including performance data, available to stakeholders in a form that is easily understandable. Projects should have clear objectives and requirements that will be measurable in order to allow the EPA and the public to evaluate the success of the project and enforce its terms. Also, the project sponsor should about the time frame within which results will be achievable.

Shifting of risk burden. In addition to the above criteria, the proposed project must be consistent with Executive Order 12898 on Environmental Justice. It must protect worker safety and ensure that no one is subjected to unjust or disproportionate environmental impacts.

Several Project XL pilot projects were initiated by government agencies, as well as private industry. Unfortunately, the Project XL program was subject to heavy criticism from environmental groups as well as from government oversight entities, and new projects are currently not being fielded.

REFERENCES

1. A good source of statutory material related to regulation of the environment can be found on the Web at http://www4.law.cornell.edu/uscode/42/. The regulations associated with the various laws and statutes are found at http://www4.law.cornell.edu/cfr/. In addition, the U.S. Environmental Protection Agency (EPA) Website contains information on environmental laws and regulations that fall under EPA jurisdiction for compliance and enforcement. The EPA home page, which contains an index and links to other sources of information, is found at http://www.epa.gov/epahome/rules.html.
2. For a summary of major pieces of legislation related to hazardous and toxic substances and wastes, see http://www.ehsgateway.com/legislation.html.
3. For a more detailed discussion of the history of the Love Canal, see http://www. essential.org/orgs/CCHW/lovcanal/lcsum.html and http://enviro.gannon.edu/ EnvRegs/eli/Love%20Canal%20Project.html.
4. For additional information regarding the incident at Bhopal, see http:// www.corpwatch.org/trac/bhopal/factsheet.html and http://www.prakash.org/issues/ environment/disasters/industrial/bhopal/envreu19941204_00.html.
5. The International Standards Organization (ISO) was established in Geneva, Switzerland, in 1947. The purpose of the ISO was (and is) to develop uniform, worldwide standards for the conduct of manufacturing and other industry, business, technical, and commercial activities. At present, 92 countries belong to the ISO (representing more than 92% of the world's industrial production). At the international level, more than 200 technical committees have been developed. The ISO has, to date, established over 8500 standards.
6. The following EPA Website provides common-sense information, links, and tools for pollution prevention and other environmental management systems approaches: http://www.epa.gov/envirosense/.
7. To view the sector notebooks online, see http://es.epa.gov/oeca/sector/index.html. For individual sector information, each listed sector is a "hot link" to more detailed information for that sector. Also, an excellent set of Web-based resources related to pollution prevention, waste minimization, and environmental management issues is maintained by the U.S. Navy. These resources are available to anyone, and are found at http://enviro.nfesc.navy.mil/p2library/.
8. See, for example, the Website for "Campus Pollution Prevention Information Resources," at http://esf.uvm.edu/labxl/Background%20Information/campusp2links.html.

5

Information Systems for Proactive Environmental Management

Steven P. Frysinger
James Madison University, Harrisonburg, Virginia

1 INTRODUCTION

Environmental computing is a very broad topic, and to some extent defies taxonomy. However, in the interest of providing an overview of this field, environmental information systems can be described generally as either *environmental management information systems (EMIS)* or *environmental decision support systems (EDSS)*. The former will be defined as systems which provide access to information, such as records and reports, while the latter include systems which provide access to tools with which to operate on information in order to arrive at an environmental management decision. There is clearly a great deal of overlap between these two definitions, and many systems might straddle the definition uncomfortably. Nonetheless, it will be useful to address this broad topic in the context of this dichotomy.

2 ENVIRONMENTAL MANAGEMENT INFORMATION SYSTEMS (EMIS)

As corporate environment, health, and safety (EH&S) organizations endeavor to become world-class operations, they must deal with a remarkable variety of processes which have evolved over many years, largely in response to regulatory events. Integration of these various processes into a cohesive and efficient management system is a necessary part of their quest to make more forward-looking, comprehensive decisions impacting EH&S stewardship. An integrated management system has the potential to significantly improve the effectiveness of the EH&S organization by facilitating awareness of hazards or opportunities across regulatory boundaries. Likewise, integration can reduce the cost of EH&S operations by encouraging the sharing of knowledge and effort across these same regulatory lines. For example, a single activity which maintains inventories of chemicals stored at company facilities might serve the needs of both Occupational Safety and Health Administration (OSHA) Hazard Communication and Super-fund Amendments and Reauthorization Act (SARA) Title III compliance.* Thus, an integrated environmental management system (used here to describe all aspects of environment, health and safety management) is a prerequisite to a world-class EH&S organization.

Integrated environmental management requires integrated environmental *information* management. While there is more to environmental management than information, there is virtually no aspect which does not depend heavily on the availability and accessibility of correct and current information. Many, if not most, EH&S processes are themselves essentially focused on information man-agement, in the form of record keeping, reporting, permitting, or training. Therefore, it is difficult to conceive of an integrated environmental management system which does not stand on an integrated environmental information man-agement system.

While integration of diverse software tools has been going on for a long time in environmental decision support systems (to be discussed later in this chapter), the integration called for by EH&S management refers to data rather than software integration. So-called enterprise resource planning (ERP) systems have the goal of spanning all aspects of a company's operations and integrating data management across organizational boundaries, but even the most widely implemented ERP systems have only recently begun to include EH&S informa-tion management. These systems typically require such an intrusive implementa-tion process that many firms decline to pursue them, preferring instead to develop their own versions based on existing business models. Unfortunately, EH&S data

---

*Regulatory references throughout this chapter are based on the laws of the United States.

management is rarely considered in this process because the EH&S organization, as a cost center, is not perceived to add value to the firm, and therefore rarely attracts such an investment. The EH&S organization is then left to manage its data on its own, even though much of the information on which it depends is in fact owned by line organizations within the company.

2.1 The Need for Integration

The many processes of the typical EH&S organization are usually supported by as many diverse environmental management information systems, many of them manual (i.e., with little or no computer support). These information systems have evolved in response to individual needs, generally without regard to inter-dependencies between processes and their information management needs.

Apart from the obvious inefficiencies which result from such circumstances, this ad-hoc structure has resulted in redundant and inconsistent databases—multiple databases store the same piece of information, and they sometimes disagree on its value. For example, several EH&S information systems may use facilities data from different databases which conflict with one another. This sort of inconsistency ultimately threatens compliance.

2.2 An Integrated Solution

There is an approach which improves the situation by developing the framework for an *integrated environmental information system (IEIS)*, an important special case of EMIS. It is important to note that the term "information system," as operationally defined here, is much broader than the computer hardware and software which might support it. It includes a data model incorporating the structure, definition, and relationships between data elements, as well as the processes and procedures by which these data are created, modified, used, and destroyed. While much of this can and should be supported by computer systems, this fact has little relevance to the conceptual definition of the information system. Once the IEIS is defined, a systems engineering activity can readily determine the design and structure of the hardware and software systems which will support it, about which more will be said later.

2.3 Conceptual Framework

The IEIS approach is predicated on the notion that one can usefully separate data from the management processes that use them. That is, most or all data of use to EH&S are descriptive of *objects*, while the various management processes undertaken by EH&S professionals are focused on these objects. An object-oriented approach to EH&S information might start with the definition of such high-level objects as employees, customers, buildings, vehicles, services, and products. Each of these can then be decomposed in a similar fashion, as appropriate, with the terminal objects described by a data structure.

The various EH&S management processes can generally be viewed as *operating on* the data objects suggested above. For instance, SARA Title III Section 312 reporting is focused (by regulation) on buildings, while OHSA training requirements are focused on employees. Furthermore, each process may be supported by one or more software applications. In general, the software applications serving EH&S processes are the agents which interact with the data required for these processes (Figure 1).

Thus, there is envisioned a clear separation between *data, processes,* and *applications*:

1. A *datum* may be used by multiple *process*es; e.g., Building Address is used for SARA Title III reporting and for OSHA accident reporting.
2. A *process* may be served by multiple *applications*; e.g., one software application might support the SARA inventory maintenance activity by site personnel, while another application is used to generate the SARA reports.
3. In some instances, *applications* may be used by multiple *process*es; e.g. the software used by site personnel to maintain chemical inventories may serve the purposes of both SARA and OSHA compliance processes.

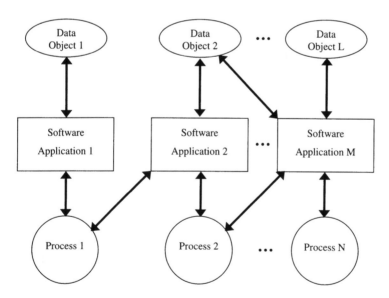

FIGURE 1 An exemplary relationship between data, processes, and software applications.

In essence, this approach addresses our need to understand this relationship between our information and our processes so that we may ensure the availability of the correct data and the correct software applications to interact with those data.

2.4 The Path to Integration

There are four essential steps to achieving an integrated environmental information system:

1. Develop an integrated data model.
2. Map the integrated data model onto corporate databases of record.
3. Define high-level requirements for the IEIS.
4. Implement the foundation of the IEIS.

While some of these can be executed concurrently, it is imperative that we recognize the precedence implicit in their ordering. As with any systems engineering activity, in this activity the *what* has to lead the *how*, rather than the other way around. It will be advantageous to look ahead to current and future system implementations to help us to achieve an understanding of requirements, but particular discipline must be applied to prevent us from erroneously finding a requirement in what is merely a habit. This discipline will be encouraged by a phased approach, in which we first define an IEIS for the set of processes as they currently exist, admitting that the model will be revisited as a result (and indeed in support of) efforts to reengineer those processes.

2.5 Model Development

The first step in the project is the development of an integrated data model which correctly describes the firm from an EH&S point of view. The initial (baseline) data model must include all data items required by the current set of EH&S processes, but must be orthogonal to these processes so that data objects and fields which are common to multiple processes occur only once in the data model, to be shared by the processes requiring them. This is critical to the identification of shared information and the elimination of redundant databases. Once such a baseline data model has been developed, it can and should be refined and revised as appropriate to reflect the ongoing reengineering of the EH&S organization's structure and processes.

2.6 Mapping the Model onto Databases

The integrated data model so developed will then be analyzed to determine the appropriate owner for each of the data categories and elements. In many cases, this will be the so-called database of record for the company, and will not be under the control of the EH&S organization. For example, much information about corporate facilities might be maintained by a real estate organization within the firm but outside of EH&S. Identifying our stake in such external databases is essential since, as customers of these databases, we will need to be recognized and have a voice in the implementation and management of the data. There may also be data items of importance to EH&S which should and could readily be maintained in these external databases; we will want to be in a position to lobby the appropriate organizations for such extensions. Furthermore, interfaces to these data sources must be engineered so that the data truly *will* be shared, rather than simply copied into yet another system, further contributing to data redundancy.

2.7   Defining IEIS Requirements

The third step is the definition of high-level requirements for the integrated information system. The integrated data model and analysis described above form the foundation for this. What must be added are the *functional requirements* for the integrated system. For example, if EH&S information must be globally accessible by EH&S leadership, this requirement should be articulated clearly.

2.8   Implementing the IEIS Foundation

The fourth step addresses the implementation of the IEIS. Implementation includes the interaction and negotiation with other organizations whose information assets have been identified as a subset of the EH&S data model in step 2. It also includes the planning and acquisition and/or development of software required to realize the IEIS from the starting position of our existing information management systems. The result of this step is *not* necessarily a single software system; in fact, this outcome is highly unlikely, given that the software to be used by individuals and groups engaged in the various processes will have to satisfy functional requirements which may be peculiar to those processes. As long as the ensemble of computer systems finally in use by the EH&S organization (a) implements the integrated data model developed in steps 1 and 2, and (b) satisfies the high-level requirements defined in step 3, then we will have achieved an integrated environmental information system and will reap the benefits thereof. This, perhaps, is the point of departure of this approach from conventional thinking about integration—we seek to achieve the benefits of integrated *information* while valuing diversity of software applications and vendors.

Once these four steps have been executed, the design and implementation of the integrated system using an appropriate combination of existing and new platforms can proceed through conventional information project management and systems engineering activities. In fact, it might be hoped that through effective communication, any ongoing procurement and development activities underway during the execution of these steps can be appropriately guided so as to minimize changes or disruption once they are complete. For example, an early intermediate result will be the identification of data common to the first key processes to be evaluated. This knowledge can surely be used during the procurement of supporting systems to anticipate the results of the integration effort.

2.9 EMIS Summary

This approach to integrating environmental management information systems into an integrated environmental information systems serves to illustrate the issues attending these systems in general. Whether this approach or some other is used, however, the critical element for proactive environmental management is that integration be achieved in the interests of eliminating compliance-threatening redundancy and removing substantial inefficiencies.

3 ENVIRONMENTAL DECISION SUPPORT SYSTEMS (EDSS)

As the complexity of our environmental management problems has increased, so has the need to apply the information management potential of computing technology to help environmental decision makers with the difficult choices facing them. Environmental information systems have already taken many forms, with most based on a relational database foundation (as described in the previous section). Such systems have helped greatly with the day-to-day operations of environmental management, such as chemical and hazardous waste tracking and reporting, but they have two critical shortcomings which have prevented them from significantly improving the lot of environmental scientists and planners tackling more strategic decisions.

Traditional environmental management information systems generally ignore the crucial spatial context of virtually all environmental management problems, and they offer little or no support for the dynamics of environmental systems, both manufacturing and otherwise. Fortunately, a relatively new category of system, called an environmental decision support system (EDSS), shows real promise in both of these areas.

3.1 What are Environmental Decision Support Systems?

Environmental decision support systems are computer systems which help humans make environmental management decisions. They facilitate "natural intelligence" by making information available to the human in a form which maximizes the effectiveness of their cognitive decision processes, and they can take a number of forms (1).

As defined here, EDSSs are focused on specific problems and decision makers. This sharp contrast with the general-purpose character of such software systems as geographic information systems (GIS) is essential if we are to put and keep EDSSs in the hands of real decision makers who have neither the time nor inclination to master the operational complexities of general-purpose systems. Indeed, it can be argued that most environmental specialists are in need of computer support which provides everything that they need, but *only* what they need. This point becomes more critical when it is understood that many important "environmental" decisions in design and manufacturing, for example, are not made by environmental specialists at all, but are instead made by professionals in other disciplines.

3.2 The Need for Environmental Decision Support Systems

The development of environmental policies and generation of environmental management decisions is currently, to a large extent, an "over-the-counter" operation. Technical specialists are consulted by decision makers (who may or may not have a technical background), to assist in gathering information and exploring scenarios. Because of the inaccessibility of data and modeling tools, decision makers must consult their technical support personnel with each new question, a time-consuming and inefficient process.

If the data and analytical tools could be placed within reach of decision makers, they would be able to consult them more readily, and would therefore be more likely to base their decisions on a technical foundation. In some instances, the availability of environmental decision support determines whether or not a product design or manufacturing process will indeed be "environmentally conscious." This is the premier reason why environmental decision support systems, of a sort described in part herein, are necessary if we are to achieve higher quality in our environmental management decisions and obtain more protection with our finite resources.

3.3 Foundations

Environmental decision support systems address a problem domain of remarkable breadth, ranging from selection of an appropriate light switch for an automobile to the determination of community risk associated with stored chemicals. The character of environmental decisions and their surrounding issues is central to the design of a successful EDSS.

3.4 The Nature of Environmental Management Decisions

To understand environmental management decisions, we must first identify the decision makers. The stereotypical image of an environmental manager is an environmentally trained business manager given the responsibility for avoiding fines and other sanctions, and perhaps pursuing "beyond compliance" goals, all within the constraints of finite (and generally tight) budgets. Indeed, many environmental decision makers fit this description.

However, these individuals also have their counterparts in the regulatory arena (such as agency compliance officers). Furthermore, critical environmental decisions are often made by market researchers, product designers, and manufacturing process developers. Naturally, the level of environmental expertise these individuals possess is highly variable. Nonetheless, all of them can and do make critical environmental decisions. It is therefore incumbent upon the toolbuilders—including EDSS architects—to craft systems and processes that will help to bridge the gap between technical expertise and the decision maker, so that the benefits of this expertise may be realized.

3.5 Characteristics of the Problem

Environmental decision makers are clearly a diverse group of people faced with a diverse group of problems. The breadth of their problem domain, in fact, defines the need for eclectic individuals with tools to match.

In general, environmental decision problems are

Spatial, in that most human activities and their environmental impacts are associated with a place having its own characteristics which influence the decision

Multidisciplinary, requiring consideration of issues crossing such seemingly disparate fields of expertise as atmospheric physics, aquatic chemistry, civil engineering, ecology, economics, geology, hydrology, toxicology, manufacturing, materials science, microbiology, oceanography, radiation physics, and risk analysis

Quantitative, because the constituent disciplines themselves are highly quantitative, and because the costs and ramifications are generally so significant, that objective metrics are desired to help mitigate controversy

Uncertain, in that while the elements are quantitative, the sparsity of data and nascent state of the constituent disciplines leaves many unknowns

Quasi-procedural, since many environmental decisions are tied to a regulatory or corporate policy framework which specifies the steps by which a decision is to be reached, and because the threat of liability dictates a defensible audit trail for the decision process

Political, reflecting the fact that environmental management is driven by public policy, influenced by such considerations as economics, social impacts, and public opinion

The diversity of these characteristics of the problem domain make effective environmental decision support extremely challenging.

3.6 Implications for Environmental Decision Support

Because of these factors, it is not practical to contemplate a generic decision framework for environmental management. Even if it were possible to capture all of the elements necessary to address the great variety of decisions to be undertaken, the system so built would be virtually unusable. Environmental managers are already confronted with a vastly complex problem space; one of the first jobs of the decision support system is to simplify this space, offering them everything that they need to make the decision at hand—but *only* those things.

Therefore, while our definition of EDSS includes the integration of multiple supporting technologies (such as simulation and GIS), we further restrict this definition to stipulate that EDSSs are focused on a particular decision problem and decision maker. Thus, they are not general-purpose tools with which anything can be done—if only you knew how to do it. Rather, they are particularly tailored to the problem facing the analyst, and offer a user interface which is optimized for this problem.

The focused nature of such EDSSs improves the user's interaction with the computer system, allowing the user to concentrate on the problem at hand and the information and tools needed to solve it. It also dictates a software architecture that facilitates the development of sibling systems embracing different decision problems with an essentially common user and data interface (2). Such a family of focused EDSS siblings offers user interface simplicity, in that the siblings share interaction style, organization, and fundamental approaches (where appropriate), while maintaining the focus each sibling has on its particular decision problem.

3.7 Task Analysis of Environmental Decision Making

The focused approach to EDSS design advocated here dictates the use of a human factors engineering technique, called *task analysis*, to support the specification of a particular EDSS for a particular problem.

As defined in the human factors community, "task analysis breaks down and evaluates a human function in terms of the abilities, skills, knowledge and attitudes required for performance of the function" (3). The EDSS designer must endeavor to understand the decision problem, and all of the factors which must be considered in solving it. In addition, the "social history" of the problem must be understood, since there will (in general) already be a number of different approaches to solving a given environmental management problem. For a system to support an analyst in arriving at a credible decision, the various competing approaches must be considered, and possibly accommodated.

A major stumbling block in task analysis is the fact that very few individuals can accurately explain the way in which they actually arrive at a particular decision. They can tell you how they think they *should* do it, and they can often develop a post-hoc analytical rationale for their decision, but people are generally unaware of the actual process by which they make decisions. Thus, other instruments must be used to understand the decision process, ranging from observation and interview up through controlled experimentation to determine the influence of different variables on decisions.

In the environmental arena, this is further complicated by the fact that there are often guidelines or regulations dictating the way in which decisions are *supposed* to be made about a particular problem. These do indeed dictate certain aspects of the process, but often leave a great deal unspecified. For example, the U.S. Resource Conservation and Recovery Act (RCRA) requires that a waste facility be monitored by a network including at least one upgradient and three downgradient wells in order to assure that no hazard to the public health results from the facility. However, though the legislature was specific about this detail, it made little effort to assist the manager in deciding where or how many (above four) wells are to be installed. Furthermore, the language of the act would suggest that certainty is required with respect to the detection of leaks, though no reasonable person would argue that this is either theoretically or economically achievable. Implicit in this example is the issue of uncertainty, which, because of its importance in environmental management, deserves further attention.

3.8 Management of Uncertainty

Uncertainty is implicit in environmental decision making. Complex technical decisions must be made regarding events—in both the past and the present— which depend on many different variables. Solutions to such problems often depend on the use of various mathematical modeling techniques. These techniques, in the main, attempt to predict the future performance of a complex system on the basis of relatively sparse empirical data. The predictions drawn from these modeling studies form the basis for the entire process to follow, including such expensive decisions as the design of a product and its associated manufacturing processes. Ultimately, the environmental effectiveness of the product throughout its life cycle, in terms of protection of human health and reduction of environmental risk, depends on these results.

However, these modeling studies are unavoidably visited by uncertainty of various types, ranging from conceptual model uncertainty—associated with the selection of assumptions necessary to choose the model(s)—to parameter uncertainty resulting from sparse empirical data, noisy measurements, and the general difficulty associated with measuring critical parameters.

3.9 Sources of Uncertainty

Uncertainty in such environmental management problems exists because of a lack of empirical data, errors in the data, incorrect models, and the general nondeterminism of nature.

The first of these, a lack of empirical data, is easy to understand; we routinely live with imperfect knowledge of the current state of systems, owing to lack of data (in a usable form). This and the second (errors in the data) are the ones typically addressed in scientific and engineering studies when the goal is to reduce uncertainty. The usual approach is to collect more data, and to attempt to reduce the measurement error in the data collected.

The third reason, the use of incorrect models, is recently receiving more attention in environmental management. As environmental managers come to accept that model building (whether mental or mathematical) is an essential part of problem solving, the disagreements as to which models are correct become more apparent. Some would argue that a model is correct to the extent that it accurately predicts the future behavior of the system; the limiting factor for environmental problems is the complexity of the system in question. And here is where an interesting human factor emerges.

As mathematical models are expanded to attempt to account for more of the fine details of the natural system under study, the mental models of the analyst become inadequate. While humans are capable of recognizing and apprehending in a *gestalt* sense the breadth of complex systems, they are ill equipped to mentally manage the myriad simultaneous details attending such systems. It can be argued that we build mathematical models precisely because we cannot manage such details mentally. Yet, as we build these models, they too become more complex than we can fully grasp, resulting in a great deal of effort and controversy associated with the development of the mathematical models. Many environmental modelers spend more time studying their models than studying the natural systems they emulate.

This problem becomes especially acute when the decision maker is not the developer of the mathematical model, because an opportunity exists for mismatch between the analyst's mental model and the quantitative mathematical model he or she is attempting to use. This results in uncertainty, both subjective (i.e., lack of confidence on the part of the analyst) and objective (i.e., a measurable variability in the decisions made by several analysts or by one analyst on several occasions). Ultimately, this uncertainty finds its way into public perception, causing the public at large to wonder how to interpret the products of science and engineering (the public's awareness of the modeling debate surrounding global warming is a good example of this).

Finally, the fourth cause of uncertainty in environmental problems arises out of the nondeterministic character of the natural environment, at least as it is currently understood. We should not expect to eliminate uncertainty entirely in solving environmental problems. Like the other three, this cause of uncertainty applies to both spatial and aspatial data, and some adaptive approaches have been proposed to help analysts arrive at accurate descriptions of the uncertain natural parameters (e.g., Ref. 4).

Unfortunately, humans tend to have some difficulty in reliably making probabilistic judgments (5). There is a tendency toward a "fish-eye" view of uncertainty, in that perception of unfamiliar issues or events is related to familiar ones, resulting in distortion not unlike the familiar cartoon maps showing "the New Yorker's view of the World." This is evident in studies examining human perception of risk, and applies to probabilistic judgments more generally.

Quantification of uncertainty has been widely acknowledged as a critical issue in risk assessment (see, for example, Ref. 6). A variety of methods for managing uncertainty have been studied (7), most of which are beyond the scope of the present chapter. One of these, which figures prominently in EDSS, involves the use of computer simulation methods to quantify the uncertainty associated with a model result, *conditioned* on the correctness and appropriateness of the model for the problem at hand.

3.10 Stochastic Analysis

In considering the uncertainty of quantitative models, one considers the output of the model to be some function of one or more input coefficients. These co-efficients become the parameters of a numerical representation of the model. The quantitative uncertainty in the modeling solution, then, results from the combined uncertainties of the input parameters.

Stochastic analysis of uncertainty is predicated on the ability to articulate the probability distributions of each uncertain parameter and then iteratively solve one or more model equations involving these parameters. To accomplish this, samples are drawn from the parameter distributions, most often employing Monte Carlo or Latin hypercube sampling methods.

To generate N Monte Carlo samples from a given probability distribution, one first produces the corresponding cumulative distribution function (CDF). The ordinate of the CDF, which ranges from zero to one, is then sampled uniformly, and the corresponding abscissa values are taken as pseudo-random samples of the target distribution.

Latin hypercube sampling, a variation on the Monte Carlo method, forces the uniform samples drawn on the ordinate to cover the entire range (zero to one) by dividing the axis into N equal-width bins. From each bin a sample is drawn, with uniform sampling within each bin. This modification helps to ensure that the tails of the target distribution are sampled, and therefore can result in convergence on the target distribution in fewer samples than the unmodified Monte Carlo method.

To solve environmental models using such stochastic methods, one solves the model equation iteratively, each time using parameter values drawn from the uncertain parameter distributions by the methods just described. The set of results of these calculations form, themselves, a distribution which aggregates the uncertainty of each of the parameters, and whose characteristics can be used to describe the model. The moments and upper and lower quantile bounds of such a calculated distribution can be employed directly in decision making based on the model. For example, if one calculates individual exposure to radionuclides using such an approach, the CDF of the distribution of results can be used to find the probability that exposure will exceed 25 mrem/year. It has been demonstrated (8) that the use of such methods can help to avoid the "creeping conservatism" which often results from the use of upper-bound parameter values alone to model risk.

4 CONTRIBUTING DISCIPLINES

Several disciplines interact with and are integrated by environmental decision support systems as defined in this chapter. This section will introduce the most prominent of these, with a special focus on the particular areas of intersection and contribution. This treatment cannot be construed as a fair representation of any of these disciplines as a whole; rather, it is intended to provide a sense of the interdisciplinary nature of EDSS, and to illuminate some of the opportunities for interdisciplinary research associated with EDSS.

4.1 Environmental Science

Environmental science is itself an interdisciplinary field, integrating biology, chemistry, mathematics, and physics in the context of environmental protection and management. There is a distinctively applied, anthropocentric orientation to environmental science; it differs from such fields as ecology in that it approaches the study of our environment with an eye toward human needs and use of the environment, and therefore addresses the science, engineering, and management practices which will help to conserve environmental resources for human benefit. This is not to imply that environmental scientists as a whole do not place value on nature in and of itself, but that their professional lives are more focused on natural *resource* protection, where the word *resource* refers to human needs and wants. This distinction is significant for the present EDSS discussion only because, as a practical matter, nearly all environmental decisions are anthropocentric. Even in the relatively rare cases where economic resources are available for "pure" ecological protection or remediation, the decisions made must necessarily consider cost/benefit as best they can in order to justify the use of the limited funds. Therefore, *worth* is an important element of virtually every practical environmental decision, and its analysis is most definitely in need of assistance from EDSS technology.

The contributions of environmental science to EDSS begin with the basics. In some instances, we are interested in the basic science involved, with no particular environmental twist, such as the solubilities of chemicals in water, the partitioning of a chemical between the vapor or aqueous phases, the chemical equilibrium of carbon dioxide and water, or the physics of radioactive decay. In others there is a distinctly environmental angle, such as the adsorption of chemicals on soil particles, or the avian toxicity of a pesticide. The line between these two cases is blurred, which is one of the reasons that the basic sciences are so readily integrated into environmental science pedagogically.

Of special interest to EDSS are environmental science's contributions in mathematical modeling of environmental processes. In this context, environmental science integrates such disciplines as geography, hydrogeology, and meteorology, along with various associated engineering disciplines, notably civil and chemical engineering. In some fields, mathematical models are employed to help discover the truth about the phenomenon under study, with the (usually optimistic) goal of arriving at *the* model which describes the way the process works. In contrast, environmental scientists develop models primarily in order to accurately predict the future (or sometimes past) behavior of the system, without suffering the delusion that the model works the same way the system does. Model fidelity—the degree to which the model reflects the way the system actually works—is usually of secondary concern in environmental science. Model robustness—the degree to which the model predicts system behavior under varying conditions consistent with the stated assumptions—is of primary concern.

The focus of environmental modeling is prediction, useful because it can help us to understand what has happened, or what will happen. Such models are central to environmental decision support systems, and in fact to environmental decision making in general. Though some environmental managers would profess to distrust models, and prefer to make predictions through some other means, they fail to realize that these other means invariably include *mental models* of the system. Mental models may not be mathematical, but they are most certainly models, and bear all of the constraints that apply to models.

These constraints can nearly all be reduced to one axiom: a model is only as good as the assumptions that accompany it. In the case of environmental models, significant assumptions are always needed in order to apply a particular model to a particular situation. Assumptions could arise in an attempt to cope with uncertainty in future events (such as the number of inches of rain that will fall next year), or in an attempt to simplify the problem to make it more tractable (such as modeling groundwater contaminant transport in two dimensions rather than three). Assumptions in environmental models are not *bad*; indeed, they are necessary. However, they must be made and validated consciously during model building, and not forgotten when the model is applied. Part of the role of EDSS in the application of environmental models is to help the decision maker to acknowledge, and to an appropriate extent participate in, the assumptions made and validated. In some systems, this is accomplished by requiring analysts to explicitly state their assumptions respecting the models to be applied.

Another multidisciplinary grouping, drawn from the health sciences, can be included in environmental science in this context, although it is not traditionally grouped together in an academic environment. Health science is here taken to include various branches of medicine, toxicology, and epidemiology. These disciplines provide crucial information regarding the ultimate human health ramifications of the systems or actions under study. For example, this would include the first phases of risk assessment, wherein the relationships between human exposure and human health effects are explored and described. Like other aspects of environmental science, this (collective) discipline also contributes models to environmental decision support. These models, both analytical and empirical, assist with such tasks as dose–response calculation and uptake prediction.

4.2 Information Systems Engineering

Information systems engineering (meant here to include computer science and its kin) is also a multidisciplinary field. Not surprisingly, information systems engineering and several of its associated technologies plays a key role in environmental decision support systems. We will explore four of these which are of particular importance to EDSS.

4.3 Geographic Information Systems

A central feature of virtually all environmental decisions is their spatial context. Geographic information systems (GIS) are computer software systems which directly target the management, analysis, and display of spatial information, and which are therefore crucial in an effective EDSS.

There are many GIS packages available, differing in the details of their design. However, some key design features are common to virtually all commercial or public-domain GIS offerings. (A more complete introduction to geographic information systems may be found in Ref. 9.) Current GISs represent spatial information as layers of two-dimensional data encoding different spatial data elements, analogous to (and in fact derived from) the traditional mapmaker's technique of drawing different map features on separate layers of transparent material. These layers can then be overlaid in whatever combination is desired to produce a map showing those features which are of interest. For example, one might overlay a property (lot/block) map onto a soils map in order to evaluate the soils present in individual lots for septic suitability analysis. These two-dimensional layers are typically managed as one of two data types, vector and raster. Early in the history of GIS, packages would use either one or the other of these two data formats, but they are now both supported in common GIS products.

The *vector* data format, as its name implies, represents spatial objects (such as building lots or soil regions) as polygons formed by sequences of vectors, or line segments, each of which is in turn represented by its endpoints (in whatever reference system, such as latitude/longitude, is convenient). Some spatial objects (such as roads or rivers) are represented simply as vector sequences which do not close into polygons. Finally, some objects (such as drinking-water wells) may be represented as a single point. While the structures discussed above represent the location of the spatial objects, they do not describe the attributes of the objects. Such attributes are typically represented in a relational database which is linked to the spatial description by an identifier field. Thus, if one selects the polygon representing a soil region—for example, by clicking the mouse within that region—the GIS would first determine the identifier of the polygon which contained the mouse pointer, and then use this identifier to extract attribute information (in this case soil classification) from the relational database. In fact, when the spatial objects are drawn on the computer screen, one or more of the attribute fields can be used to determine such drawing options as line color or type, or polygon fill color or pattern. In this way a color-coded soils map can be displayed, at the same time that the information used to produce it is available to other computer software. Foremost among the virtues of the vector approach to spatial data representation is the fact that the points (which are the building blocks of all types of spatial objects) can be expressed with a level of precision limited only by the computer's number representation. (Of course, this has no bearing on the accuracy of the data so represented.)

The *raster* data format takes an entirely different approach to spatial data storage. Data layers are represented as regular matrices, with the (normally square) cell dimensions determining the resolution of the layer. The name *raster* is related to the raster display of modern cathode-ray tube (CRT) displays, which are composed of rows and columns of pixels. However, there is no actual correspondence between a GIS raster layer and a CRT's pixels: the data in one cell of a GIS raster layer can be drawn using one or more CRT pixels. In a raster representation of a soils map layer, each cell of the raster contains a value corresponding to the soil category within that cell. If the cell dimension is, for example, 30 m, then the soil category assigned to the cell is that of the soil which dominates the 30 m × 30 m area represented. It is obviously quite a simple matter to display a color-coded soils map by mapping a raster's cell values onto the video memory's pixel values through a color lookup table. This results in display operations which are somewhat faster than can be achieved with a vector (polygon) display. Alternatively, the raster layer's cells can contain key values providing connectivity to a relational database, similar to vector systems, although this approach is used less often. In any case, the precision of the spatial representation using raster data structures is limited by the data storage available for each raster. If one wanted 10-m rather than 30-m resolution (supposing one had corresponding information resolution), the space required to store the layer would increase by a factor of 9.

The chief advantage of a raster data structure is the ease with which one can perform calculations oriented toward the intersection of two or more layers. For example, if one defines septic-suitable areas as those which have a sandy loam soil *and* a slope of less than 10%, one can produce a new layer by performing a cell-by-cell comparison of the soils layer with a slope layer (which itself could be produced by analyzing an elevation layer).

Such calculations are common in natural resource management, which has resulted in raster-oriented GISs dominating these fields. On the other hand, in areas where precise locations are important (such as tax maps or pipeline location), vector-oriented GISs have dominated. Since most GIS packages have migrated into a hybrid orientation, supporting both data structures and conversions between them, one no longer has to make the choice when purchasing the software, and can choose the structure appropriate for the problem at hand.

As was hinted during the foregoing discussion, GIS technology includes more than the simple storage and display of map layers. A critical component of GIS is the analytical suite which permits calculations, comparisons, and manipulation of data layers to produce either new derived layers or simple answers. For example, given a soils layer, any competitive GIS can very simply report the area represented by a particular soil type, either as a percentage of the whole or in such units as acres, hectares, or square meters. Likewise, most GIS packages permit more sophisticated spatial statistics, such as the generation of rasters by interpolation of contour maps, or conversely the generation of contours from rasters. Such analytical capabilities differentiate GIS packages from more simple mapping packages.

These analytical capabilities have increasingly permitted GIS technology to be the basis for decision making in many contexts, not the least of these being environmental management. GIS capability is now a standard in nearly all organizations undertaking environmental analyses, with the useful side effect that many sites of interest have already compiled significant repositories of GIS data pertinent to their problems. However, GIS largely remains an over-the-counter operation. Because of relatively complicated user interfaces, exacerbated by rather breathtaking secondary memory (disk) storage requirements for GIS data, most organizations maintain something along the lines of a GIS department or group. Decision makers, if they recognize that they have a problem which can be addressed by GIS methods, approach this group with the problem description and enter the group's service queue. For some problems, this sort of specialist attention is necessary. GIS groups tend to be staffed by individuals with considerable knowledge of cartography and the tricks necessary to manipulate map data without corrupting it. On the other hand, a good deal of GIS capability is in principle within the grasp of workers from other fields, but the tools and/or data themselves are not available. In integrating GIS technology into environmental decision support systems, we attempt to address the latter problem, not the former. For the subset of GIS-tractable problems which can be approached by the non-GIS specialist, integration into a decision tool addressing their larger problem will solve the batch-oriented, over-the-counter bottleneck which more often than not results in GIS methods not being used where they might otherwise be put to good effect. Another way to think of this is to consider that EDSS can bring some elements of GIS to decision makers in such a way that they need not know it is GIS.

4.4 Computer Data Representation

While the geographic information systems technology just described goes a long way toward providing display capabilities for environmental management problems, it does not satisfy all such needs. First of all, GIS displays are overwhelmingly two-dimensional in nature, with a strong bias toward representing data in map format, or "plan view." Many GIS packages also provide a "2.5-dimensional" representation capability wherein map layers containing elevation information in the raster cells are drawn as surfaces from a user-specified perspective. While often useful, such displays are not by themselves adequate.

For many environmental management problems, true three-dimensional displays are helpful. For example, when evaluating the behavior of a modeled airborne contaminant plume, the analyst should be able to navigate about (and through) the three-dimensional plume in order to get a better feel for its shape and character; contour plots fail to communicate this information. Computing and displaying such volumetric renderings rests squarely within the domain of information systems engineering. The algorithms required to efficiently draw, shade, and cast virtual light upon three-dimensional objects drawn on a two-dimensional computer screen are the result of considerable research in the field of computer graphics. Many of these algorithms have been known for quite some time, but the ability to use them to generate very sophisticated volumetric displays in near-real-time is relatively new, especially on common desktop computing hardware. These tools have begun to play an important role in environmental decision support, and will be integrated into EDSS platforms with increasing frequency.

More recently, however, advances in personal computing have included the development and widespread dissemination of what has been called multimedia technology. This suite of computer capabilities has added photographic and motion-picture display to the more conventional computer graphics world, and has also added high-fidelity sound-generation capability to the platform. The ability to include photographs (such as site familiarization photos) and videos (such as a sequence capturing the removal of a well core) has the potential to greatly enhance the information delivery potential of environmental decision support systems. The audio capabilities have an obvious use in delivery of speech (such as in online help or cooperative work situations), but also support the use of sound to represent quantitative information which cannot readily be accommodated by available visual display channels (e.g., Refs. 10 and 11).

4.5 Supercomputing and Networking

A third area of information systems engineering which has the potential to significantly impact EDSS design relates to the execution of computationally intensive models. Historically, such computations have been relegated to a segment of computer technology called supercomputers. Supercomputers may be operationally defined as computers which are both fast and expensive enough that few of them are in existence. This rather awkward definition is necessary to account for the fact that current personal computers offer a level of computational throughput which would have been considered supercomputing 25 years ago. It is pointless to attempt to define supercomputers in absolute performance terms, because the technology advances so rapidly as to render such boundaries obsolete in very few years.

Nonetheless, it may be presumed that no matter how fast individual workstations become, there will be still faster computers which are few in number but which are made available to a wide population. In this work, such super-computing technology is considered in combination with digital networks because high-bandwidth data networks have made it possible to consider linking supercomputers with personal workstations in such a way that the interactive user need not be aware that computations have been "contracted out."

In some sense, this sort of approach would represent a distribution of the EDSS architecture across multiple, remotely located machines. This view is especially appropriate if one distributes the data or code repository functions as well. For example, one might keep national meteorological data in a disk farm associated with a National Oceanic and Atmospheric Administration (NOAA) supercomputing facility, which might also store and maintain modeling codes that have been submitted to a quality assurance process. An individual EDSS being used to evaluate potential emissions from a factory might make use of these data and codes, as well as the supercomputer power, to solve a local air-modeling problem. Avoiding the need to distribute the data and codes saves a considerable amount of space (which would have been redundantly consumed on every similar EDSS platform), and also reduces the risk of data (or code) contamination.

In any case, the environmental models currently in use already stretch even high-end workstation capabilities to the point that analysts might wait several days for a single iteration of a model to execute. As computer throughput increases, more iterations of the Monte Carlo simulation will be executed, and more complicated models employed. Although individual workstations can satisfy many environmental management computation requirements, there will be a need for supercomputer access in environmental decision support for quite some time.

4.6　Expert Systems

Finally, information systems engineering offers the environmental decision support system a technology to help capture and deliver the knowledge of experts in particular problem domains. EDSS is predicated on the notion that *human* intelligence is needed to make responsible environmental management decisions. *Artificial* Intelligence (AI) might therefore seem anachronistic in this work. However, although expert systems research has indeed grown out of AI research, the connection stops there. Expert systems offer the possibility of providing advisors to environmental analysts, for example, to help them choose the assumptions and parameters of their conceptual model of the problem (4). In this sense, the interactive user has the benefit of aggregated advice from many experts who would otherwise be unavailable, but still has the last word. This area of research in EDSS is the most prospective, and much work remains to be done before it can be claimed that expert systems technology has contributed substantially. Nonetheless, there is great potential for a productive collaboration.

4.7　Decision Science

The term *decision science* is used here to refer collectively to the various fields of investigation which attempt to provide quantitative (or at least controlled qualitative) structure to the decision-making process. This includes subdisciplines ranging from statistics and geostatistics, through operations research and linear programming optimization, to classical and Bayesian probability theory. While such formal decision methods are only sparingly applied in current environmental decision frameworks, it can be expected that this will increase in the future, if for no other reason than they provide some accountability for the decision process and remove some of the air of subjectivity from it.

There is a formalism associated with decision science, the terms of which are fairly intuitive. To begin with, a *decision* itself is a choice between alternatives. These alternatives are compared according to some *criteria*, the measurable evidence on which the decision is to be based. A criterion can be a *factor* which enhances or detracts from the suitability of an alternative, or a *constraint* which limits the alternatives under consideration. In order to combine criteria for evaluation and action, one employs *decision rules*. These include procedures for aggregating criteria into a single index, along with an algorithm for comparing alternatives according to this index. Decision rules can be *choice functions* (sometimes called objective functions) or *choice heuristics*. The former provide a mathematical method for alternative comparison, typically involving some form of optimization. The latter provide an algorithm or procedure to be followed, sometimes with a stopping rule to indicate when the procedure should terminate and the solution either taken or the search abandoned. For example, if one seeks to fit a linear equation to a set of data points, one can solve the conventional linear regression equation which sets a derivative to zero to solve for the minimum cumulative squared error. This would be a choice function. Alternatively, one can solve the equation iteratively while varying the coefficients according to some prescription, stopping either when this same error metric is "small enough" (but not necessarily a minimum) or when the number of iterations has exceeded one's patience. This choice heuristic *might* result in the same solution as the choice function, but in examples such as this one it probably will not. On the other hand, there may not be unique analytical solutions to the problem at hand, leaving heuristic approaches the only game in town.

There is usually a specific *objective* of the decision at hand, and the decision rules are structured in the context of this objective. When there are multiple criteria which must be considered in the decision, this is termed a *multicriteria evaluation*, in which some method for combining the criteria must be selected. More complicated is the *multiobjective* case, in which there are multiple objectives which may be complementary or may conflict.

While a great many techniques are available from decision science, two are commonly employed in environmental decision making: linear programming and decision trees.

4.8 Linear Programming

Linear programming methods are usually associated with operations research. They are typically applied to optimization and resource allocation problems where there are linear relationships between problem parameters, both objectives and constraints. The linear equations describing the constraints associated with decision variables are solved simultaneously to define a solution space or feasible region (in as many dimensions as there are variables). The linear objective function is then evaluated to determine its minimal or maximal value (for cost functions or benefit functions, respectively). If this optimal value, plotted in the space of the decision variables, is contained within the feasible region defined by the constraints, then an optimal, feasible solution has been found. Given this structure, linear programming solutions strongly resemble conventional (multiple) linear regression methods, solved either graphically or iteratively. These methods are frequently used in optimization problems such as cost/benefit analysis for monitoring or remediation systems, or allocation of monitoring wells along a site perimeter.

Vogel (12) cites an example of this form of systems analysis applied to a so-called conjunctive use problem in which the best balance of water supply sources (surface and groundwater) is sought, with the goal that the total system yield under coordinated use exceeds the sum of the yields under uncoordinated use. Given an annual water demand of K, the decision maker seeks to find optimal values of groundwater withdrawal (G) and surface water withdrawal (S) such that $G + S \geq K$. Naturally, there are various constraints on both ground and surface water usage (for example, there are maximum yields from each source), which taken together form the feasible region. If the objective function of this problem can be described linearly (i.e., there is some linear combination of G and S whose coefficients describe the normalized benefits of each source of water), then the family of curves (straight lines) representing this function under various coefficient values can be plotted on the decision axes overlying the feasible region. The selection of the optimal coefficients can then be made graphically.

For more realistic problems there are many constraint equations and terms in the objective function, preventing the use of graphical methods. However, a variety of techniques have been developed to evaluate such systems mathematically. Even in cases where the exact optimal solution is intractable, linear programming has the potential to identify a range of solutions in the neighborhood of the optimal solution.

4.9 Decision Trees

Decision trees are associated with a decision analytic method which accounts for both expected value and uncertain events. A hierarchical graphical structure is used to describe the structure of the decision problem. Nodes (vertices) in the tree are either decision nodes or chance nodes, depending on whether the branches result from the decision maker's choice or some uncertain event, respectively. Every decision tree has as its root a decision node, which is the first decision under consideration. Every branch in the tree eventually terminates in a "leaf" representing the outcome of that particular path through the tree, with its associated probabilities and expected value. Folding back the path probabilities and expected values of chance nodes (by multiplication), one can arrive at expected values for the decision nodes, and make an optimal decision based on this value. However, to do this one requires some metrics for expected value of each outcome, and probabilities for each branch from each chance node. Furthermore, the decision and chance alternatives must be finite: one is selecting from a particular set of decision alternatives, rather than adjusting an operating point on a continuum.

5 APPLICATIONS OF EDSS

The foregoing has provided a foundation for environmental decision support systems in general. Though the technology has seen most of its application in natural resource management and environmental remediation, there are many opportunities to bring the power of EDSS to bear on problems in industry. Three examples will serve to illustrate this point.

5.1 Integrated Factory Decision Support

Environmental compliance within a manufacturing environment is an information-intensive pursuit, and can be facilitated by the integration of information systems and repositories (13). The value of such integration can best be illustrated in the breach. When changes are made in the configuration of a manufacturing area, such as the movement of solvent baths from one location to another, the failure to include environmental compliance managers in the decision process can result in permit violations (for example, the movement of Volatile Organic Compound (VOC) sources from one vent stack to another can require air permit modifications). In general, layout of the factory floor can affect environmental performance, as well as compliance, so that one must consider environmental ramifications when attempting to develop and/or modify the manufacturing layout.

Geographic information systems, and their cousins the computer-aided design (CAD) systems, have been involved in facilities management for some time, but their use in support of environmental management is relatively new (14). Combining the spatial plant design data with relational data describing such domains as materials inventory provides the basis for integrated decision making—integrating environmental management with overall plant management functions. Combining these with decision tools and simulation capabilities allows managers to make superior decisions about plant layout, and improve their compliance record.

Beyond physical plant arrangements, this sort of information system integration can also go a long way toward reducing the cost of regulatory compliance. For example, in the United States the Emergency Preparedness and Community Right-to-Know Act (EPCRA) requires annual reporting of the quantities and whereabouts of hazardous materials, with the intent of ensuring the safety of emergency responders in the event of fire or other disaster. This simple and arguably worthwhile requirement can result in a great deal of expense to a company whose information management and decision tools are not integrated. It is typical for such companies to issue annual inventory surveys to plant personnel, who then must physically locate and record such materials so that the regulatory reports can be completed. A far better alternative is to integrate the purchasing, storeroom, and environmental compliance software systems so that the flow of materials into and within the facility is generally known at any time. This not only permits the decision support system to easily produce the annual reports required for the EPCRA, but also allows regular review of materials movements and usage, which in turn can facilitate such other tasks as tracking of air emissions calculated by mass balance.

This example illustrates an EDSS emphasizing GIS and relational database technology, and especially the integration of these technologies across organizational and functional boundaries within the operation.

5.2 Risk Management Planning

In the United States, provisions of the Clean Air Act require owners or operators of a stationary air pollution source with more than a threshold quantity of a regulated substance to submit a Risk Management Plan (RMP). Among other things, this plan must describe the accidental release prevention and emergency response policies at the source, the regulated substances handled, and the worst-case release scenarios and alternative release scenarios, including administrative controls and mitigation measures to limit the distances for each reported scenario.

This regulation provides a natural application for environmental decision support systems. To plan for a release of chlorine from tank cars on a siding, for example, one must evaluate the dynamics of material movement in the siding area, including variations in quantity of chlorine present at any one time. Then atmospheric transport models must be used to predict, for each of a variety of weather conditions, where the toxic gas is likely to go, and in what concentrations. Since weather prediction and atmospheric models are attended by a great deal of uncertainty, quantitative means must be employed to manage this uncertainty. Monte Carlo simulation, as described previously, can help to quantify the uncertainty given historical data on which to base probability distributions.

This example illustrates an EDSS based on modeling and simulation, with substantial support provided by GIS technology.

5.3 Design for Environment

A third example of the use of environmental decision support systems in the industrial context involves supporting the decision-making process engaged in by product and process designers intending to minimize the environmental impact of their product. Designing for environment (DFE) requires the availability of a great deal of information regarding alternative materials, components, and processes available for consideration by the designer. Such information is notoriously difficult to find, and when available its applicability to different situations is quite variable. To support the designer adequately, the system must make this information available for ready access, but it must also help the user to select only the information appropriate to the problem at hand, and perhaps also assist in the actual design decisions.

By integrating an expert system with a highly descriptive relational database, the EDSS can meet this need. Expert assistance (even if delivered by computer) is very appropriate in this situation, especially if one considers that the designer very likely has little training in environmental issues.

6 EDSS SUMMARY

Environmental decision support systems, defined here as a class of information systems integrating several technologies in support of improved environmental decision quality, have served well in a variety of applications in natural resource management and environmental remediation. They offer similar benefits to the industrial environmental manager prepared to invest in their deployment. While not turnkey, off-the-shelf solutions, such systems, once developed, can earn their keep by helping to solve problems which might otherwise be intractable.

7 CONCLUSION

The foregoing has described a wide range of information systems designed to help improve the quality of environmental management. The integrated approach to environmental management information systems, while based largely on conventional relational database technology, still offers the prospect of real risk reduction and performance improvement when information maintenance and management is required. For more active decision-making processes, the described framework for environmental decision support systems offers the opportunity to put scientifically sound tools into the hands of real decision makers. The key to both of these approaches in *integration*.

REFERENCES

1. G. Guariso and H. Werthner, *Environmental Decision Support Systems*. Chichester, U.K.: Ellis Horwood, 1989.
2. S. P. Frysinger, An Open Architecture for Environmental Decision Support. *Int. J. Microcomput. Civil Eng.,* vol. 10, no. 2, pp. 119–126, 1995.
3. R. W. Bailey, *Human Performance Engineering*. Prentice-Hall, London, 1982.
4. A. S. Heger, F. A. Duran, S. P. Frysinger, and R. G. Cox, Treatment of Human-Computer Interface in a Decision Support System. *IEEE International Conference on Systems, Man, and Cybernetics,* pp. 837–841, 1992.
5. R. M. Hogarth, *Judgement and Choice*. New York: Wiley, 1987.
6. National Research Council, *Issues in Risk Assessment*. Washington, DC: National Academy Press, 1993.
7. M. G. Morgan and M. Henrion, *Uncertainty*. Cambridge, U.K.: Cambridge University Press, 1990.
8. T. E. McKone and K. T. Bogen, Predicting the Uncertainties in Risk Assessment. *Environ. Sci. Technol.,* vol. 25, no. 10, pp. 1674–1681, 1991.

9. S. Aronoff, *Geographic Information Systems: A Management Perspective.* Ottawa: WDL Publications, 1989.

10. S. P. Frysinger, Applied Research in Auditory Data Representation, In E. J. Farrell (ed.), *Extracting Meaning From Complex Data—Proceedings of the SPIE/SPSE Symposium on Electronic Imaging*, 1990.

11. G. Kramer, *Auditory Display: Sonification, Audification, and Auditory Interfaces.* Proceedings of the 1992 International Conference on Auditory Display. Addison-Wesley, 1994.

12. R. M. Vogel, Resource Allocation. In R. A. Chechile and S. Carlisle (eds.), *Environmental Decision Making: A Multidisciplinary Perspective*. New York: Van Nostrand Reinhold, pp. 156–175, 1991.

13. S. P. Frysinger, New Approaches to Environmental Information and Decision Support Systems. National Association for Environmental Management's *Environmental Management Forum*, Dallas, TX, October 28–31, 1997.

14. W. J. Douglas, *Environmental GIS: Applications to Industrial Facilities.* Lewis Publishers, Boca Raton, FL, 1995.

6

European Policies for Waste Management

Ingo F. W. Romey and Marc Obladen
University of Essen, Essen, Germany

1 INTRODUCTION

With the treaty establishing the European Community and with the ratification of the Treaty of Amsterdam, the European Community receives wide legislative rights. The following states are members of the European Community: Austria, Belgium, Denmark, Finland, France, Germany, Greece, Ireland, Italy, Luxembourg, Netherlands, Portugal, Spain, Sweden, and the United Kingdom of Great Britain and Northern Ireland.

In order to carry out their tasks and in accordance with the provisions of this treaty, the European Parliament, acting jointly with the Council and the Commission, make regulations and issue directives, take decisions, and make recommendations or deliver opinions.

> A regulation shall have general application. It shall be binding in its entirety and directly applicable in all Member States.

A directive shall be binding, as to the result to be achieved, upon each Member State to which it is addressed, but shall leave to the national authorities the choice of form and methods.

A decision shall be binding in its entirety upon those to whom it is addressed. Recommendations and opinions shall have no binding force.

By the Treaty of Amsterdam the members of the European Community are obligated to pursue a common environmental policy. Community policy on the environment shall contribute to pursuit of the following objectives:

Preserving, protecting, and improving the quality of the environment
Protecting human health
Prudent and rational utilization of natural resources
Promoting measures at the international level to deal with regional or worldwide environmental problems

Further Community policy on the environment shall aim at a high level of protection, taking into account the diversity of situations in the various regions of the Community. It shall be based on the precautionary principle and on the principles that preventive action should be taken, that environmental damage should as a priority be rectified at the source, and that the polluter should pay. In this context, harmonization measures answering environmental protection requirements shall include, where appropriate, a safeguard clause allowing Member States to take provisional measures, for noneconomic environmental reasons, subject to a Community inspection procedure.

Without prejudice to certain measures of a Community nature, the Member States shall finance and implement the environment policy.

Without prejudice to the principle that the polluter should pay, if a measure based on the provisions of paragraph 1 involves costs deemed disproportionate for the public authorities of a Member State, the Council shall, in the act adopting that measure, lay down appropriate provisions in the form of temporary derogation, and/or financial support from the Cohesion Fund of the European Community.

Exemplary discussed should be the newest and most important decisions dealing with the universal treatment of wastes and the waste management strategy.

2 EUROPEAN DIRECTIVES ON WASTE MANAGEMENT

2.1 European Parliament and Council Directive on Packaging and Packaging Waste

The Directive on packaging and packaging waste aims to harmonize national measures concerning the management of packaging and packaging waste in order, on the one hand, to prevent any impact thereof on the environment of all Member States as well as of third countries or to reduce such impact, thus providing a high level of environmental protection, and, on the other hand, to ensure the functioning of the internal market and to avoid obstacles to trade and distortion and restriction of competition within the Community. This Directive lays down measures aimed, as a first priority, at preventing the production of packaging waste and, as additional fundamental principles, at reusing packaging, at recycling and other forms of recovering packaging waste and, hence, at reducing the final disposal of such waste. It covers all packaging placed on the market in the Community and all packaging waste, whether it is used or released at industrial, commercial, office, shop, service, household, or any other level, regardless of the material used. This Directive shall apply without prejudice to existing quality requirements for packaging such as those regarding safety, the protection of health and the hygiene of the packed products or to existing transport requirements.

In this context, "packaging" shall mean all products made of any materials of any nature to be used for the containment, protection, handling, delivery, and presentation of goods, from raw materials to processed goods, from the producer to the user or the consumer. "Nonreturnable" items used for the same purposes shall also be considered to constitute packaging. Generally recommendable is the prevention of packaging. Therefore it shall be ensured that preventive measures are implemented. Such other measures may consist of national programs or similar actions adopted, if appropriate in consultation with economic operators, and designed to collect and take advantage of the many initiatives taken within Member States as regards prevention. Furthermore, all Member States should encourage reuse systems of packaging, which can be reused in an environmentally sound manner, in conformity with the Treaty.

In order to comply with the objectives of this Directive, Member States shall take the necessary measures to attain the following targets covering the whole of their territory. No later than five years between 50% as a minimum and 65% as a maximum by weight of the packaging waste will be recovered. Moreover, within this general target, and with the same time limit, between 25% as a minimum and 45% as a maximum by weight of the totality of packaging materials contained in packaging waste will be recycled, with a minimum of 15% by weight for each packaging material. At least no later than 10 years from the date by which this Directive must be implemented in national law, a percentage of packaging waste will be recovered and recycled, which will have to be determined by the Council.

Member States shall, where appropriate, encourage the use of materials obtained from recycled packaging waste for the manufacturing of packaging and other products.

Member States which have, or will, set programs going beyond the above-mentioned targets and which provide to this effect appropriate capacities for recycling and recovery, are permitted to pursue those targets in the interest of a high level of environmental protection, on condition that these measures avoid distortions of the internal market and do not hinder compliance by other Member States with the Directive.

Member States shall take the necessary measures to ensure that systems are set up to provide on the one hand the collection and return of used packaging and packaging waste from the consumer, other final user, or from the waste stream in order to channel it to the most appropriate waste management alternatives. On the other hand, the reuse or recovery including recycling of the packaging and/or packaging waste collected, in order to meet the objectives laid down in this Directive, should be ensured. These systems shall be open to the participation of the economic operators of the sectors concerned and to the participation of the competent public authorities.

For waste management, a marking and identification system of waste packaging is essential. Therefore, the Council shall decide no later than two years after the entry into force of this Directive on the marking of packaging. To facilitate collection, reuse, and recovery (including recycling), packaging shall indicate for purposes of its identification and classification by the industry concerned the nature of the packaging material(s) used. To that end, the Commission will specify the numbering and abbreviations on which the identification system is based and shall specify which materials shall be subject to the identification system in accordance with the same procedure. Packaging shall bear the appropriate marking either on the packaging itself or on the label. It shall be clearly visible and easily legible. The marking shall be appropriately durable and lasting, even when the packaging is opened.

In addition, the Commission will promote, as appropriate, the preparation of European standards relating to the essential requirements, in particular, the preparation of European standards relating to:

> Criteria and methodologies for life-cycle analysis of packaging
> The methods for measuring and verifying the presence of heavy metals and other dangerous substances in the packaging and their release into the environment from packaging and packaging waste
> Criteria for a minimum content of recycled material in packaging for appropriate types of packaging
> Criteria for recycling methods
> Criteria for composting methods compost produced
> Criteria for the marking of packaging

With this Directive it will be ensured that the sum of concentration levels of lead, cadmium, mercury, and hexavalent chromium present in packaging or packaging components shall not exceed the following:

600 ppm by weight in 1998
250 ppm by weight in 1999
100 ppm by weight in 2001

Regarding effective packaging management, European information system should be introduced. Thus, Member States shall take the necessary measures to ensure that databases on packaging and packaging waste are established, where not already in place, on a harmonized basis in order to help Member States and the Commission to monitor the implementation of the objectives set out in this Directive. To this effect, the databases shall provide in particular information on the magnitude, characteristics, and evolution of the packaging and packaging waste flows (including information on the toxicity or danger of packaging materials and components used for their manufacture) at the level of individual Member States. In order to harmonize the characteristics and presentation of the data produced and to make the data of the Member States compatible, Member States shall provide the Commission with their available data in standard formats which shall be adopted by the Commission. Member States shall require all economic operators involved to provide competent authorities with reliable data on their sector.

Regarding an information system for users of packaging, Member States shall take measures to ensure that users of packaging, including in particular consumers, obtain the necessary information about:

The return, collection, and recovery systems available to them
Their role in contributing to reuse, recovery, and recycling of packaging and packaging waste
The meaning of markings on packaging existing on the market

In pursuance of the objectives and measures referred to in this Directive, Member States shall include in the waste management plans a specific chapter on the management of packaging and packaging waste. Acting on the basis of the relevant provisions of the Treaty, the Council shall adopt economic instruments to promote the implementation of the objectives set by this Directive. In the absence of such measures, the Member States may, in accordance with the principles governing Community environmental policy, inter alia, the polluter pays principle, and the obligations arising out of the Treaty, adopt measures to implement those objectives.

At least, this Directive specifies essential requirements on the composition and the reusable and recoverable, including recyclable nature of packaging.

Packaging shall be so manufactured that the packaging volume and weight be limited to the minimum adequate amount to maintain the necessary level of safety, hygiene, and acceptance for the packed product and for the consumer. Furthermore, they shall be designed, produced, and commercialized in such a way as to permit reuse or recovery, including recycling, and to minimize impact on the environment when packaging waste or residues from packaging waste management operations are disposed of. The manufacturing has to avoid the presence of noxious and other hazardous substances and materials as constituents of the packaging material or of any of the packaging components or minimized with regard to their presence in emissions, ash, or leachate when packaging or residues from management operations or packaging waste are incinerated or landfilled.

The physical properties and characteristics of the packaging shall enable a number of trips or rotations in normally predictable conditions of use and offer the possibility of processing the used packaging in order to meet health and safety requirements for the workforce. At least, the properties should fulfil the requirements specific to recoverable packaging when the packaging is no longer reused and thus becomes waste.

Furthermore, packaging must be manufactured in such a way as to enable the recycling of a certain percentage by weight of the materials used in the manufacture of marketable products, in compliance with current standards in the Community. The establishment of this percentage may vary, depending on the type of material of which the packaging is composed. Packaging waste processed for the purpose of energy recovery shall have a minimum inferior calorific value to allow optimization of energy recovery. Packaging waste processed for the purpose of composting shall be of such a biodegradable nature that it should not hinder the separate collection and the composting process or activity into which it is introduced, whereas biodegradable packaging waste shall be of such a nature that it is capable of undergoing physical, chemical, thermal, or biological decomposition such that most of the finished compost ultimately decomposes into carbon dioxide, biomass, and water.

2.2 Council Directive on Landfill of Waste

The aim of the Directive on landfill of waste is, by way of stringent operational and technical requirements on the waste and landfills, to provide for measures, procedures, and guidance to prevent or reduce as far as possible negative effects on the environment, in particular the pollution of surface water, groundwater, soil, and air, and on the global environment, including the greenhouse effect, as well as any resulting risk to human health, from landfilling of waste, during the whole life cycle of the landfill.

In principle, each landfill shall be classified in one of the following classes:

Landfill for hazardous waste
Landfill for nonhazardous waste
Landfill for inert waste

Member States shall set up a national strategy for the implementation of the reduction of biodegradable waste going to landfills. This strategy should include measures to achieve the following targets by means of, in particular, recycling, composting, biogas production, or materials/energy recovery. It ensures that long-term biodegradable municipal waste going to landfills must be reduced to 35% of the total amount (by weight) of biodegradable municipal waste produced in 1995. Furthermore, the Directive shall take measures in order that liquid waste, waste which, in the conditions of landfill, is explosive, corrosive, oxidizing, highly flammable, or flammable, as well as hospital and other clinical wastes arising from medical or veterinary establishments, are not accepted in a landfill.

The landfill permit shall state at least the class of the landfill and the list of defined types and the total quantity of waste which are authorized to be deposited in the landfill.

Member States shall take measures to ensure that all of the costs involved in the setting up and operation of a landfill site, including as far as possible the cost of the financial security and the estimated costs of the closure and after-care of the site for a period of at least 30 years shall be covered by the price to be charged by the operator for the disposal of any type of waste in that site. Regarding existing landfill sites, it should be ensured that landfills which have been granted a permit may not continue to operate unless the landfill sites will fulfil the standard of state of the art landfill sites within eight years.

The directive sets up general requirements for all classes of landfills. The landfill can be authorized only if the characteristics of the site indicate that the landfill does not pose a serious environmental risk.

The location of a landfill must take into consideration the existence of groundwater, coastal water, or nature protection zones in the area, the geological and hydrogeological conditions in the area, as well as the risk of flooding, subsidence, landslides, or avalanches on the site. Therefore the water control and leachate management must control water from precipitations entering into the landfill body, prevent surface water and/or groundwater from entering into the landfilled waste, and collect contaminated water and leachate.

Where the geological barrier does not naturally meet the conditions of permeability and thickness, it has to be completed artificially and reinforced by other means giving equivalent protection. An artificially established geological barrier should be no less than 0.5 m thick.

Recommended parameters for a check study are ph, TOC, phenols, heavy metals, fluoride, AS, and oil/hydrocarbons. In order to reduce global warming, the methane-containing landfill gas shall be collected from all landfills receiving biodegradable waste and the landfill gas must be treated and used. If the gas collected cannot be used to produce energy, it must be flared.

2.3   European Parliament and Council Directive on Incineration of Waste

Having regard to the Treaty establishing the European Community and a proposal of the Commission, the Council of the European Union and the European Parliament adopted a new directive on the incineration of waste in July 1999.

In accordance with the principle of subsidiary, it was considered that the objective of reducing emissions from incineration and co-incineration plants cannot be achieved effectively by Member States acting individually. Unconcerted action offers no guarantee of achieving the desired objective, and regarding the need to reduce emissions across the Community, it is more effective to take action at the level of the Community. This Directive confines itself to minimum requirements for incineration and co-incineration plants.

The Directive is a consequence of the fifth Environment Action Program: Towards Sustainability—a European Community program of policy and action in relation to the environment and sustainable development which sets as an objective "no exceedance ever of critical loads and levels" of certain pollutants such as nitrogen oxides (NO_x), sulfur dioxide (SO_2), heavy metals, and dioxins, while in terms of air quality the objective is that "all people should be effectively protected against recognized health risks from air pollution." That program further sets as an objective a "90% reduction of dioxin emissions of identified sources by 2005 (1985 level)" and "at least 70% reduction from all pathways of cadmium (Cd), mercury (Hg), and lead (Pb) emissions in 1995."

The purpose of the incineration plants established and operated in accordance with this Directive is to reduce the pollution-related risks of waste through a process of thermal treatment, especially oxidation, to reduce the quantity and volume of the waste and to produce residues that can be recycled or disposed of safely. The co-incineration of waste in plants not primarily intended to incinerate waste should not be allowed to cause higher emissions of polluting substances in that part of the exhaust gas volume resulting from such co-incineration and should therefore be subject to appropriate limitations.

The aim of this Directive is to prevent or, where that is not practicable, to reduce as far as possible negative effects on the environment, in particular the pollution of air, soil, surface water, and groundwater, and the resulting risks to human health, from the incineration and co-incineration of waste and, to that end, to set up and maintain appropriate operating conditions and emission limit values for waste incineration and co-incineration plants within the Community.

Incineration plants shall be operated in order to achieve a level of incineration such that the total organic carbon (TOC) of the slag and bottom ashes is less than 3% or their loss upon ignition is less than 5% of the dry weight of the material. If necessary, appropriate techniques of waste pretreatment shall be used. All incineration plants shall be designed, equipped, built, and operated in such a way that the gas resulting from the process is raised, after the last injection of combustion air, in a controlled and homogeneous fashion and even under the most unfavorable conditions, to a temperature of at least 850°C, as measured near the inner wall of the combustion chamber, for at least 2 s. If hazardous wastes with a content of more than 1% of halogenated organic substances, expressed as chlorine, are incinerated, the temperature has to be raised to at least 1100°C.

Any heat generated by the incineration or co-incineration process shall be recovered as far as possible.

The directive determines following air emission limit values.

2.3.1 Air Emission Limit Values

Daily average values:

Total dust	10 mg/m^3
Gaseous and vaporous organic substances, expressed as total organic carbon	10 mg/m^3
Hydrogen chloride (HCl)	10 mg/m^3
Hydrogen fluoride (HF)	1 mg/m^3
Sulfur dioxide (SO$_2$)	50 mg/m^3
Nitrogen monoxide (NO) and nitrogen dioxide (NO$_2$), expressed as nitrogen dioxide for existing incineration plants with a capacity exceeding 3 tonnes per hour or new incineration plants	200 mg/m^3
Nitrogen monoxide (NO) and nitrogen dioxide (NO$_2$), expressed as nitrogen dioxide for existing incineration plants with a capacity of 3 tonnes per hour or less	400 mg/m^3

Until 1 January 2007, the emission limit value for NO$_x$ does not apply to plants incinerating hazardous waste only.

Half-hourly average values:

Total dust	30 mg/m^3
Gaseous and vaporous organic substances, expressed as total or organic carbon	20 mg/m^3
Hydrogen chloride (HCl)	60 mg/m^3
Hydrogen fluoride (HF)	4 mg/m^3
Sulfur dioxide (SO$_2$)	200 mg/m^3
Nitrogen monoxide (NO) and nitrogen dioxide (NO$_2$), expressed as nitrogen dioxide for existing incineration plants with a capacity exceeding 3 tonnes per hour or new incineration plants	400 mg/m^3

Until 1 January 2007, the emission limit value for NO_x does not apply to plants incinerating hazardous waste only.

All average values over the sample period of a minimum of 30 mins and a maximum of 8 h.

Cadmium and its compounds, expressed as cadmium (Cd)	Total	Total
Thallium and its compounds, expressed as thallium (Tl)	0.05 mg/m^3	0.1 mg/m^3
Mercury and its compounds, expressed as mercury (Hg)	0.05 mg/m^3	0.1 mg/m^3
Antimony and its compounds, expressed as antimony (Sb)		
Arsenic and its compounds, expressed as arsenic (As)		
Lead and its compounds, expressed as lead (Pb)		
Chromium and its compounds, expressed as chromium (Cr)	Total	Total
Cobalt and its compounds, expressed as cobalt (Co)	0.5 mg/m^3	1 mg/m^3
Copper and its compounds, expressed as copper (Cu)		
Manganese and its compounds, expressed as manganese (Mn)		
Nickel and its compounds, expressed as nickel (Ni)		
Vanadium and its compounds, expressed as vanadium (V)		

These average values also cover gaseous and the vapor forms of the relevant heavy-metal emissions as well as their compounds. Until 1 January 2007 these average values shall apply to existing plants for which the permit to operate has been granted before 31 December 1996, and which incinerate hazardous waste only.

Average values shall be measured over a sample period of a minimum of 6 h and a maximum of 8 h. The emission limit value refers to the total concentration of dioxins and furans calculated using the concept of toxic equivalence in accordance with Annex I:

Dioxins and furans	0.1 ng/m^3

The following emission limit values of carbon monoxide (CO) concentrations shall not be exceeded in the combustion gases (excluding the start-up and shut-down phases):

50 mg/m^3 of combustion gas determined as daily average value

150 mg/m^3 of combustion gas of at least 95% of all measurements determined as 10-min average values or 100 mg/m^3 of combustion gas of all measurements determined as half-hourly average values taken in any 24-h period.

Exemptions may be authorized by the competent authority for incineration plants using fluidized bed technology, provided that the authorization foresees an emission limit value for carbon monoxide (CO) of not more than 100 mg/m^3 as a hourly average value.

2.3.2 Determination of Emission Limit Values for the Co-incineration of Waste

The limit value for each relevant pollutant and carbon monoxide in the exhaust gas resulting from the co-incineration of waste shall be calculated as follows:

$$\frac{V_{waste} \cdot C_{waste} + V_{proc} \cdot C_{proc}}{V_{waste} + V_{proc}} = C$$

where

V_{waste} is exhaust gas volume resulting from the incineration of waste only, determined from the waste with the lowest calorific value specified in the permit and standardized at the conditions given by this Directive.

C_{waste} is emission limit value set in Annex V for plants intended to incinerate wastes only (at least the emission limit values for the pollutants and carbon monoxide).

V_{proc} is exhaust gas volume resulting from the plant process, including the combustion of the authorized fuels normally used in the plant (wastes excluded), determined on the basis of oxygen contents at which the emissions must be standardized as laid down in Community or national regulations. In the absence of regulations for this kind of plant, the real oxygen content in the exhaust gas without being thinned by addition of air unnecessary for the process must be used. The standardization at the other conditions is given in this Directive.

C_{proc} is emission limit value as laid down in the tables of this Annex for certain industrial sectors or in case of the absence of such a table or such value, emission limit values of the relevant pollutants and carbon monoxide in the flue gas of plants which comply with the national laws, regulations, and administrative provisions for such plants while burning the normally authorized fuels (wastes excluded). In the absence of these measures the emission limit values laid down in the permit are used. In the absence of such permit values the real mass concentrations are used.

C is total emission limit value as laid down in the tables of this Annex for certain industrial sectors and certain pollutants or in case of the absence of such a table or such values total emission limit values for CO and the relevant pollutants replacing the emission limit values as laid down in specific Articles of this Directive. The total oxygen content to replace the oxygen content for the standardization is calculated an the basis of the content above respecting the partial volumes.

2.3.3   Special Provisions for Large Combustion Plants

C_{proc} for solid fuels expressed in mg/Nm^3 (O$_2$ content 6%):

Pollutant	50–100 MWth	100–300 MWth	>300 MWth
SO$_2$ General case	850	850–200 (linear decrease from 100 to 300 MWth)	200
Indigenous fuels	or rate of desulfurization $\geq$ 90%	or rate of desulfurization $\geq$ 92%	or rate of desulfurization $\geq$ 95%
NO$_x$	400	300	200
Dust	50	30	30

Until 1 January 2007 and without prejudice to other Community legislation, the emission limit value for NO$_x$ does not apply to plants co-incinerating hazardous waste only.

C_{proc} for biomass expressed in mg/Nm^3 (O$_2$ content 6%):

Pollutant	50–100 MWth	100–300 MWth	>300 MWth
SO$_2$	200	200	200
NO$_x$	350	300	300
Dust	50	30	30

Until 1 January 2007 and without prejudice to other Community legislation, the emission limit value for NO_x does not apply to plants co-incinerating hazardous waste only.

C_{proc} for liquid fuels expressed in mg/Nm^3 (O_2 content 3%):

Pollutant	50–100 MWth	100–300 MWth	>300 MWth
SO_2	850	850 to 200 (linear decrease from 100 to 300 MWth)	200
NO_x	400	300	200
Dust	50	30	30

Until 1 January 2007 and without prejudice to other Community legislation, the emission limit value for NO_x does not apply to plants co-incinerating hazardous waste only.

2.3.4 *C*: Total Emission Limit Values

C expressed in mg/Nm^3 (O_2 content 6%). All average values over the sample period of a minimum of 30 min and a maximum of 8 h:

Pollutant	C
Cd + Tl	0.05
Hg	0.05
Sb + As + Pb + Cr + Co + Cu + Mn + Ni + V	0.5

C expressed in ng/Nm^3 (O_2 content 6%). All average values measured over the sample period of a minimum of 6 h and a maximum of 8 h:

Pollutant	C
Dioxins and furans	0.1

C for solid fuels expressed in mg/Nm^3 (O_2 content 6%); C for biomass (as defined in Council Directive 88/609/EEC as amended) expressed in mg/Nm^3 (O_2 content 6%); C for liquid fuels expressed in mg/Nm^3 (O_2 content 3%):

Pollutant	C
HCl	10
HF	1

3 CONCLUSION

The Council of Europe and the European Parliament legislate the general frame-work of environment. The European Treaty binds the Member States to put Community laws into national legislation. Regarding the precision of the guide-lines, the elaboration of the directives will differ more or less in each Member State. Furthermore, different time scales may exist to take into account the different standards of environmental legislation in the Member States as well as their specific situations. Nevertheless, the Community legislation will be harmo-nized in the future, and separate treatment more and more abolished. Although the principle of subsidiary guarantees the diversity of the European Union, a common EU legislative gains more importance determined by the tempo of the European integration.

Current European environmental legislation has already reached a very high standard in environmental protection. Potential new member states for the Euro-pean Union will be constrained to adopt stronger environmental legislation.

REFERENCES

Further information is available at http://europa.eu.int/pol/env/index_en.htm.

7

Energy Conservation

K. A. Strevett, C. Evenson, and L. Wolf
University of Oklahoma, Norman, Oklahoma

1 INTRODUCTION

A large proportion of our current pollution problems is the result of energy technologies that rely on combustion of carbon-based fuels. Included in these problems are emissions of greenhouse gases, acid rain precursors (oxides of sulfur and nitrogen), and carbon monoxide; formation of photochemical oxidants; releases to the biosphere of raw and refined petroleum products; and mining-related pollution. Obviously, then, decreasing our consumption of carbon-based energy will result in decreases in the amounts of these pollutants entering the biosphere.

Global warming poses the threat of an environmental impact that is global and, at least on a time scale of centuries, irreversible. Over the very long term of two to three centuries, temperatures could rise by as much as 10 to 18°C. While it is impossible at this point to predict accurately all the effects of global warming, its consequences are potentially so threatening to human and ecosystem health that humans have an ethical obligation to do something about it (1).

It is obvious that strategies for reducing consumption of energy derived from combustion of carbon-based fuels are among the most important means of preventing global pollution. After a look at energy demands, this chapter discusses several such energy conservation strategies, the fuels currently being used

99

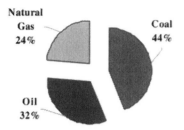

FIGURE 1   Percent contribution of coal, oil, and natural gas to global energy markets.

to supply these demands, and a survey of the environmental impacts of some of the pollutants produced by these fuels.

2   ENERGY SUPPLIES AND DEMANDS

Coal, oil, and natural gas supply about 95% of global energy. Coal dominates energy markets, accounting for about 44% of fossil energy consumption. Oil accounts for about 32% of fossil fuel supply, while natural gas contributes 24% (Figure 1).

Coal is the most abundant fossil fuel worldwide, with current reserves expected to last more than 200 years. "Conventional" oil production is expected to peak between 2010 and 2020, resulting in a switch to "unconventional"* sources and a possible increase in price (2). The total ultimately recoverable natural gas resources in the world are estimated to amount to about 80% as much energy as the recoverable reserves of crude oil. At current usage rates, gas reserves represent approximately a 60-year supply (3).

Although developed countries account for less than 20% of the world's population, these countries use more than two-thirds of the commercial energy supply, consuming 78% of the natural gas, 65% of the oil, and about 50% of the coal produced each year (Figure 2). The United States and Canada, for example, account for only about 5% of the world's population, but consume about one-quarter of the available energy (3). Carlsmith et al. (1990, as cited in Ref. 4) estimated that 36% of U.S. energy consumption is in commercial and residential buildings; industry accounts for another 36% and transportation for the remaining 28%.

---

*Oil is considered unconventional if it is not produced from underground hydrocarbon reservoirs by means of production wells, and/or it requires additional processing to produce synthetic crude. It includes such sources as oil shales, oil sands-based synthetic crudes and derivative products, and liquid supplies derived from coal, biomass, or gas (2).

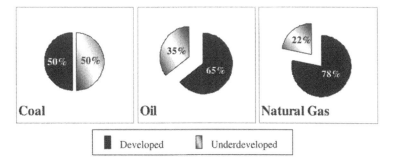

FIGURE 2 Comparison of coal, oil, and natural gas consumption in developed and less developed countries.

In November 1998, the World Energy Outlook (2) predicted 65% growth in world energy demand and 70% growth in CO_2 emissions between 1995 and 2020, without policy changes. The Outlook estimates that fossil fuels will meet 95% of additional global energy between 1995 and 2020 and that two-thirds of the increase in energy demand and energy-related CO_2 emissions over this period could occur in China and other developing countries. The market share of gas is expected to increase, while that of oil will decline slightly and the share of coal will remain stable. By 2020, global electricity generation is predicted to have increased by nearly 88% over 1995 rates. While electricity generation from energy sources other than carbon-based fuels and hydropower is growing fast, it is expected to represent less than 1% of world electricity generation by 2020 without policy changes.

3 NONRENEWABLE ENERGY SOURCES

3.1 Coal

Coal is fossilized plant material preserved by burial in sediments and altered by geological forces that compact and condense it into a carbon-rich fuel. Its advantage lies in its abundance of supply. The environmental effects of burning all the remaining coal, however, could be catastrophic. Coal is the worst offender among fossil fuels in terms of CO_2 per unit of energy generated. The supply of coal is enough to permit atmospheric carbon buildup of severalfold (4). In addition, the burning of coal is a primary source of acid rain precursors. Pollution associated with the mining of coal is discussed later.

Industrialized countries generate between 20% and 30% of their energy from coal; in the case of China, the figure is nearly 75% (5). In the United States, the relative contribution of coal declined from a peak of about 75% of total energy supply in 1910 to about 17% in 1973 and increased again to about 23% in 1989. In 1989, about 86% of domestic coal consumption was accounted for in electric power production (6).

3.2 Petroleum

Petroleum, like coal, is derived from organic molecules created by living organisms millions of years ago and buried in sediments where high pressures and temperatures concentrated and transformed them into energy-rich compounds. Petroleum has represented a relatively inexpensive source of fuel for transportation and provides the chemical industry with feedstocks, e.g., for the production of plastics. However, its use results in emissions of carbon dioxide, carbon monoxide, and acid rain precursors, and in the formation of photochemical oxidants. In addition, aquatic and terrestrial systems may become polluted by unintentional releases of raw and refined petroleum.

3.3 Natural Gas

Natural gas is a combustible mixture of methane (CH_4) and other hydrocarbons formed during the anaerobic decomposition of organic matter. It is the least polluting of the fossil fuels, releasing only a little more than half as much CO_2 as coal. Important disadvantages of natural gas are its limited supply, difficulty of storage in large quantities, and difficulty of transport across oceans. It can be transported across land via pipelines; however, leaks of methane from these pipelines represent a significant contribution to global warming. Furthermore, such pipeline networks are prohibitively expensive for developing countries. As a result, much of the natural gas produced in conjunction with oil pumping is simply burned (flared off), representing a terrible waste of a valuable resource (3).

4   SOURCES AND ENVIRONMENTAL IMPACTS OF POLLUTANTS

The production and/or consumption of carbon-derived energy result in release to the biosphere of a variety of pollutants. These include gaseous pollutants [carbon dioxide, acid rain precursors (nitrogen oxides and sulfur dioxide), and carbon monoxide], photochemical oxidants, unintentional releases of raw and refined petroleum, mining-related pollution (i.e., acid mine drainage), methane, and thermal pollution.

4.1   Gaseous Pollutants

4.1.1   Carbon Dioxide

Carbon dioxide is responsible for 55% of global warming. The two primary anthropogenic sources of atmospheric CO_2 are fossil fuel burning (~77%) and deforestation (~23%). Cline (4) has estimated that if human sources of atmospheric carbon were immediately reduced by about 43%, warming could be held to about 2.5°C.

Atmospheric CO_2 concentration was more or less stable near 280 ppm for thousands of years until about 1850, and has increased significantly since then (Figure 3) (Schimel et al., 1995, as cited in Ref. 7). Since the beginning of the industrial era, about 40% of all CO_2 released through the burning of fossil fuel has been absorbed by sinks; the remainder has remained in the atmosphere (1). The human-caused increase in atmospheric CO_2 already represents nearly a 30% change relative to the preindustrial era (7); annual global emissions of CO_2 have increased 10 times this century (8). At the current rate of increase in concentrations of CO_2 and other heat-trapping gases in the atmosphere, greenhouse gas concentrations will be equivalent to double the preindustrial CO_2 concentration by 2050 (National Academy of Sciences, 1992, as cited in Ref. 1). Ultimately, this could increase the average global temperature by about 1–5°C, with a likely figure of 2.5°C. According to Cline (4), we are already committed to about 1.7°C of warming from the existing accumulation of greenhouse gases, and warming could increase by 10°C or more if nothing is done to alter likely fossil fuel consumption patterns. The historic record suggests that the average global surface temperature has already risen approximately 0.3–0.6°C since the nineteenth century (1).

Natural gas releases slightly less than half the amount of CO_2 released during the combustion of coal, with petroleum in between. Coal and natural gas each accounts for about 27% of U.S. fossil fuel supply, but coal accounts for about

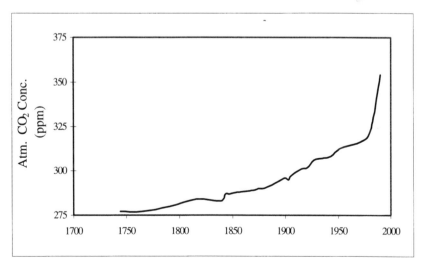

FIGURE 3   Historical increase in global CO_2 emissions. (*Sources*: Refs. 35–37.)

one-third of U.S. CO_2 emissions. In the United States, electric utilities account for about one-third of all CO_2 releases, with transportation activities adding approximately an additional third. Globally, oil consumption accounts for nearly half of total CO_2 emissions and much of its air pollution (6).

4.1.2 Nitrogen Oxides

Nitrogen oxides (NO_x) are responsible for about 35% of acid rain, and are a precursor of O_3 pollution (Figure 4). Of all U.S. air pollutants, oxides of nitrogen have been the most difficult to control. They are formed when ambient diatomic nitrogen (N_2) is heated to temperatures > 1200°F, and their dominant sources are the internal combustion engine and power plants (Figure 5) (1). The 900 million tons of coal burned annually in the United States are responsible for about one-third of all this country's NO_x emissions (3).

$$2NO + O_2 \rightarrow 2NO_2$$

$$2NO_2 + H_2O \rightarrow HNO_2 + HNO_3$$

There are various ways of reducing nitrogen oxide emissions including combustion control and the use of catalysts (9). Our best option for reducing this pollutant, however, is through reduced burning of fossil fuels and forests.

4.1.3 Sulfur Dioxide

Sulfur dioxide (SO_2) is responsible for about 60% of acid rain (Figure 4). At least two-thirds of the sulfur oxides in the United States are emitted from coal-fired power plants. Much of the coal burned in the United States has a high sulfur content—2% or more. Most of the remaining SO_2 emissions are accounted for by industrial fuel combustion and industrial processes such as petroleum refining, sulfuric acid manufacturing, and smelting of nonferrous metals (Figure 5) (10).

4.1.4 Carbon Monoxide

Carbon monoxide (CO) is the result of incomplete combustion. CO inhibits respiration in animals by binding irreversibly to hemoglobin. About half the CO released to the atmosphere each year is the result of human activities. In the United States, two-thirds of the CO emissions are created by internal combustion engines in transportation (3).

4.2 Photochemical Oxidants

Photochemical oxidants are products of secondary atmospheric reactions driven by solar energy—e.g., splitting of an O_2 or NO_2 molecule, freeing an oxygen atom which reacts with another O_2 to form ozone (O_3). O_3 is the result of atmospheric chemistry involving two precursors, nonmethane hydrocarbons (HCs) and NO_x, which react in the presence of heat and sunlight (Figure 6) (11).

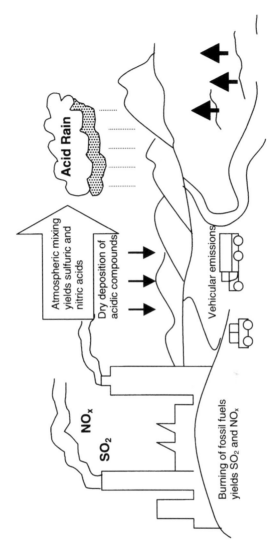

FIGURE 4   NO$_x$ and SO$_2$ contributions to acid rain formation.

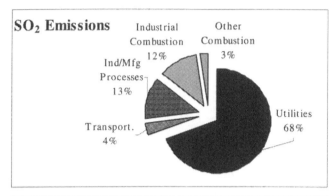

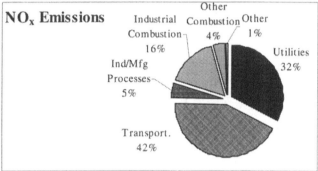

FIGURE 5 Percent contribution to SO_2 and NO_x emissions in the United States of various industries. (*Source*: Ref. 34.)

This ground-level O_3 is a pollutant that can have harmful effects on human health, while O_3 present in the upper atmosphere protects the earth from harmful ultraviolet radiation. Figure 6 demonstrates the dynamic interactions between HCs and NO_x, which are produced from combustion, and atmospheric oxygen. In addition to forming O_3, NO_x can also remove ground-level O_3. This removal is often temporary, however, as O_3 is re-formed through other reactions.

Ground-level O_3 is a respiratory irritant that causes health concerns at very low concentrations because its very low solubility in water means it tends not to be removed by the mucous in the upper respiratory tract and penetrates deeper into the lungs. There is evidence that exposure to O_3 accelerates the aging of lung tissue and increases susceptibility to respiratory pathogens. Human exposure to O_3 can produce shortness of breath and, over time, permanent lung damage (12). Costs of the health effects of O_3 in the United States are estimated to be about $50 billion per year. In addition, O_3 causes more damage to plants than any other pollutant (1). O_3 concentrations rise with temperature and are therefore expected to be exacerbated by global warming. If cloud cover decreases as a result of global

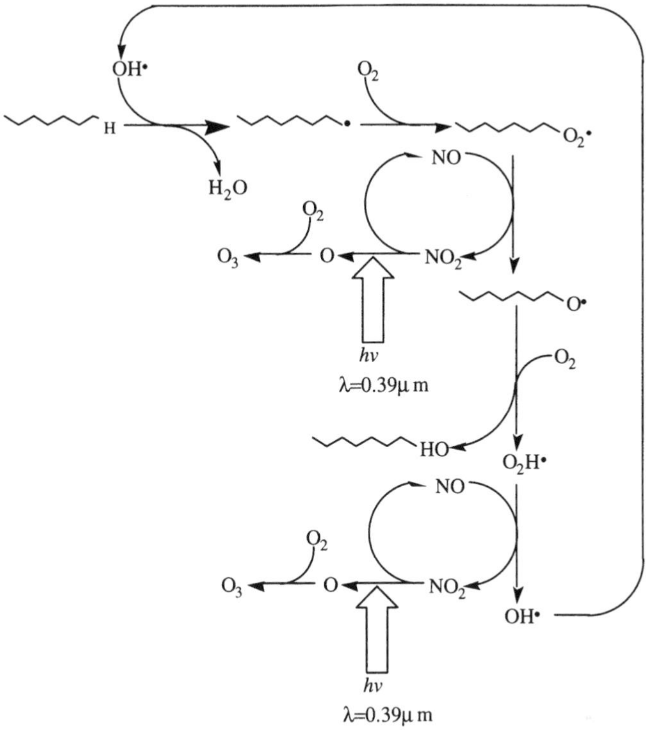

FIGURE 6 The release of hydrocarbons and NO during combustion results in the conversion of NO to NO_2. Increased formation of NO_2 increases the production of O_3.

warming, thus permitting increased penetration of sunlight, O_3 concentrations will be further increased.

4.3 Raw and Refined Petroleum Spills and Leaks

Crude oil spills such as that of the *Exxon Valdez* are probably the most widely known examples of this type of energy-related pollution. In addition, it has been estimated that about 11 million gallons of gasoline are lost each year by leaking underground storage tanks (3).

4.4 Mining-Related Pollution

Acid mine drainage is one of the most common and damaging problems in the aquatic environment. Many waters flowing from coal mines and draining from the waste piles that result from coal processing and washing have low microbial densities due to the highly acidic nature of these waters. Acidic mine water results from the presence of sulfuric acid produced in a series of microbially mediated reactions that begin with the oxidation of pyrite, FeS_2 (13). Often, mining operations result in surface waters infiltrating into the subsurface voids, especially after the mine is exhausted and pumping ceases. In some areas of Appalachia, large underground impoundments of water have filtered into coal mines. These waters have become very acidic and, when they are returned to the surface via pumping or by subsurface flows, their low pH value devastates the aquatic systems they infiltrate (14).

Another impact of underground mining is the waste materials that are a by-product of any mining operation. Gaining access to the vein or seam of coal, as well as transporting the coal to the surface, requires large amounts of waste materials to be removed to the surface. These waste materials, or tailings, are often piled up in large mounds in close proximity to the mine. The composition of many tailings can contain toxic minerals such as mercury, lead, or iron sulfide. Water percolating through these waste materials often produces water quality problems downstream from the tailings similar to those associated with subsurface water flows from within the mines. In addition to the sterile conditions on tailings mounds themselves, rain water running off the tailings often is so acidic as to kill both the vegetation in the immediately affected lands and the aquatic life in streams and rivers receiving these waters. Many lands and streams within the Appalachian coalfield areas of western Pennsylvania, West Virginia, eastern Kentucky, and eastern Ohio are devastated by the acidic waters resulting from coal mining operations. The enactment of environmental legislation limits the damage associated with active mining operations, but the land degradation associated with past mining has left a filthy legacy of degraded landscapes (14).

Surface mining is usually favored over underground mining for primarily economic reasons. It is virtually impossible to prevent land degradation when surface mining occurs. First, in some operations, huge depressions result. Second, the overburden (extracted soil, subsoil, and unconsolidated earth and rocks) must be stored and then replaced systematically in their original order after the mineral is removed. Even under optimal conditions, which rarely occur, restoration usually results in a landscape that is less productive than it was prior to mining. Subsurface groundwater flow is always disturbed, and revegetation is often slow. Restoration is further complicated when toxic materials are leached from the overburden during its storage. These conditions often occur in coal mining operations, which have disturbed about 2.3 million acres in the United States (14).

The area affected by mining can be three to five times more widespread than the area actually exploited (15). Even when increased acidity is not considered, mining-related soil erosion alone can impact natural waters significantly. Added nutrients may increase aquatic productivity, resulting in eutrophication. Lower levels of dissolved oxygen associated with eutrophication may render the water uninhabitable by other aquatic organisms. On the other hand, suspended sediments may reduce light penetration, *reducing* productivity and therefore available fish food. Sediments may also interfere with salmon and trout spawning and reduce survival of their eggs. Young fish may be more susceptible to predation when sediments fill or cover hiding places (14). Species that stalk their prey visually may be unable to survive in murky water.

4.5 Methane

Methane is responsible for about 20% of the greenhouse effect, and concentrations have already risen to more than double preindustrial estimations. Concentrations continue to rise at about 0.9% annually (4). The majority of anthropogenic methane is the result of non-energy-related human activities such as ruminant livestock and cultivation of rice (from which about half the world's population derive about 70% of their calories), and decomposition of organic matter in landfills. However, leaks in natural gas pipelines contribute about 21% of anthropogenic methane, and the burning of coal adds an additional 6%. Other energy-related sources of methane include coal mines, natural gas leaks, gas associated with oil production, and the creation of new wetlands when forests are flooded following construction of hydroelectric dams.

4.6 Thermal Pollution

When coal is burned to generate electricity at a power plant, some of the coal's energy content is lost to coolant water, which is then discharged into rivers or lakes. Since an inverse relationship exists between water's temperature and its oxygen-holding capacity, the water's dissolved oxygen concentration can be diminished to a point below which some aquatic organisms may be able to survive.

5 POLLUTION PREVENTION THROUGH DECREASED FOSSIL FUEL CONSUMPTION

Carbon dioxide can be considered an inevitable product of fossil fuel combustion; therefore, CO_2 emissions can be reduced only through reduced consumption of fossil fuels. It is important to note that emissions of every other pollutant discussed in Section 4 will be reduced as an additional benefit of reducing fossil fuel consumption and thereby CO_2 emissions.

5.1 Imposition of a Tax on Traditional Energy Sources

Internal costs are the expenses, monetary or otherwise, that are borne by those who actually use a resource. External costs are the expenses, monetary or otherwise, borne by someone other than the individuals or groups who use a resource (3). As an example, according to Tenenbaum (27), a 1990 study at Pace University concluded that the true cost of an unscrubbed coal plant was 11.6 cents per kilowatt hour (kWh), double the 5.8 cents that utilities were charging.

Cline (4) has produced an extensive analysis of the economic effects of global warming. One strategy for reducing dependence on fossil energy sources is the imposition on these sources of a tax large enough to "internalize" the costs associated with fossil fuels, such as sea-level rise (estimated by Cline to amount to about $7 billion annually in the United States*), agricultural losses ($18 billion), curtailed water supply due to reduced runoff ($7 billion), forest loss (>$3 billion, considering only lumber value), increased electricity demand for additional cooling ($11 billion), exacerbation of urban O_3 problems ($4 billion), increased mortality from heat waves ($6 billion, valued at lifetime earning potential), losses of leisure activities associated with winter sports (ski industry $1.5 billion), increased hurricane ($750 million) and forest fire damage, and species loss. Cline estimates total damage from CO_2-equivalent doubling the amount to about $61 billion,[†] or about 1.1% of the Gross Domestic Product (GDP). Intangible losses such as species loss and decline in human quality of life could raise the total to about 2% of GDP. If CO_2 doubling results in a temperature increase of 4.5°C rather than 2.5°C, the corresponding damage could be as high as 4% of GDP, with even greater losses in some other countries such as low-lying island nations.

Some of the revenue derived from the tax could be channeled toward improvements in public transportation, development and/or subsidization of more environmentally benign energy sources, and research directed toward efficiency improvements. Cline (4) suggests that some of the revenue be channeled to developing countries "to secure their participation in international abatement. . . . The importance of including developing countries in international measures for restraining and reducing emissions, and the political and equity considerations that seriously limit the amount of growth these countries can be expected to sacrifice to help avoid global warming, strongly point to the need to channel some of the revenue from a carbon tax from industrial countries to assist developing countries that are prepared to take measures to reduce deforestation and configure future energy development along lines that minimize carbon emissions."

---

*Figures are in 1990 dollars and are based on a doubling of CO_2-equivalent resulting in an approximate temperature increase of 2.5°C; concentrations of more than double preindustrial levels obviously would result in even higher costs.

†In contrast, Tenenbaum (27) cites a 1991 report that says the external costs of energy *currently* range from $100 billion to $300 billion in the United States.

5.2 Establishing Emissions Caps and Trading Programs

Establishing emissions caps and trading programs would be similar to the imposition of limits on sulfur emissions established by the 1990 Clean Air Act Amendments (CAAA); a brief discussion of these limits is therefore warranted.

The CAAA established an absolute cap on sulfur dioxide (SO_2) emissions by electrical utilities of 8.95 million tons after an initial reduction of 10 million tons; it is assumed that this cap is sufficient to protect ecosystem health. Under the technology-forcing regulatory approach of the past, each utility would have been required to install a technology that reduced emissions by an amount sufficient to achieve the 10-million-ton reduction. Economists have argued that this approach results in higher control costs than necessary. Different utilities are likely to incur different control costs due to age and technological differences in their facilities (i.e., one utility may have a much lower per-ton incremental cost for emission reduction than another).

The 1990 Clean Air Act Amendments provides for the issuance of permits to utilities equal to 30–50% of their emissions 10 years earlier. Utilities whose per-ton incremental costs for emission reduction are low can reduce emissions beyond the level required for permit compliance and then sell surplus permits. In turn, utilities whose incremental costs are high can reduce their control costs by purchasing permits from utilities whose incremental costs are low. The end result is achievement of the desired level of SO_2 emissions reduction without imposing unreasonable economic burdens on utilities while, at the same time, providing an economic incentive for industries to reduce their SO_2 emissions.

A similar program could be developed and used for carbon emissions. The cap for carbon emissions could be based on the degree to which countries would like to limit global warming. For instance, freezing global carbon emissions at the current level of about 6 billion tons (gigatons, or GtC) would limit warming to about 5°C (1). Capping emissions at 4 GtC would limit warming to 2.5°C. Carbon emissions could by reduced by as much as 20–25% through energy efficiency improvements and substitution of non-carbon energy sources (both of which are discussed later) at zero net economic cost with significant economic benefit to those companies involved in this trading program (4).

According to Cline (4), it is widely believed that a system of tradable permits can be applied globally, on a country-by-country basis, in much the same manner as would a carbon tax. If a country has a quota allocation that is small relative to its demand, its firms could bid to purchase quotas from other countries. Other countries could sell a portion of their quotas at a price that would equal or exceed the cost of reducing their overall carbon emissions. Thus, global carbon emissions could be reduced through an economic incentive program that would reward countries that reduce their overall carbon emissions.

Booth (1), on the other hand, believes that permits issued on an individual basis rather than by country would be more effective:

> Permits could be domestically distributed annually on a per person basis equal in amount to existing emissions initially, and then reduced by 3.6 percent of the initial amount each year over a phase-in period of approximately 25 years to arrive at a 90 percent total reduction. Individuals who don't need the full allocation for their own energy consumption could sell their surplus permits at the going market price. Such a system would tend to redistribute income away from industries and high-income families who are heavy consumers of energy to low-income families who tend to consume less energy. Because of the potential to sell surplus permits, the public resistance to a permit system would be less than to a carbon tax. The rising price of permits over time would provide the incentive needed for increased energy conservation and to shift to non-fossil fuel energy sources. As in the case of acid rain control, a marketable permit system for carbon emissions control results in control being achieved at the lowest possible cost (1). p. 23

Either of the above strategies would constitute impetus for increases in efficiency and other conservation measures. Both taxes and tradable permits minimize overall abatement costs by allocating the cutbacks to the countries where marginal costs of emissions reductions are the lowest. A major difference between the two strategies is that, with tradable permits, it is possible to specify the exact cutback in emissions (4). Cline (4) believes the best strategy to be reliance on nationally set carbon or greenhouse gas taxes during an initial phase-in period and then, in a subsequent phase, to set the taxes at an internationally agreed rate while each individual nation would continue to collect them. If such taxes failed to achieve satisfactory progress toward global emission targets, it would then be appropriate to shift to an international system of tradable permits.

5.3 Elimination of Subsidies

5.3.1 International Subsidies

For some years, the World Bank (33) has been drawing attention to the fact that electricity is sold in developing countries at, on average, only 40% of the cost of its production. A recent study pointed out:

> Such subsidies waste capital and energy resources on a very large scale. Subsidizing the price of electricity is both economically and environmentally inefficient. Low prices give rise to excessive demands and, by undermining the revenue base, reduce the ability of utilities to provide

and maintain supplies. Developing countries use about 20 percent more electricity than they would if consumers paid the true marginal cost of supply. Underpricing electricity also discourages investment in new, cleaner technologies and more energy efficient processes (16). p. 12

Shah and Larsen (1991, as cited in Ref. 4) estimated that nine large developing and Eastern European countries (China, Poland, Mexico, Czechoslovakia, India, Egypt, Argentina, South Africa, and Venezuela) spend a combined $40 billion annually in subsidization of fossil fuels (with China's* $15.7 billion the largest). The former Soviet Union spends more than twice this amount—$89.6 billion annually—on fossil fuel subsidies. The removal of these subsidies would eliminate an estimated 157 million tons of carbon annually from the developing group and 233 million tons from the former Soviet Union alone. These cutbacks would represent about 8% of global carbon emissions (or about 6% if deforestation emissions are included).

Prices that cover production costs and externalities are likely to encourage efficiency, mitigate harmful environmental effects, and create an awareness conducive to conservation. Subsidized energy prices, on the other hand, are one of the principal barriers to raising energy efficiency in developing countries, where it is only 50–65% of what would be considered best practice in the developed world. Studies indicate that with the present state of technology a saving of 20–25% of energy consumed would be achieved economically in many developing countries with existing capital stock. If investments were made in new, more energy-efficient capital equipment, a saving in the range of 30–60% would be possible (9).

5.3.2   U.S. Subsidies

According to Ackerman (30), two studies have attempted to measure federal energy subsidies. The Department of Energy's Energy Information Administration identifies subsidies worth $5–$13 billion annually, while the Alliance to Save Energy, an energy conservation advocacy group, estimates energy subsidies at $23–$40 billion annually (in 1992 dollars). Ackerman also states that several provisions of the tax code are, effectively, subsidies to the oil and gas industry and that, depending on one's view of a local tax controversy, the total subsidy to oil and gas production alone might be as much as $255 million, almost 5% of sales in 1990.

---

*China accounts for 11% of global carbon emissions, excluding emissions from deforestation. Seventy percent of China's energy comes from coal (4).

5.4 Increases in Energy Efficiency

Primary energy is defined as the energy recovered directly from the Earth in the form of coal, crude oil, natural gas, collected biomass, hydraulic power, or heat produced in a nuclear reactor from processed uranium. Generally, primary energy is not used directly but is converted into secondary energy (9). The process of energy conversion and transformation results in part of the energy being wasted as heat. Energy efficiency considerations focus on the following factors:

> The efficiency of original extraction and transportation
>
> The primary energy conversion efficiency of central power plants, refineries, coal gasification plants, etc.
>
> The secondary energy conversion efficiency into storage facilities, distribution systems and transport networks (e.g., of electricity grids)
>
> Efficiency of final energy conversion into useful forms such as light and motion (9)

For the world as a whole, the overall efficiency with which fuel energy is currently used is only around 3–3.5% (17). According to Orr (32), a Department of Energy study showed that U.S. energy consumption could be reduced by 50% with present technologies with a net positive economic impact. The United States did indeed reduce the energy intensity of its domestic product by 23% between 1973 and 1985 (18).

5.4.1 The Industrial Sector

The industrial sector in the more advanced industrial countries is the most efficient energy user. It is easier to be efficient when operating on a larger scale and when energy is an explicit element of operating costs. Profit margins mandate careful cost analysis, and in industries where energy costs comprise a significant portion of total costs, managers are more alert to opportunities for savings (9). According to the Office of Technology Assessment (1991, as cited in Ref. 2) four sectors—paper, chemicals, petroleum, and primary metals—account for three-fourths of the energy used in manufacturing. More than half the energy consumed by industry in the leading industrial countries is as fuel for process heat, and over one-fifth (gross) is in the form of electricity for furnaces, electrolytic processes, and electric motors. Most process heat is delivered in the form of steam, with an overall efficiency variously estimated to be between 15% and 25%. The biggest users of process heat are the steel, petroleum, chemicals, and paper and pulp industries (9).

Potential for improvements does exist. In general, sensors and controls, advanced heat-recovery systems, and friction-reducing technologies can decrease energy consumption (5). Many efficiency measures are specific to each industry. For instance, the World Energy Council (9) offers several options for improving efficiency in the chemical industry, including the use of biotechnology and catalysts (Table 1).

In the paper industry, automated process control, greater process speeds, and high-pressure rollers can boost efficiencies significantly (5). According to Carlsmith et al. (1990, as cited in Ref. 4), electric arc furnaces using scrap are much more energy efficient for steel production than are traditional techniques and could increase their share of output from 36% to 60%. According to Cline (4), these authors also estimate that by 2010, direct reduction or smelting of ore for making iron would reduce energy requirements in steelmaking by 42% with a net cost savings. Even greater opportunities exist for improving energy efficiency in developing countries: for example, China and India use four times as much energy as Japan does to produce a ton of steel (5).

In aluminum production, energy efficiency can be increased by improved design of electrolytic reduction cells, recycling, and direct casting. Other examples of improvements in industrial processes include low-pressure oxidation in industrial solvents, changes in paper-drying techniques (as well as paper recycling), and shifting from the wet to the dry process in cement making (4).

Co-generation, the simultaneous production of both electricity and steam or hot water, represents a great opportunity for improving energy efficiency in that the net energy yield from the primary fuel is increased from 30–35% to 80–90%. In 1900, half the United States' electricity was generated at plants that also provided industrial steam or district heating. However, as power plants became larger, dirtier, and less acceptable as neighbors, they were forced to move away from their customers. Waste heat from the turbine generators became an unwanted pollutant to be disposed of in the environment. In addition, long transmission lines, which are unsightly and lose up to 20% of the electricity they carry, became necessary. By the 1970s, co-generation had fallen to less than 5% of our power

TABLE 1 Options for Improving Efficiency in Chemical Industry

Options	Benefits
Biotechnology	Speed reaction times
	Reduce necessary temperatures and pressures
Catalysts	Improve yields and reaction times
	Reduce necessary temperatures and pressures
Separation and concentration	Improve product purity
Waste heat management	Reduce necessary temperatures and pressures

supply, but interest in this technology is being renewed, and the capacity for co-generation has more than doubled since the 1980s.

5.4.2 Buildings

In developed countries, buildings are the largest or second-largest consumers of energy. In the United States, buildings account for about 75% of all electricity consumption (19) and about 35% of total primary energy consumption (3); most of this is for heating and cooling. Electricity generation alone produces more than 25% of energy-related carbon dioxide emissions (20). Building improvements could therefore have a major impact on overall energy consumption and carbon emissions.

In a "typical" North American house, the average efficiency of insulation is about 12% compared with the ideal. As a result, the overall energy efficiency of air cooling systems has been estimated to be barely 5%, and the overall energy efficiency for space heating is less than 1%. These figures do not take into account avoidable losses through heating or cooling unoccupied rooms (9).

Building design is one of the simplest yet most effective ways to take advantage of solar energy. Buildings can incorporate either passive or active solar technologies. Passive solar heating and cooling function with few or no mechanical devices; primarily they involve designing the form of landscape and building in relation to each other and to sun, earth, and air movement (19). In general, passive technologies use a building's structure to capture sunlight and store heat, reducing the requirements for conventional heating and lighting. Heating can be cut substantially by the use of one or several technologies in the building's design (Table 2). When included in a building's initial design, these methods can save up to 70% of heating costs (21). Orr (32) points out that it is cheaper and less risky by far to weatherize houses than it is to maintain a military presence in the Persian Gulf at a cost of $1 billion or more each month.*

Cooling needs also may be reduced by passive means; one strategy is the reduction of internal heat gains. Another passive strategy for reducing cooling needs is by reduction of external heat gains. Several technologies that can be used to reduce internal and external heat gains are listed in Table 2. Also, it is important

---

*Nearly one-quarter of all jet fuel in the world, about 42 million tons per year, is used for military purposes. The Pentagon is considered to be the largest consumer of oil in the United States and perhaps in the world. One B-52 bomber consumes about 228 liters of fuel per minute; one F-15 jet, at peak thrust, consumes 908 liters of fuel per minute. It has been estimated that the energy the Pentagon uses up annually would be sufficient to run the entire U.S. urban mass transit system for almost 14 years. Further, it has been estimated that total military-related carbon emissions could be as high as 10% of emissions worldwide, and that between 10% and 30% of all global environmental destruction can be attributed to military-related activities (28).

TABLE 2 Technologies for Increasing a Building's Energy Efficiency

Area for improving energy efficiency	Technology
Heating	Heat-circulation systems using natural convective forces
	Heat pumps
	Solar-thermal collectors
	Insulated windows and shutters
	Special window glazings
	Heat-storing masses built into structure
	Building orientation
	Draft proofing
	Superinsulation of structure
Cooling	Fluorescent lighting over incandescent
	Lower-wattage bulbs
	Landscaping that provides maximum shade
	Window shades
	Reflective or tinted window coatings
	Insulated windows
	Light-colored roofs
	Ventilation by natural convection
	Ground absorption of heat

to trade in old, wasteful for newer, more efficient ones; the payback period may be as little as two to three years (3).

One measure proposed in several developed countries is to require all houses to be subject to an energy efficiency survey that would lead to an energy efficiency rating which would have to be disclosed to prospective buyers when the house is sold (9).

5.4.3 Lighting

About 40–50% of the energy consumed in a typical house is used for heating and cooling, with an additional 5–10% used for lighting. Lighting is the least efficient common use of energy: about 95% of the energy used in an average lighting system dissipates as heat (19). Incandescent bulbs have an efficiency of about 4% in converting electricity to visible radiant energy. In contrast, the efficiencies of fluorescent lights is typically around 20%, and can be as high as 35% (9). According to Lovins and Lovins (1991, as cited in Ref. 4), a 15-W compact fluorescent bulb emits the same amount of light as a 75-W incandescent bulb and lasts 13 times as long. Further, over its lifetime, it can save enough coal-fired electricity to reduce carbon emissions by 1 ton with a net savings. The National Academy of Science (1991, as cited in Ref. 4) contends that the replacement of an average of just 2.5 heavily used interior incandescent bulbs and one exterior bulb by compact fluorescent lights would reduce average household lighting energy requirements by 50%. Why, then, do we continue to use incandescents?

Lack of awareness
Easy commercial availability or promotion
High first cost
High replacement cost in the event of breakage
Cost and inconvenience of retrofitting new lighting systems to existing domestic buildings, where rewiring and new sockets, holders, and appliances may be needed (9)

5.4.4 Government's Role

MacNeill (31) contends that, in order to make steady gains in energy efficiency, governments must institute politically difficult changes in at least three areas:

1. Countries must consider "conservation pricing," i.e., taxing energy during periods of low real prices to encourage increases in efficiency.
2. Stricter regulations should demand steady improvement in the efficiency of appliances and technologies, and in building design, automobiles, and transportation systems. [In the United States, efficiency standards for appliances were adopted in 1986. For refrigerators, the biggest users of electricity in most households, the energy efficiency of new models almost tripled from 1973 to 1993 (22).]
3. Institutional innovation will be necessary to break utility-supply monopolies and to reorganize the energy sector so that energy services can be sold on a competitive, least-cost basis.

In addition, governments should excise policies that retard the development of new and renewable energy resources, particularly those that serve as substitutes for fuelwood.

5.4.5 Caution

As a final word on the issue of efficiency, it is worthwhile to quote Cline (4):

In reaching the overall conclusion that some 20 percent to 25 percent of carbon emissions in the United States might be eliminated at zero cost by a move to "best practices," it is important that there not be a misguided inference that dealing with the greenhouse problem will be cheap over the longer term. . . . Serious action to curb global warming would involve emissions restraints over a period of two to three centuries. . . . Whether the first step is low-cost (or even no-cost) is significant but of limited help in gauging the eventual costs.

The central point is that a one-time gain from elimination of inefficiencies would shift the entire curve of baseline emissions downward but still leave future emissions far above present levels. Consider the period through the year 2100 . . . a central baseline estimate calls for approximately 20 GtC of global carbon emissions by that year . . . an aggressive program to limit global warming would mean restricting emissions to approximately 4 GtC annually. Suppose the engineering approach is correct that, 20 percent of emissions can be eliminated for free. Such gains would still leave emissions at 16 GtC in the year 2100, far above the 4-GtC ceiling needed to substantially curb the greenhouse effect. The remaining cutbacks would have to be achieved through more costly industrial reductions in energy availability beyond those achievable through costless efficiency gains. In short, the "best practices" school provides a basis for expecting that addressing the global warming problem may be less costly than otherwise might be thought, but it by no means warrants the conclusion that action will be costless over the longer term (4).

5.5 Energy Conservation in Transportation

Transport activities account for about 30% of the energy used by final consumers, and about 20% of the gross energy produced (9). About 98% of the total comes from petroleum products refined into liquid fuels, and the remaining 2% is provided by natural gas and electricity (3). Movement of people takes about 70% of the total, and movement of freight about 30%. Within this sector, road transport accounts for the largest proportion, over 80% in industrialized countries, with air transport next, at 13% (9). According to the United Nations Fund for Population Activities (29), the world car fleet increased by seven times between 1950 and 1980 while human population only doubled during that period. Fifteen percent of the world's oil is consumed by automobiles and light trucks in the United States alone (Office of Technology Assessment, 1991, as cited in Ref. 4).

About 75% of all freight in the United States is carried by trains, barges, ships, and pipelines, but because they are very efficient, they use only 12% of all transportation fuel (3). The rapid increase in road transport in recent years is a major contributor to the rise in oil demand. Further, motor vehicles are believed to be responsible for 14% of all CO_2 derived from fossil fuel combustion (9), along with their contribution to acid rain and other forms of air pollution such as O_3.

The Reagan administration relaxed automobile efficiency standards that had already been met by Chrysler. If the regulations had been left in place, the amount of gasoline saved in a decade or so would have been equivalent to the entire amount of oil estimated to underlie the Arctic National Wildlife Refuge (23).

Gasoline prices in Europe and Japan are double or triple the U.S. price because governments there impose levies that force consumers to consider and internalize the full costs of their behavior (5). The gradual imposition of a significantly higher gasoline tax, until the cost of gasoline in the United States is comparable to that in Europe, would create a powerful incentive for people to drive smaller, more fuel-efficient cars and use energy-efficient alternative forms of transportation. Highways and bridges would last longer, and emissions would be reduced, attenuating global warming and acid rain. This would, of course, necessitate improvement of public transportation to accommodate people who could no longer afford to drive to work; some of the gas-tax funds could be set aside for this. In the United States, mass transport accounts for only 6% of all passenger travel; in Germany the figure is over 15% and in Japan it is 47% (9).

Another possibility for internalization of the many hidden costs of driving would be the implementation of an insurance program based on the average number of miles a driver travels. This would link a portion of drivers' insurance programs to the number of miles they drive and collect payments at the gas pump (12). Ledbetter and Ross (11) provide the details of such an arrangement:

The price of gasoline at the pump could include a charge for basic, driving-related automobile insurance that would be organized by state governments and auctioned in blocks to private insurance companies. All registered drivers in the state could automatically belong. Supplementary insurance above that provided by the base insurance purchased at the pump could be independently arranged, as we presently do for all our insurance. For example, owners of expensive cars, or people who desire higher levels of liability coverage, could purchase supplemental insurance. Drivers with especially bad driving records could be required to purchase supplemental liability insurance. Below are some of the advantages of such an arrangement.

Insurance costs become much more closely tied to the amount of driving alone. The more miles a person drives, the more insurance he or she pays. Since accident exposure is closely correlated with miles driven, the proposed system would be more fair than the present system, in which people who drive substantially less than the average miles per year are given only small discounts, and people who drive substantially more than the average don't pay any additional premium.

If insurance were part of the cost of gasoline, a person could not drive without paying for insurance. Uninsured motorists would be brought into the system, substantially lowering the cost of driving for insured motorists: in California for example, uninsured motorists increase premiums for insured motorists by about $150 per year.

The apparent cost of gasoline at the pump would rise substantially, roughly 50 cents to a dollar per gallon. Such a price rise would encourage the purchase of more fuel-efficient vehicles and help slow the growth in vehicle miles of travel. For consumers, the increase in the price of fuel would be offset by a decrease in the annual insurance premium motorists would pay directly to insurance companies, resulting in no net increase in driving costs.

Unlike a gasoline tax, this system would not be regressive: many low-income persons drive substantially less miles per year than their higher-income counterparts. They would, therefore, see a substantial drop in the money they pay for auto insurance (11).

5.5.1 Efficiency Issues in Transportation

The efficiency of a motor vehicle is a function of several factors (Table 3). Typically, about 80% of the fuel used in a representative vehicle traveling over a mix of urban, rural, and highway routes is unproductive energy spent in overcoming internal friction in auxiliary items and in thermodynamic losses in the engine (9). Improvements in vehicle design and alternative fuels can have a major impact in improving efficiency and reducing emissions. However, much of the forward momentum achieved in the decade prior to 1985 has slowed in response to downward oil price movements and apparent consumer preferences (9).

The inherent efficiency of the internal combustion engine began to approach its limits in the 1960s. Engines built since then range from 34% efficiency for spark-ignition automobile-type engines under optimum load/speed conditions to about 42% for large marine-type and direct-injection diesels. The difference is attributable to the higher compression ratios, lower throttle losses and improved direction injection achievable in large diesels.

In practice, however, optimum load/speed conditions are never achieved. The energy efficiency of a vehicle operating in traffic, with variable speeds and

TABLE 3 Factors Affecting a Motor Vehicle's Fuel Efficiency

Factor	Components
Design	Weight
	Efficiency
	Frictional losses
	Aerodynamics
Use	Effectiveness of use in transporting materials and people
Typical operational cycle	Length of journey
	Traffic conditions

loads, is at least 30% lower. Short journeys, when the engine is cold at start-up and never warms up sufficiently for optimal fuel combustion, create suboptimal fuel use and high emissions. Stop/start conditions in heavy traffic also cause relatively high fuel use and emissions (9).

Engine efficiency is further reduced, often by an additional 30% or so, by the carrying of oil pumps, air pumps, fuel pumps, electrical systems, heating, air conditioning, and other related equipment. Friction and viscosity losses in the vehicle's drive train—e.g., in automatic transmissions, which alone can reduce engine efficiency by 10–15%, cut efficiency still further. As a result, the average thermodynamic efficiency of the motor vehicle is only between 10% and 17%.

Nevertheless, significant improvements in automobile fuel economy have been achieved in recent years. The biggest gains have been made by cutting down on excess weight in the body, improving aerodynamics, and improving tires. Still, the "payload efficiency" of a medium-sized car is only about 10%, while that of fully loaded commercial aircraft is around 30–35%. Heavy-duty trucks, freight trains, and ships also achieve greater payload efficiencies than cars (9).

Raising the average fuel efficiency of the U.S. car and light truck fleet by 1 mpg would cut oil consumption about 295,000 bbl per day. In one year, this would equal the total amount the Interior Department hopes to extract from the Arctic National Wildlife Refuge in Alaska (3). Increased fuel efficiency can be supplemented by savings from transportation management, including increased mass transit, carpooling, and improved maintenance (including proper tire inflation) (4).

5.6 Increased Exploitation of Natural Gas

Increased exploitation of natural gas in preference to coal or oil as an interim measure has the potential to slow global warming as non-hydrocarbon primary energy sources are developed and put into place. Natural gas provides about one-fifth of global commercial energy and is our most efficient "traditional" energy source. Only about 10% of its energy content is lost in shipping and processing, since it moves by pipelines and usually needs very little refining. Ordinary gas-burning furnaces are about 75% efficient, and high-economy furnaces can be as much as 95% efficient (3). It generates fewer pollutants than any other traditional fuel and less CO_2 as well: 42% less than coal and 30% less than oil (5).

According to Gibbons et al. (5), some analysts feel that the most promising future option for electric power generation is the aeroderivative turbine, which is based on jet engine designs and burns natural gas. With additional refinement, this technology could raise conversion efficiency from its present 33% to more than 45%.

North America has a pipeline network for delivering natural gas to market. However, most countries cannot afford a pipeline network, and much of the natural gas that comes out of the ground in conjunction with oil pumping is simply burned (flared off), a terrible waste of a valuable resource (3).

Natural gas is quite easy to ship through pipelines as long as it is going from one place to another on the same continent. The problem is that much of the gas is in Russia or the Middle East, while the markets are in Europe, Japan, or North America. One way of shipping gas across oceans is to liquefy it by cooling it below its condensation point (−140°C). Liquefied natural gas (LNG) has only 1/600 the volume of the gaseous form, and is therefore economical to transport by tanker ship. However, if a very large LNG tanker had an accident and blew up, it would release as much energy as several Hiroshima-sized atomic bombs (3).

5.7 Increased Exploitation of Passive Technologies

Because most paved surfaces, and the surfaces of most buildings, tend to retain and release more heat than is true of vegetated areas, and because heating and air conditioning equipment releases/generates a great deal of heat, urban areas typically are several degrees warmer than vegetated areas. For example, an early study of this subject showed downtown St. Louis to be 13°F warmer in the winter and 9°F warmer in June than the large, tree-canopied Forest Park, 5 miles away. Tree cover can moderate this "heat island effect," helping to control micro-climate in three different ways:

1. Absorption and reflection of solar radiation. A tree in full leaf intercepts between 60% and 90% of the radiation that strikes it, depending on the density of its canopy. Clusters of trees spaced closely together can therefore reduce ambient summer temperature significantly. Placed directly adjacent to buildings on the east, west, and south sides, they can reduce incoming solar radiation in the summer and, if deciduous, allow most of it to pass through in the winter, when a deciduous tree intercepts only 25–50%.

2. Creation of a "still zone" under the canopy. Around the edges of a tree canopy is a band of air turbulence where the cooler air within and the warmer outside air meet and mix. This turbulent zone appears to form a containing frame for the still, cool air beneath the canopy.

3. Release of cooling water vapor from their leaf surfaces through evaporation and transpiration (19).

A study of a mobile home in Florida showed that well-placed plantings could reduce cooling costs by more than 50% (Hutchinson et al., 1983, as cited in Ref. 19). Calculations of electrical energy saved by tree planting suggest that this is one of the most cost-effective means of reducing the heat island effect and thus electrical energy consumption (19). According to McPherson (1990, as cited in Ref. 19), about 97% of the total carbon conserved annually by a tree is in reduced power-plant emissions resulting from reduction in electrical energy use rather than in carbon dioxide absorbed.

6 POLLUTION PREVENTION VIA CHOOSING REPLACEMENTS FOR FOSSIL FUELS

6.1 Introduction

Despite potentially significant technological improvements in efficiency and decreases in environmental impact, some of the inefficiencies and pollutants associated with traditional energy sources cannot be avoided. Uneven distribution of resources can increase transportation costs, which can amount to 25% or more of the cost of crude oil, for example (9). Indeed, about 75% of the original energy in crude oil is lost during distillation into liquid fuels, transportation of that fuel to market, storage, marketing, and combustion in vehicles (3). For this reason, alternative energy sources such as solar, geothermal, and wind should receive much more attention.

In the United States, "renewable" energy sources account for about 7.5% of total consumption. The vast majority of this energy comes from two sources that have reached commercial maturity: hydroelectric power and biofuels (24). Currently, biofuels, primarily wood, account for about 4% of the U.S. energy supply. More than 6% of all homes burn wood as their principal heating fuel. The paper and pulp industry burns wood scraps to provide heat and electricity to run its operations. Wood and other biofuels are also used to generate a small amount of electricity by utilities (6).

Worldwide, potentially sustainable or renewable energy resources, including solar, biomass, hydroelectric, and other, less developed types of power production, currently provide less than 3% of total energy use (3). As of 1990, traditional biomass (e.g., fuelwood, crop residues, and dung) accounted for 60% of total available renewable energy, and large-scale hydropower for another 30% (9). About half of all wood harvested in the world annually is used for fuelwood; many countries use fuelwood (including charcoal) for more than 75% of their nonmuscle energy. About 40% of the world's total population depend on firewood and charcoal as their primary energy source. In some African countries, such as Rwanda and Sudan, firewood demand is already 10 times the sustainable yield of remaining forests (3). These figures illustrate the enormity of the potential for environmentally benign energy sources such as solar to replace not only fossil fuels but also traditional renewables which also cause environmental harm.

6.2 Biomass

As recently as 1850, wood supplied 90% of the fuel used in the United States. Wood now provides less than 1% of energy in the United States, but in many of the world's poorer countries, wood and other biomass fuels provide up to 95% of all energy consumed. Approximately half of all wood harvested annually is for fuel. About 40% of the world's population depend on firewood and charcoal as their primary energy source; however, about three-fourths of these lack an adequate, affordable supply (3).

In wood-burning power plants, pollution-control equipment is easier to install and maintain than in individual home units. Wood burning contributes less to acid precipitation than does coal, as wood contains little sulfur and burns at lower temperatures than coal, resulting in the production of fewer nitrogen oxides. However, unless trees cut for fuel are replaced with seedlings, wood burning results in a net increase in atmospheric CO_2.

Inefficient and incomplete burning of wood in stoves and fireplaces produces smoke laden with fine ash and soot and hazardous amounts of carbon monoxide and hydrocarbons. The U.S. Environmental Protection Agency (EPA) ranks wood burners high on a list of health risks to the general population, and standards are being considered to regulate the use of woodstoves nationwide. Highly efficient and clean-burning woodstoves are available but expensive (3).

6.3 Hydroelectric Dams

As of 1987, hydroelectric dams in the United States provided the energy equivalent of about 71 large power plants, about 10–14% of U.S. electricity, or about 3% of total energy supply, depending on year-to-year rainfall patterns. Of the pollutants associated with fossil fuel energy, methane is the only one that results from the damming of rivers. However, large dams have drowned out some of the most beautiful stretches of American rivers, flooded agricultural lands, forests, and areas of historical and geological value, and resulted in the dislocation of communities and loss of wildlife (6). Dam failure can cause catastrophic floods and thousands of deaths. Sedimentation often fills reservoirs rapidly and reduces the usefulness of the dam for either irrigation or hydropower (3).

6.4 Synthetic Fuels

Methanol would provide little reduction in greenhouse gases if made from natural gas (Office of Technology Assessment, as cited in Ref. 4). Synthetic fuels derived from coal or oil shales would result in the release of even more CO_2 than coal because the conversion processes require so much energy (6). The use of compressed natural gas brings the potential for leaks of methane that could largely offset the lesser carbon content of natural gas when compared to oil.

Ethanol (grain alcohol) and methanol (wood alcohol) are produced by anaerobic digestion of plant materials (Figure 7). Ethanol is unlikely ever to play an important energy role in our transportation future: 8% of the U.S. corn crop would replace only 1% of U.S. gasoline. Further, making ethanol from corn requires almost as much energy as the ethanol contains; therefore, it offers little if any global-warming benefit (6). Among biomass fuels, synthetic natural gas or methanol produced from woody biomass hold the largest potential for reducing greenhouse gases (a reduction of 60–70% from that emitted by vehicle fuels used at present), so long as the feedstock were offset with replacement biomass growth (Office of Technology Assessment, as cited in Ref. 4).

6.5 Tides

The stormy coasts where waves are strongest are usually far from major population centers that need the power. In addition, the storms that bring this energy can destroy the equipment intended to exploit it (3). Even if the technology for capture of tidal energy were available, only a minute fraction could, even theoretically, be harnessed for useful purposes (25). France operates a tidal generating station on the Rance Estuary that is designed to produce 240 MW of electricity but that usually only generates 62 MW (26).

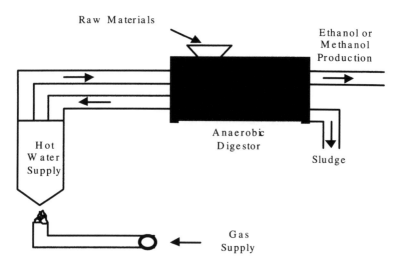

FIGURE 7 Production of ethanol and methanol through anaerobic digestion of plant material.

6.6   Nuclear Power

Nuclear power provides about 17% percent of the world's electricity (5) and about 5% of total energy needs, led by Western Europe with about 11% reliance on nuclear plants (3). Although the United States has the world's largest nuclear power program, it provides only about 7.5% of our energy needs. In the United States, at least, the management and operation of existing plants must improve significantly, and existing unresolved safety problems must be convincingly solved. The design of new reactors must be simplified and incorporate more passive shut-down safety features. Further, there must be tangible progress in solving the problems of storing radioactive wastes (6).

However, even if all these requirements were met, the potential contribution of nuclear energy to solving the global energy–climate problem would be limited for several reasons:

It is unlikely that a significant number of safer new reactors can be designed, approved, constructed, operated, and "debugged" in a relatively short period of time—say, less than 20 years. They will therefore be unable to make a significant contribution to meeting the world's energy needs during the next 20–40 years.

Because of their inherent cost and complexity, nuclear plants are unlikely to be deployed in poor, developing countries. Such facilities demand a high level of sophisticated and expensive support to be safely constructed and operated, a condition unlikely to occur in most of the developing world.

Unless the world suddenly embraces peaceful solutions to its age-old ethnic and boundary problems, the prospect of nations using nuclear materials to build weapons clandestinely will grow with nuclear plant deployment (6).

Of the nuclear plants that have been decommissioned so far, the costs of tearing them down have been about two to ten times the costs of building them in the first place.

We may never reach a breakeven point where we get back more energy from nuclear plants than we put into them, especially considering the energy that may be required to decommission nuclear plants and guard their waste products in secure storage for thousands of years (3).

The raw materials required for nuclear fuels result in the same disturbances of the landscape as other mined minerals.

Denmark has never permitted atomic power plants to be constructed within its boundaries, and Sweden has a policy of decommissioning all its existing plants. In the United States, few plants currently are under construction because it has become so costly due to required environmental safeguards and the inevitable litigation of nuclear opponents. Further, in most parts of the United States, it is so politically risky that no new plans for nuclear power plant construction are currently in existence (14).

6.7 Geothermal Energy

Geothermal energy is heat contained below the earth's surface, either in rock or in trapped hot water or steam. Geothermal power offers a number of environmental advantages. When compared with other alternative energy sources, geothermal plants are reliable; the Department of Energy reports that they have a 65% "capacity factor" (the ratio of actual output to the output that would result if the plant ran full-tilt, full time). This is comparable with the capacity factors of new coal or gas turbine plants. In contrast, the capacity factor of wind and solar thermal plants is about 21%. Geothermal energy produces no ash, no scrubber waste, and no radioactive waste. Although geothermal energy sometimes produces toxic waste from the dissolved or suspended chemicals naturally found deep in the earth, these materials tend to be more easily disposed of than those from other energy sources; virtually all U.S. generating plants simply reinject them into the reservoir (27).

Geothermal power, however, suffers from resource, technological, and economic constraints. The only type of geothermal energy that has been widely developed is hydrothermal energy, which consists of trapped hot water or steam (24). The total geothermal energy of the world's volcanoes and hot springs is only about 2% of today's global commercial energy use. This energy flux can be utilized in hyperthermal areas such as Iceland (where most buildings are heated by geothermal steam) or in the "Ring of Fire" surrounding the Pacific Ocean, where 18 nations (including the western United States) currently generate geothermal electricity (27). Hydrothermal cannot, however, be of more than local or regional importance (25).

Creating a geothermal plant is expensive, because developers usually must bore holes a mile or so deep through hard rock. Even though geothermal plants need no fuel, making operating costs extremely low, the capital cost still amounts to about $3000 per kilowatt, in contrast for about $824 per kilowatt for an efficient gas turbine plant. Innovations such as new drilling technologies promise to cut expenses (27).

Other problems associated with the use of geothermal power include the following:

> Geothermal facilities are very large-scale plumbing pipes with an abundance of giant pipes, huge valves, and specialized fittings. Some plants need mufflers and sound blankets to reduce drilling and generating noise, and they usually emit a plume of steam.

The rotten-egg stench of hydrogen sulfide released from underground can often be overpowering (27).

The geothermal heat conducted by rocks is two and one-half times today's commercial energy use (25). Preliminary estimates of the cost of electricity derived from hot dry rock (HDR) suggest that it might be relatively cheap, at least in areas where the earth becomes at least 144°F warmer with each mile of depth and drilling costs are thus somewhat less formidable. Conditions like these reportedly are found under 40,000 square miles of the lower 48 states, primarily in Nevada, Oregon, and California (27). In view of the potentially catastrophic effects of global warming, as well as the other environmental problems associated with traditional energy sources, HDR-derived energy deserves serious study.

6.8 Wind Farms

Wind farms are large-scale public utility efforts to take advantage of wind power. In 1990, wind machines in California generated enough electricity to meet the annual residential needs of a city the size of San Francisco, or more than 1% of California's total electrical needs. There are enough windy sites in California to meet about 20% of existing electricity demand. Advanced wind machines could supply energy to the United States in amounts far in excess of the nation's total present energy demand.

The towers, roads, and other structures on a wind farm actually take up only about one-third as much space as would be consumed by a coal-fired power plant or solar thermal energy system to generate the same amount of energy over a 30-year period. The land under windmills is more easily used for grazing or farming than is a strip-mined coal field or land under solar panels. Further, wind power generates many more jobs per unit energy produced than do most other technologies, even though its total cost is generally lower (3).

An obvious limitation of wind farms is the necessity of locating them in windy areas. The best sites are in the Great Plains and include North and South Dakota, Kansas, and Montana (6). Seacoasts also offer great potential for siting wind farms. Opponents have complained of visual and noise pollution. While most wind farms are too far from residential areas to be heard or seen, they do interrupt the view in remote places and destroy the sense of isolation and natural beauty. They can also pose a hazard to birds that fly into the whirling blades.

6.9 Solar Energy

Of all the available forms of energy, renewable or nonrenewable, solar has the greatest potential for providing clean, safe, reliable power. The supply is inexhaustible: the solar energy falling on the earth's continents is more than 2000 times the total annual commercial energy currently being used by humans (25).

Solar technologies can be broadly grouped into two categories:

1. Active technologies—solar thermal power plants, solar ponds, wind turbines, and photovoltaic cells (Figure 8)
2. Passive technologies—natural materials or absorptive structures with no moving parts that simply gather and hold heat (Figure 8) (3)

Low-temperature thermal collectors can provide heat for domestic hot water, space heating, and industrial purposes (e.g., supplying hot water for car washes). According to Cunningham and Saigo (3), water heating consumes 15% of the U.S. domestic energy budget.

Active solar systems generally pump a heat-absorbing, fluid medium (air, water, or an antifreeze solution) through a relatively small collector rather than passively collecting heat in a stationary medium such as masonry (Figure 8) (3). There are three main types of solar-thermal collectors: the parabolic trough, the parabolic dish, and the central receiver.

Parabolic trough and parabolic dish units are modular and relatively small, so that systems can be sized to suit almost any application. Central-receiver systems generally are much larger. In all three, sunlight striking reflectors is collected and used to heat a fluid that is piped to a central location. The heat can be used directly to produce steam for industrial processes or to drive turbines that generate electricity (21).

Photovoltaic cells are elegantly simple devices that generate electricity directly from sunlight without going through the process of thermal–electric conversion. These cells are made of silicon or other semiconductor materials; they have no moving parts and therefore are quiet and reliable. Photovoltaic cells require no maintenance, have the potential for long life, produce no pollution, and consume no water in generating electricity. They can convert 20% or more of the sunlight striking them into electricity; practical efficiencies in the 30–40 range are possible (21).

In the United States, the land area of the lower 48 states intercepts about 47,000 quadrillion BTUs of direct sunlight per year, about 600 times total U.S. primary energy use. At a solar collection efficiency of 15%, readily achievable using present photovoltaic cells, significantly less than 1% of the land area of the lower 48 states would be required to meet all our energy needs. This can be compared to the 20% of U.S. lands devoted to croplands, or the 31% to pastures. Moreover, many of the solar cells could be placed on the walls and roofs of existing structures, reducing the area of land needed (6).

If the entire present U.S. electrical output came from central tower solar steam generators, 780 square miles of collectors would be needed. This is less land, however, than would be strip-mined in a 30-year period if all our energy came from coal or uranium. Further, we can put solar collectors wherever we choose (such as lands unsuited for agriculture, grazing, or habitation), whereas

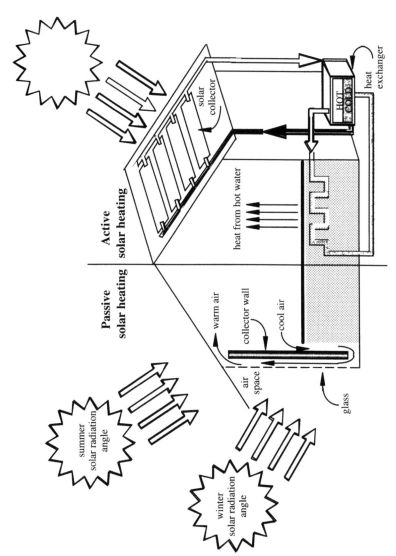

FIGURE. 8 Passive and active solar systems. In a passive system, the length of roof overhang is based on the latitude of the winter and summer sun. Natural air convection circulates heated air between outer glass wall and collector wall. The collector wall is designed to be ~40 cm thick to collect and store solar heat. In the active system, water is passed through solar collector panels, the heated water is then pumped into the house, where heat is radiated from the hot water and then recirculated into the system.

strip-mining occurs wherever coal or uranium exist, regardless of other values associated with the land (3).

What happens when the sun goes down? One solution is hybrid energy systems that run on 75% solar energy and 25% natural gas.

Unlike nuclear power, solar presents no problems of safety, disposal of radioactive wastes, or danger that nuclear materials will fall into the wrong hands. However, although solar schemes probably have the smallest environmental impact of all current forms of energy, care must be taken with size of concentrators and with the films and silicon used in photovoltaics (9).

It must be remembered that net yields and overall conversion efficiencies are not the only considerations when different energy sources are compared. The yield/cost ratio and conversion-cycle efficiency is much higher for coal burning, for example, than for photovoltaic electrical production, making coal appear to be a better source of energy than solar radiation. However, solar energy is free, renewable, and nonpolluting; therefore, if we use solar energy to obtain electrical energy, it does not matter how efficient the process is, as long as we get more out of it than we put in (3).

6.10   Costs of Renewables

Geothermal energy is currently the least expensive renewable energy source. It is closely followed by wood, hydroelectric, wind, and solar energy (Table 4). Most solar technologies have high initial costs while providing savings down the road in the form of lower fuel costs. For example, a solar water heater may cost $2500 to purchase and install. A solar power plant may cost $2500–$3000 per kilowatt of capacity, while a conventional power plant costs between $400 and $1200 per kilowatt of capacity. The difference is that solar technologies cost very little to operate, whereas the major cost associated with conventional technologies is usually fuel, which will be paid later. Unfortunately, our tax system gives

TABLE 4   Cost Comparison of
Renewable Energy Sources

Energy source	Cost (cents/kWh)
Geothermal	4.5–6.5
Hydroelectric	5
Wood	5
Wind	7.5
Solar-thermal	8
Solar (photovoltaic)	30

advantages to the conventional plants, which can deduct the high, ongoing fuel costs as operating expenses (21).

Commercialization of solar technologies could be greatly accelerated with market incentives such as solar-energy tax credits, regulations requiring that cost-effective passive and active solar technologies be included in new buildings, and increased federal funding for solar-energy research and development. Currently, the great majority of subsidies go to established energy sources—fossil fuels and nuclear power (21). U.S. federal funding for renewable energy sources fell from $1.3 billion in 1980 to $135 million in 1990 (in 1990 dollars; Office of Technology Assessment, 1991, as cited in Ref. 4). By recognizing the environmental and social costs of energy technologies, federal, state, and local governments can help provide a "level playing field" for solar technologies and play a decisive role in influencing energy choices (21).

When a homeowner or community invests independently in solar or wind generation, what should be done about energy storage when electricity production exceeds use? Many private electricity producers believe the best use for excess electricity is in cooperation with the public utility grid. When private generation is low, the public utility runs electricity through the meter and into the house or community. When the wind generator or photovoltaic systems overproduce, the electricity runs back into the grid and the meter runs backward. Ideally, the utility reimburses individuals for this electricity, for which other consumers pay the company. The 1978 Public Utilities Regulatory Policies Act required utilities to buy power generated by small hydro, wind, cogeneration, and other privately owned technologies at a fair price. Not all utilities comply yet, but some—notably in California, Oregon, Maine, and Vermont—are purchasing significant amounts of private energy (3).

7 CONCLUSION

Significant reductions in all pollutants that result from petroleum combustion (CO_2, CO, acid rain precursors, photochemical oxidants, unintentional petroleum releases) could be achieved by the imposition of a gasoline tax that would encourage the use of public transportation and fuel-efficient vehicles. Reductions in motor vehicle-related urban runoff represent an additional pollution benefit of reducing the use of fossil fuels for transportation. The benefits of freezing carbon emissions via a permitting system could have similar benefits to those of a carbon tax, possibly with less public resistance.

At present, the possibilities for alternative transportation fuels appear rather limited. The primary drawback for the use of ethanol is the relatively low energy value obtained through its use as compared to the energy required for its production. Synthetic natural gas or methanol produced from woody biomass may be our most attractive options in this area; obviously, however, employment of these does not eliminate CO_2 emissions.

Elimination of fossil fuel subsidies is vital to the creation of incentive for increases in efficiency, a conservation measure whose potential is enormous. A dramatic increase in exploitation of passive technologies and especially non-carbon energy sources is essential. Of all the available forms of energy, solar has the greatest potential for providing clean, safe, reliable power. Wind farms also represent a significant potential means of producing energy with minimal environmental impacts, and geothermal energy deserves thorough investigation as well. During a period of transition to noncarbon energy sources, increased exploitation of natural gas represents a means of reducing CO_2 emissions significantly.

REFERENCES

1. D. E. Booth, *The Environmental Consequences of Growth*. London and New York: Routledge, 1998.
2. International Energy Agency, *World Energy Outlook, 1998 Edition*. Paris: Organization for Economic Co-operation and Development, 1998.
3. W. P. Cunningham and B. W. Saigo, *Environmental Science*. Dubuque, IA: William C. Brown, 1997, 1992.
4. W. R. Cline, *The Economics of Global Warming*. Washington, DC: Institute for International Economics, 1992.
5. J. H. Gibbons, P. D. Blair, and H. L. Gwin, Strategies for Energy Use. *Sci. Am.*, vol. 261, no. 3, pp. 136–143, 1989.
6. J. J. MacKenzie, J.J. Energy and Environment in the 21st Century: The Challenge of Change. In J. Byrne and D. Rich (eds.), *Energy and Environment: The Policy Challenge*, New Brunswick, NJ: Transaction, 1992.
7. P. M. Vitousek, H. A. Mooney, J. Lubchenco, and J. M. Melillo, Human Domination of Earth's Ecosystems. *Science*, vol. 277, pp. 494–499, 1997.
8. A. Whyte, The Human Context. In H. Coward (ed.), *Population, Consumption and the Environment*, pp. 41–59. Albany: State University of New York Press, 1995.
9. World Energy Council (WEC), *Energy for Tomorrow's World*. New York: St. Martin's Press, 1993.
10. L. W. Canter, Environmental Impact Assessment, 2nd edition, p. 480, New York: McGraw-Hill, 1995.
11. M. Ledbetter and M. Ross, Light Vehicles: Policies for Reducing Their Energy Use and Environmental Impacts. In New Brunswick, NJ: Transaction, 1992. *Energy and Environment: The Policy Challenge*, J. Burne and D. Rich (eds.), pp. 187–233.
12. Union of Concerned Scientists, Assessing the Hidden Costs of Fossil Fuels (briefing paper). Cambridge, MA: Union of Concerned Scientists, 1993.
13. S. E. Manahan, *Environmental Chemistry*. Chelsea, MI: Lewis, 1991.
14. D. L. Johnson and L. A. Lewis, *Land Degradation: Creation and Destruction*. Cambridge, MA, and Oxford, U.K. Blackwell, 1995.
15. Committee on Interior and Insular Affairs to Accompany HR 11500, Surface Mining

Control and Reclamation Act of 1974, HR93-1072. Washington, DC: U.S. House of Representatives, 30 May 1974.

16. World Bank, *Energy Efficiency and Conservation in the Developing World*. New York: World Bank, 1993.

17. R. U. Ayres, The Energy Policy Debate: A Case of Conflicting Paradigms. *WEC J.*, vol. 111, p. 57, July 1992.

18. W. D. Ruckelshaus, Toward a Sustainable World. Sci. Am., vol. 261, no. 3, pp. 166–174, 1989.

19. J. T. Lyle, *Regenerative Design for Sustainable Development*. New York: Wiley, 1994.

20. M. P. C. Munasinghe, Sustainable Energy Development: Issues and Policy. In P. R. Kleindorfer, H. C. Kunreuther, and D. S. Hong (eds.), *Energy, Environment and the Economy*, pp. 3–42. Brookfield, VT/Cheltenham, U.K.: Edward Elgar, 1996.

21. Union of Concerned Scientists, *Solar Power: Energy for Today and Tomorrow*. Cambridge, MA: Union of Concerned Scientists, 1992.

22. E. Hirst and J. Ito, Justification of Electric-Utility Energy-Efficiency Programs. Oak Ridge National Laboratory Report ORNL/CON-419, August 1995.

23. P. Ehrlich, and A. Ehrlich, *The Population Explosion*. New York: Simon & Schuster, 1990.

24. Union of Concerned Scientists, *Cool Energy: The Renewable Solution to Global Warming*. Cambridge, MA: Union of Concerned Scientists, 1991.

25. D. Abrahamson, Climatic Change and Energy Supply: A Comparison of Solar and Nuclear Options. In J. Byrne and D. Rich (eds.), *Energy and Environment: The Policy Challenge*, p. 430. New Brunswick, NJ: Transaction, 1992.

26. E. D. Enger and B. F. Smith, *Environmental Science: A Study of Interrelationships*. Dubuque, IA: William C. Brown, 1995.

27. D. Tenenbaum, Tapping the Fire. *Technol. Rev.*, vol. 2, pp. 39–47, 1995.

28. M. Renner, *Assessing the Military's War on the Environment. State of the World 1991*. New York: Norton, 1991.

29. United Nations Fund for Population Activities, *State of the World Population 1990*.

30. F. Ackerman, *Why Do We Recycle?* Washington, DC/Covelo, CA: Island Press, 1997.

31. J. MacNeill, Strategies for Sustainable Economic Development. *Sci. Am.*, vol. 216, no. 3, pp. 155–165, 1989.

32. D. W. Orr, *Ecological Literacy*. Albany, NY: SUNY Press, 1992.

33. World Bank, *World Bank Development Report 1999*. Hong Kong: Asia 2000, 1999.

34. U.S. Environmental Protection Agency, Acid Rain Program—Overview, EPA 430/F-92/019. Washington, DC: EPA, 1992.

35. H. Friedli, H. Lötscher, H. Oeschger, U. Siegenthaler, and B. Stauffer, Ice Core Record of the $^{13}C/^{12}C$ Ratio of Atmospheric CO_2 in the Past Two Centuries. *Nature*, vol. 324, no. 20, pp. 237–238, 1986.

36. A. Neftel, H. Oeschger, and B. Stauffer, Evidence from Polar Ice Cores for the Increase in Atmospheric CO_2 in the Past Two Centuries. *Nature*, vol. 315, no. 2, pp. 45–48, 1985.

37. National Academy of Sciences, Committee on Science, Engineering, and Public Policy, *Policy Implications of Greenhouse Warming: Mitigation, Adaptation, and the Science Base*. Washington, DC: National Academy Press, 1992.

8

Fundamentals of Heat Transfer

René Reyes Mazzoco
Universidad de las Américas–Puebla, Cholula, Mexico

1 HEAT TRANSFER MECHANISMS

1.1 Conduction

Conduction heat transfer is explained through the molecular motion in the solid's structure. Heat is transferred from one molecule to the adjacent molecule by means of vibrational motion. This basic description points out that heat transfer through a solid takes place entirely by conduction, and also states that it occurs to a limited extent in liquids and gases because of their molecular mobility.

The mathematical formulation of conduction heat transfer was proposed by Joseph Fourier while solving heat transfer problems in metal casting and template. The first step for this formulation is the recognition that the amount of heat transferred, q(W), from one point of a metal piece to another point of the same medium (continuum) is proportional to the temperature difference between those two points. The evaluation of the temperature difference through the derivative in any direction (s) makes the measurement independent of any two specific points and the distance between them:

$$q \propto \frac{dT}{ds} \tag{1}$$

For most cases of conduction heat transfer in solids it is possible to change the proportionality relation to an equality defining the proportionality constant called thermal conductivity, k(W/m K). Thus, the calculation of the amount of heat transferred per cross-sectional surface area, A(m^2), is obtained from Fourier's first law:

$$\frac{q}{a} = -k\frac{dt}{ds} \tag{2}$$

1.1.1 Measurement of Thermal Conductivity

The thermal conductivity is a physical property characteristic of the medium and is defined similarly for the three states of the matter. The expression for the calculation of the amount of heat transferred by conduction is also used to define the thermal conductivity, and one procedure for its measurement, that is, k, is calculated after measuring the linear temperature difference between two points, the amount of heat transferred, and the cross-sectional area associated with the heat flow trajectory:

$$k = \frac{q/a}{-dT/ds} \tag{3}$$

The procedure for calculating values of k is especially relevant for the evaluation of this property in mixtures of contaminated media for which no information is normally available. For this evaluation it is easier to calculate the thermal diffusivity, α (where $\alpha = k/\rho C_p$, m^2/s) of the mixture while solving Fourier's second law in cylindrical coordinates:

$$\frac{\partial T}{\partial t} = \alpha\left(\frac{\partial T}{\partial r^2} + \frac{1}{r}\frac{\partial T}{\partial r}\right) \tag{4}$$

Thus, a cylinder with a substance of unknown thermal conductivity is placed in a bath at constant temperature to obtain the initial and boundary conditions associated with the solution of Eq. (4):

$T(r, 0)$ = initial sample temperature

$T(0, t)$ = measurement at the cylinder's center

$T(R, t)$ = constant bath temperature

The value of α can be obtained from the solution of Eq. (4) with these boundary conditions. The values of the density, ρ, and the constant-pressure heat capacity, C_p, are normally available in the literature, or are measured experimentally.

Thermal conductivity prediction models are available for gases and liquids in several references (e.g., Ref. 1).

1.1.2 Effect of Shape on Calculation of Conduction Heat Transfer

Evaluation of the integral between two specific points in the direction of heat transfer allows for the calculation of the macroscopic amount of heat. For an object with constant cross-sectional area in the direction of heat flow, integration of Eq. (2) gives:

$$\frac{q}{A} = k\frac{T_1 - T_2}{s_2 - s_1} \tag{5}$$

Instead, for a cylindrical object, heat flow in the direction of the radius finds a constantly changing cross-sectional area. Thus, integration of Eq. (2) with $A = 2\pi rL$ gives

$$q = \frac{T_1 - T_2}{ln(r_2/r_1)/2\pi kl} \tag{6}$$

which involves a logarithmic distance (radius) difference instead of the thickness of the medium involved.

1.1.3 Combined Resistances

A common problem in heat transfer design is the combination of several layers of solid to provide heat insulation, or layers of solid and fluid as in heat exchanger design. For a rectangular geometry, two combined resistances to heat transfer can be expressed as

$$\frac{q}{A} = \frac{T_1 - T_2}{(L_1/k_1) + (L_2/k_2)} \tag{7}$$

where L and k are the thickness and the thermal conductivity of components (1) and (2), respectively, and T_1 and T_2 are the temperatures at the external surfaces of the combined wall. Each additional layer of material increases by one the resistances to heat transfer added in the denominator of Eq. (7).

The effect of geometry on the combined resistances to heat transfer can be obtained by integrating Eq. (2) for a double-layered cylinder:

$$q = \frac{T_2 - T_1}{[ln(r_2/r_1)/(2\pi k_1 L)] + [ln(r_3/r_2/(2\pi k_2 L)]} \tag{8}$$

1.1.4 Relative Magnitude of Values of Thermal Conductivity

The values of the thermal conductivity depend on the phase of the material considered:

$$k_{gas} < k_{liquid} < k_{solid}$$

Thus, solid materials are good heat conductors, while for heat insulation trapped gases are the best option. Good electric conductor metals are the best selection for heat conductors. The design of heat insulation follows the criteria for air entrapment in fabrics or ceramics that could be resistant to high temperatures.

1.2 Convection

Heat convection is described as heat transport in fluid eddies promoted by the flow derived from a mechanical device, a pump or fan (forced convection), or a density difference (natural convection). The mechanism is associated with the definition of the convective heat transfer coefficient, $h(W/m^2 \, °C)$:

$$h = \frac{q}{A(T_2 - T_1)} \tag{9}$$

As the turbulent flow process carrying the heat cannot be fully described, the temperature difference is considered at two points (1) and (2) in the direction of heat transfer. It is not possible to describe this process through a differential equation, and Eq. (9) is a definition for h that is related to the specific geometry associated to the surface area, A, and the flow conditions.

The convective heat transfer coefficient can be calculated for design purposes from experimental information gathered in the open literature. Experiments have been carried out under geometry, flow range, and similar thermophysical properties conditions that can be encountered in process applications. The information has been grouped in terms of flow conditions and thermophysical properties involved.

Flow conditions are described through the Reynolds number (Re) for forced convection. The Reynolds number relates the momentum convection associated to the flow velocity, v, to the momentum diffusivity associated to ν, the kinematic viscosity ($\nu = \mu/\rho$), μ is Newtonian viscosity (kg/ms). At low Reynolds numbers, implying low flow velocity, momentum diffusivity dominates, and the fluid displacement is in the laminar flow condition. When the flow velocity is high relative to the kinematic viscosity, the Reynolds number is high, indicating turbulent flow conditions.

$$Re = \frac{Lv}{\nu} \tag{10}$$

L is the flow characteristic length; for internal flow in circular pipes, L is the internal diameter.

The Grashof number (Gr) describes flow conditions for natural convection and is used instead of the Reynolds number.

$$\text{Gr} = \frac{g\beta(T_w - T_{\text{infinity}})L^3}{v^2} \tag{11}$$

Here g is the acceleration of gravity, T_w is the solid wall temperature, T_{infinity} is the fluid bulk temperature, L is the heat transfer characteristic length, and β is the volume coefficient of expansion:

$$\beta = \frac{(\rho_{\text{infinity}} - \rho)}{\rho(T - T_{\text{infinity}})} \tag{12}$$

ρ_{infinity} is the fluid bulk density.

In natural and forced convection, the Prandtl number describes the influence of thermophysical properties in the calculation of the convective heat transfer coefficient, normally to the $1/3$ power.

$$\text{Pr} = \frac{v}{\alpha} = \frac{C_p \mu}{k} \tag{13}$$

The Nusselt number, Nu, is the ratio of heat convection to diffusion associated to the heat transfer characteristic length, L:

$$\text{Nu} = \frac{hL}{k} \tag{14}$$

From the exact analysis of the boundary layer between the fluid and the solid wall transferring heat, the correlation in forced convection among Nusselt, Reynolds, and Prandtl numbers is

$$\text{Nu} = 0.664 \, \text{Re}^{1/2} \, \text{Pr}^{1/3} \tag{15}$$

This theoretical correlation has very limited application, and the dependence of these dimensionless numbers on the geometry makes experimentation necessary to calculate correlations for each geometry. The correlation results are normally reported with the same mathematical formulation:

$$\text{Nu} = c_0 \, \text{Re}^n \, \text{Pr}^m \tag{16}$$

For natural convection, the analysis of the boundary layer provides the correlation of the important dimensionless numbers:

$$\text{Nu} = C(\text{Gr Pr})^m \tag{17}$$

1.3 Radiation

For practical conditions, radiation emitted (or received) by surface is calculated from an equation that involves the effect of the area, A_{12}, the emissivity, ε_1, of the emitting surface involved, and a view factor, F_{12}, that describes the effect of the relative positions of the two surfaces involved on the amount of radiation exchanged. The formulation of the exchanged radiation is

$$q = \sigma \varepsilon_1 A_{12} F_{12}(T_1^4 - T_2^4) \tag{18}$$

All practical terms in Eq. (18) are measured experimentally and are reported in several references (e.g., Ref. 2)

2 HEAT ACCUMULATION

Heat accumulation is described through the heat capacity. The specific property normally used to achieve this calculation is the constant-pressure heat capacity, C_p (J/kg °C). The total amount of material that stores heat should be expressed in the mass or molar terms used for the C_p. The heat stored is then a function of the temperature change in the total mass considered:

$$C_p = \frac{q}{m(dT/dt)} \tag{19}$$

The temperature variation with time allows the evaluation of the heat flow accumulated.

2.1 Sensible Heat

The variation of the temperature in a fluid medium defines sensible heat. The calculation of the amount of sensible heat is obtained from Eq. (19). The heat capacity should correlate the fluid and phase considered.

2.2 Latent Heat

The process where a change of phase takes place requires the addition of latent heat. The latent heat is used to change phase in a fluid without a change in the medium temperature. The evaluation of the latent heat is necessary to measure the amount of heat required for phase change. Latent heat values and prediction correlations are available in Ref. 1.

3 EXPERIMENTAL MEASUREMENT AND PREDICTION OF HEAT TRANSFER THERMOPHYSICAL PROPERTIES

3.1 Constant-Pressure Heat Capacity, C_p

Measurement of the C_p requires the evaluation of temperature change in a fixed mass of material due to a heat flow from the surroundings according to Eq. (19).

3.2 Thermal Conductivity, k

For the measurement of k, Fourier's first law is normally used to define the parameters involved in the evaluation. The heat flux in Eq. (2) is determined from the heat flow and the body geometry while the temperature gradient is measured directly.

3.3 Convective Heat Transfer, h

The convective heat transfer coefficient is experimental measured of forced- and natural-convection conditions. h is part of Nu, while flow conditions are represented by Re or Gr, and the thermophysical properties form Pr.

Normally, the values of h are obtained from reported correlations. If it is necessary to evaluate h for conditions not previously studied, the information is gathered and analyzed according to Eq. (16) or (17).

3.4 Thermophysical Properties of Mixtures in Pollution Control

Mixtures of contaminated media normally require the experimental evaluation of the thermophysical properties. In some cases, due to nonavailability of the experimental data, correlations for calculating the thermophysical properties are limited.

4 HEAT TRANSFER DESIGN

Process efficiency is defined at the design stage. Design algorithms for heat transfer equipment can be found in several classic references (e.g., Ref. 3) and are still used for designing heat transfer equipment. Several software options are also available for efficient heat transfer equipment design; software the description can be obtained from demos downloaded from an Internet search (any search engine) on "Heat Exchangers."

The basic equation for heat exchange design is

$$q = U_o A_o \Delta T_{\text{LM}} F \tag{20}$$

where U_o is the overall heat transfer coefficient and includes all the heat transfer conductances around the solid wall transferring heat. For a flat wall transferring heat,

$$U_o = \frac{1}{1/h_{\text{inside}} + G/k + 1/h_{\text{outside}}} \tag{21}$$

where h_{inside} is the inside convective heat transfer coefficient, G is the wall thickness, and h_{outside} is the outside heat transfer coefficient.

The driving force for heat transfer in a heat exchanger is the logarithmic mean temperature difference:

$$\Delta T_{LM}F = \frac{[(T_{outlet} - t_{inlet}) - (T_{inlet} - t_{outlet})]F}{\ln[(T_{outlet} - t_{inlet})/(T_{inlet} - t_{outlet})]} \tag{22}$$

T is the hot fluid temperature, t is the cold fluid temperature, and F is the efficiency factor adapted for each configuration of shell and tube, plate exchangers, and direct-contact heat exchangers (4).

From the calculation of the amount of heat transferred, including the temperature changes involved and the overall heat transfer conductance, the area for heat exchange is determined. Several heat transfer equipment can be used to accomplish the heat exchange between the media in a given process condition.

4.1 Heat Transfer Design and Good Engineering Practices

Design defines the efficiency of the operation of a process. Once the optimized design is utilized, it is necessary to maintain good engineering practices. These practices should include pollution control and waste minimization.

Heat transfer equipment is subjected to fouling and corrosion, which are among the major hurdles for the operation. Fouling increases heat transfer resistance and waste of energy. Good engineering practices include the use of fouling suppressants in heat transfer fluids and periodic cleaning of the exchanger walls.

For water as the cooling or heating medium in industrial operations there are several standard techniques for keeping fouling low. Water in cooling-water circuits has to be treated to keep salts and dirt content low. Common treatments include the addition of coagulants for sedimentation of some salts and particles; addition of biocides, to prevent microbial growth that is another source of fouling; and the addition hardness suppressants such as polyphosphates; among others. Although the materials used in heat transfer fluids treatment are a source of solid waste, handling its final deposition should follow normal procedures. Fouling prevention is not considered a polluting operation. Fouling prevention by-products can be integrated to cement kiln operations when feasible, in order to eliminate waste generation.

Corrosion protection of heat transfer surfaces is a suggested practice for pollution control and waste minimization. In order to prevent corrosion, begin with the analysis of the appropriate combination of materials and fluids. For the operating equipment, passive and active cathodic protection are recommended.

4.2 Innovations for Efficient Heat Use

Efficient energy use is a direct way to reduce pollution and minimize wastes from industrial sources. The ongoing research in energy efficiency and resulting innovations highlight the intensity of scientific activity in this field.

New approaches to increase heat transfer efficiency include the following.

1. Fluidized bed combustion is the choice for eliminating solids in solid-waste management schemes. In general, direct contact between the materials increases heat transfer efficiency. Direct contact reduces the heat transfer resistances due to the wall in conventional equipment, and increases the convective heat transfer coefficients due to the higher contact velocities between the materials and fluids.

2. To increase the efficiency in steam generation, direct-contact heat exchangers make use of residual heat from combustion gases to preheat the feed streams to the boiler. Thermal recovery is a possibility from direct-contact heat exchangers and heat pumps. Rotary drums recover heat from a residual discharge in an steam generator and transport it to preheat the inlet streams to the generator.

3. Heat pipes are a promising technology for increasing residual heat usage as heat pumps. Heat pipes use capillary pressure as the driving force for condensing and evaporating the working fluid, thus eliminating the necessity for pump and compressor in the power cycle. The understanding of heat pipe operation is related to the evaluation of convective heat transfer coefficients for change-of-phase heat transfer.

4. Plate heat exchangers are now available for almost any process condition, including high-pressure and corrosivity conditions. Enhanced heat transfer surfaces improve energy management, reducing wastes. Improved surfaces increase the convective heat transfer coefficients for heating–cooling operations, and change-of-phase heat transfer.

5. Co-generation in chemical and petrochemical processes makes use of the process integration gained from the use of simulation and pinch-point techniques to increase energy usage.

5 CONCLUSIONS

The understanding of heat transfer fundamentals is a basic step toward the proposition of improved industrial solutions in terms of energy wastes minimization.

Clear fundamental concepts make the use of design software straightforward. This is the approach to equipment design that produces the best results for waste minimization.

Heat transfer innovations are improving energy handling in industrial processes, reducing pollution and wastes. This research field is active in fundamentals such as enhanced heat transfer or heat pipe development.

REFERENCES

1. R. C. Reid, J. M. Prausnitz and B. E. Poling, *The Properties of Gases and Liquids*, 4th ed. New York: McGraw-Hill, 1987.
2. J. P. Holman, *Heat Transfer*, 8th ed. New York: McGraw-Hill, 1997.
3. D. Q. Kern, *Process Heat Transfer*. New York: McGraw-Hill, 1950.
4. O. Levenspiel, *Engineering Flow and Heat Exchange*, 2nd ed. New York: Plenum Press, 1992.

9

Macroscopic Balance Equations

Paul K. Andersen and Sarah W. Harcum
New Mexico State University, Las Cruces, New Mexico

The prevention of waste and pollution requires an understanding of numerous technical disciplines, including thermodynamics, heat and mass transfer, fluid mechanics, and chemical kinetics. This chapter summarizes the basic equations and concepts underlying these seemingly disparate fields.

1 MACROSCOPIC BALANCE EQUATIONS

A balance equation accounts for changes in an extensive quantity (such as mass or energy) that occur in a well-defined region of space, called the *control volume* (CV). The control volume is set off from its surroundings by boundaries, called *control surfaces* (CS). These surfaces may coincide with real surfaces, or they may be mathematical abstractions, chosen for convenience of analysis. If matter can cross the control surfaces, the system is said to be *open*; if not, it is said to be *closed*.

1.1 The General Macroscopic Balance

Balance equations have the following general form:

$$\frac{dX}{dt} = \sum_{CS,i}(\dot{X})_i + (\dot{X})_{gen} \tag{1}$$

where X is some extensive quantity. A dot placed over a variable denotes a rate; for example, $(\dot{X})_i$ is the flow rate of X across control surface i. The terms of Eq. (1) can be interpreted as follows:

$\frac{dX}{dt}$ = rate of change of X inside the control volume

$\sum_{CS,i}(\dot{X})_i$ = sum of flow rates of X across the control surfaces

$(\dot{X})_{gen}$ = rate of generation of X inside the control volume

Flows into the control volume are considered positive, while flows out of the control volume are negative. Likewise, a positive generation rate indicates that X is being a created within the control volume; a negative generation rate indicates that X is being consumed in the control volume.

The variable X in Eq. (1) represents any *extensive* property, such as those listed in Table 1. Extensive properties are additive: if the control volume is

TABLE 1   Extensive Quantities

Quantity	Flow rate (CS i)	Generation rate
Total mass, m	$\dot{m}_i$	$\dot{m}_{gen} = 0$ (conservation of mass)
Total moles, N	$\dot{N}_i$	$\dot{N}_{gen}$
Species mass, m_A	$(\dot{m}_A)_i$	$(\dot{m}_A)_{gen}$
Species moles, N_A	$(\dot{N}_A)_i$	$(\dot{N}_A)_{gen}$
Energy, E	$\dot{E}_i = \dot{Q}_i + \dot{W}_i + \dot{m}_i\hat{E}_i$ $\dot{E}_i = \dot{Q}_i + \dot{W}_i + \dot{N}_i\tilde{E}_i$	$\dot{E}_{gen} = 0$ (conservation of energy)
Entropy, S	$\dot{S}_i = \dot{Q}_i/T_i + \dot{m}_i\hat{S}_i$ $\dot{S}_i = \dot{Q}_i/T_i + \dot{N}_i\tilde{S}_i$	$\dot{S}_{gen} \geq 0$ (second law of thermodynamics)
Momentum, $\mathbf{p} = m\mathbf{v}$	$\dot{\mathbf{p}}_i = \dot{m}_i\mathbf{v}_i$	$\dot{\mathbf{p}}_{gen} = \mathbf{F}$ (Newton's second law of motion)

Notes:

$\dot{Q}_i \equiv$ heat transfer rate through CS i

$\dot{W}_i \equiv$ work rate (power) at CS i

$\hat{E}_i \equiv$ energy per unit mass of stream i; $\tilde{E}_i \equiv$ energy per unit mole of stream i

$\hat{S}_i \equiv$ entropy per unit mass of stream i; $\tilde{S}_i \equiv$ entropy per unit mole of stream i

$T_i \equiv$ absolute temperature of CS i

$\mathbf{F} \equiv$ net force acting on control volume

subdivided into smaller volumes, the total quantity of X in the control volume is just the sum of the quantities in each of the smaller volumes. Balance equations are not appropriate for *intensive* properties such as temperature and pressure, which may be specified from point to point in the control volume but are not additive.

It is important to note that Eq. (1) accounts for overall or gross changes in the quantity of X that is contained in a system; it gives no information about the distribution of X within the control volume. A differential balance equation may be used to describe the distribution of X (see Section 2.2).

Equation (1) may be integrated from time t_1 to time t_2 to show the change in X during that time period:

$$\Delta X = \sum_{CS,i} (X)_i + (X)_{gen} \tag{2}$$

1.2 Total Mass Balance

Material is conveniently measured in terms of the mass m. According to Einstein's special theory of relativity, mass varies with the energy of the system:

$$\dot{m}_{gen} = \frac{1}{c^2} \frac{dE}{dt} \qquad \text{(special relativity)} \tag{3}$$

where $c = 3.0 \times 10^8$ m/s is the speed of light in a vacuum. In most problems of practical interest, the variation of mass with changes in energy is not detectable, and mass is assumed to be *conserved*—that is, the mass-generation rate is taken to be zero:

$$\dot{m}_{gen} = 0 \qquad \text{(conservation of mass)} \tag{4}$$

Hence, the mass balance becomes

$$\frac{dm}{dt} = \sum_{CS,i} (\dot{m})_i \tag{5}$$

1.3 Total Material Balance

The quantity of material in the CV can be measured in moles N, a mole being 6.02×10^{23} elementary particles (atoms or molecules). The rate of change of moles in the control volume is given by

$$\frac{dN}{dt} = \sum_{CS,i} (\dot{N})_i + (\dot{N})_{gen} \tag{6}$$

where $(\dot{N})_i$ is the *molar flow rate* through control surface i and $(\dot{N})_{gen}$ is the *molar generation rate.* In general, the molar generation rate is not zero; the determination of its value is the object of the science of chemical kinetics (Section 4.2).

1.4   Macroscopic Species Mass Balance

A *solution* is a homogenous mixture of two or more chemical species. Solutions usually cannot be separated into their components by mechanical means. Consider a solution consisting of chemical species A, B, ... For each of the components of the solution, a mass balance may be written:

$$\frac{dm_A}{dt} = \sum_{CS,i} (\dot{m}_A)_i + (\dot{m}_A)_{gen}$$

$$\frac{dm_B}{dt} = \sum_{CS,i} (\dot{m}_B)_i + (\dot{m}_B)_{gen} \tag{7}$$

$$\vdots$$

Conservation of mass requires that the sum of the constituent mass generation rates be zero:

$$(\dot{m}_A)_{gen} + (\dot{m}_B)_{gen} + \ldots = 0 \quad \text{(conservation of mass)} \tag{8}$$

1.5   Macroscopic Species Mole Balance

The macroscopic species mole balances for a solution are

$$\frac{dN_A}{dt} = \sum_{CS,i} (\dot{N}_A)_i + (\dot{N}_A)_{gen}$$

$$\frac{dN_B}{dt} = \sum_{CS,i} (\dot{N}_B)_i + (\dot{N}_B)_{gen} \tag{9}$$

$$\vdots$$

In general, moles are not conserved in chemical or nuclear reactions. Hence,

$$(\dot{N}_A)_{gen} + (\dot{N}_B)_{gen} + \ldots = \dot{N}_{gen} \tag{10}$$

1.6   Macroscopic Energy Balance

Energy may be defined as the capacity of a system to do work or exchange heat with its surroundings. In general, the total energy E can expressed as the sum of three contributions:

$$E = K + \Phi + U \tag{11}$$

where K is the *kinetic energy*, Φ is the *potential energy*, and U is the *internal energy*.

Energy can be transported across the control surfaces by heat, by work, and by the flow of material. Thus, the rate of energy transport across control surface i is the sum of three terms:

$$\dot{E}_i = \dot{Q}_i + \dot{W}_i + \dot{m}_i \hat{E}_i \tag{12}$$

where $\dot{Q}_i$ is the *heat transfer rate*, $\dot{W}_i$ is the *working rate* (or *power*), $\dot{m}_i$ is the *mass flow rate*, and $\hat{E}_i$ is the *specific energy* (energy per unit mass).

Energy is conserved, meaning that the energy generation rate is zero:

$$\dot{E}_{\text{gen}} = 0 \quad \text{(conservation of energy)} \tag{13}$$

Therefore, the energy of the control volume varies according to

$$\frac{dE}{dt} = \sum_{\text{CS},i} (\dot{Q} + \dot{W} + \dot{m}\hat{E})_i \tag{14}$$

The energy flow rate can also be written in terms of the molar flow rate $\dot{N}_i$ and the molar energy $\tilde{E}_i$. Hence, the energy balance can be written in the equivalent form

$$\frac{dE}{dt} = \sum_{\text{CS},i} (\dot{Q} + \dot{W} + \dot{N}\tilde{E})_i \tag{15}$$

1.7 Entropy Balance

Entropy is a measure of the unavailability of energy for performing useful work. Entropy may be transported across the system boundaries by heat and by the flow of material. Thus, the rate of entropy transport across control surface i is given by

$$\dot{S}_i = \frac{\dot{Q}_i}{T_i} + \dot{m}_i \hat{S}_i \tag{16}$$

where $\dot{Q}_i$ is the heat transfer rate through control surface i, T_i is the absolute temperature of the control surface, $\dot{m}_i$ is the mass flow rate through the control surface, and $\hat{S}_i$ is the specific entropy or entropy per unit mass. In terms of the molar flow rate and the molar entropy $\tilde{S}_i$, the entropy transport rate is

$$\dot{S}_i = \frac{\dot{Q}_i}{T_i} + \dot{N}_i \tilde{S}_i \tag{17}$$

According to the second law of thermodynamics, entropy may be created—but not destroyed—in the control volume. The entropy generation rate therefore must be non-negative:

$$\dot{S}_{gen} \geq 0 \quad \text{(second law of thermodynamics)} \tag{18}$$

Processes for which the entropy generation rate vanishes are said to be *reversible*. Most real processes are more or less irreversible.

In terms of mass flow rates, the entropy balance is

$$\frac{dS}{dt} = \sum_{CS,i} \left(\frac{\dot{Q}}{T} + \dot{m}\hat{S} \right)_i + \dot{S}_{gen} \tag{19}$$

If molar flow rates are used instead, the entropy balance is

$$\frac{dS}{dt} = \sum_{CS,i} \left(\frac{\dot{Q}}{T} + \dot{N}\tilde{S} \right)_i + \dot{S}_{gen} \tag{20}$$

1.8 Macroscopic Momentum Balance

The momentum **p** is defined as the product of mass and velocity. Because velocity is a vector—a quantity having both magnitude and direction—momentum is also a vector. Momentum can be transported across the system boundaries by the flow of mass into or out of the control volume:

$$\dot{\mathbf{p}}_i = (\dot{m}\mathbf{v})_i \tag{21}$$

According to Newton's second law of motion, momentum is generated by the net force **F** that acts on the control volume:

$$\dot{\mathbf{p}}_{gen} = F \quad \text{(Newton's second law)} \tag{22}$$

Hence, the momentum balance takes the form

$$\frac{dmv}{dt} = \sum_{CS,i} (\dot{m}v)_i + \mathbf{F} \tag{23}$$

Because this is a vector equation, it can be written as three component equations. In Cartesian coordinates, the momentum balance becomes

$$x \text{ momentum:} \quad \frac{dmv_x}{dt} = \sum_{CS,i} (\dot{m}v_x)_i + F_x$$

y momentum: $\qquad \dfrac{dmv_y}{dt} = \displaystyle\sum_{\text{CS},i} \left(\dot{m}v_y\right)_i + F_y$ (24)

z momentum: $\qquad \dfrac{dmv_z}{dt} = \displaystyle\sum_{\text{CS},i} \left(\dot{m}v_z\right)_i + F_z$

2 DIFFERENTIAL BALANCE EQUATIONS

As noted previously, macroscopic balance equations account only for overall or gross changes that occur within a control volume. To obtain more detailed information, a macroscopic control volume can be subdivided into smaller control volumes. In the limit, this process of subdivision creates infinitesimal control volumes described by differential balance equations.

2.1 General Differential Balance Equation

The general macroscopic balance for the extensive property X is Eq. (1):

$$\frac{dX}{dt} = \sum_{\text{CS},i} (\dot{X})_i + (\dot{X})_{\text{gen}}$$ (1)

Division by the system's volume V yields

$$\frac{d}{dt}\left(\frac{X}{V}\right) = \sum_{\text{CS},i} \frac{\dot{X}_i}{V} + \frac{\dot{X}_{\text{gen}}}{V}$$

The differential or microscopic balance equation results from taking the limit as $V \to 0$:

$$\frac{\partial[X]}{\partial} = -\nabla \cdot (\mathbf{X}) + [\dot{X}]_{\text{gen}}$$ (25)

where $[X]$ is read as "the concentration of X" and $\mathbf{X}$ is "the flux of X." The terms of this equation can be interpreted as follows:

$\dfrac{\partial}{\partial t}[X]$ = rate of change of the concentration of X

$-\nabla \cdot (\mathbf{X})$ = net influx of X

$[X]_{\text{gen}}$ = generation rate of X per unit volume

The flux $\mathbf{X}$ is the rate of transport per unit area, where the area is oriented perpendicular to the direction of transport. In Cartesian (x, y, z) coordinates, $\mathbf{X}$ may be defined as

$$\mathbf{X} = \frac{\dot{X}_x}{A_x}\mathbf{i} + \frac{\dot{X}_y}{A_y}\mathbf{j} + \frac{\dot{X}_z}{A_z}\mathbf{k}$$

Here, A_x, A_y, and A_z are the areas perpendicular to the x, y, and z directions, respectively; $(\mathbf{i}, \mathbf{j}, \mathbf{k})$ are the (x, y, z) unit vectors. In Cartesian coordinates, the del operator ∇ takes the form

$$\nabla = \frac{\partial}{\partial x}\mathbf{i} + \frac{\partial}{\partial y}\mathbf{j} + \frac{\partial}{\partial z}\mathbf{k}$$

The form of the del operator in other coordinate systems may be found in texts on fluid mechanics and transport phenomena (1–4).

Table 2 shows the concentrations, fluxes, and volumetric generation terms for the extensive quantities considered in this chapter.

2.2 Differential Total Mass Balance

Assuming conservation of mass ($\rho_{gen} = 0$), the differential mass balance can be written as

TABLE 2 Concentrations, Fluxes, and Volumetric Generation

Quantity	Flux	Volumetric generation rate
Total mass, $[m] = \rho$	$\mathbf{m} = \rho\mathbf{v}$	$\dot{\rho}_{gen} = 0$ (conservation of mass)
Total moles, $[N] = c$	$\mathbf{N} = c\mathbf{v}$	$\dot{c}_{gen}$
Species mass, $[m_A] = \rho_A$	$\mathbf{m}_A = \rho_A\mathbf{v} + \mathbf{j}_A$	$(\dot{\rho}_A)_{gen}$
Species moles, $[N_A] = c_A$	$\mathbf{N}_A = c_A\mathbf{v} + \mathbf{J}_A$	$(\dot{c}_A)_{gen}$
Energy, $[E] = e$	$\mathbf{E} = \mathbf{q} + \sigma \cdot \mathbf{v} + \mathbf{m}\hat{E}$ $\mathbf{E} = \mathbf{q} + \sigma \cdot \mathbf{v} + \mathbf{N}\tilde{E}$	$\dot{e}_{gen} = 0$ (conservation of energy)
Entropy, $[S] = s$	$\mathbf{S} = \mathbf{q}/T + \mathbf{m}\hat{S}$ $\mathbf{S} = \mathbf{q}/T + \mathbf{N}\tilde{S}$	$\dot{s}_{gen} \geq 0$ (second law of thermodynamics)
Momentum, $[\mathbf{p}] = \rho\mathbf{v}$	$\mathbf{P} = \mathbf{mv} = \rho\mathbf{vv}$	$\mathbf{f} = -\nabla \cdot \sigma + \mathbf{b}$ (second law of motion)

Notes:

$\mathbf{b} \equiv$ body force per unit volume

$\mathbf{j}_A \equiv$ diffusive mass flux of species A

$\mathbf{J}_A \equiv$ diffusive molar flux of species A

$\mathbf{f} \equiv$ total force per unit volume

$\mathbf{q} \equiv$ heat flux

$\sigma \equiv$ material stress

$\mathbf{v} \equiv$ fluid velocity

$$\frac{\partial \rho}{\partial t} = -\nabla \cdot (\mathbf{m}) \tag{26}$$

where $\mathbf{m}$ is the mass flux. It is more common to write the mass flux in terms of the density and velocity, $\mathbf{m} = \rho\mathbf{v}$. Hence,

$$\frac{\partial \rho}{\partial t} = -\nabla \cdot (\rho\mathbf{v}) \tag{27}$$

Equation (27) is called the *continuity equation*; it is one of the basic equations of fluid mechanics.

2.3 Differential Total Material Balance

The differential material balance is

$$\frac{\partial c}{\partial t} = -\nabla \cdot (\mathbf{N}) + \dot{c}_{gen} \tag{28}$$

where $\mathbf{N}$ is the molar flux. In the absence of chemical or nuclear reactions, $\dot{c}_{gen} = 0$.

2.4 Differential Species Balances

Consider a solution consisting of component species A, B, ... In general, a chemical species in such a solution may be transported by convection and by diffusion. *Convection* is transport by the bulk motion of the solution. The convective flux of species A is the product of the mass concentration ρ_A and the solution velocity $\mathbf{v}$:

$$\rho_A \mathbf{v} = \text{convective (mass) flux of A}$$

Diffusion is the transport of a species resulting from gradients of concentration, electrical potential, temperature, pressure, and so on. The diffusive flux of species A is denoted by $\mathbf{j}_A$:

$$\mathbf{j}_A = \text{diffusive (mass) flux of A}$$

The overall material flux is the sum of the convective and diffusive fluxes:

$$\mathbf{m}_A = \rho_A \mathbf{v} + \mathbf{j}_A \tag{29}$$

A differential mass balance may be written for each of the components of the solution:

$$\frac{\partial \rho_A}{\partial t} = -\nabla \cdot (\rho_A \mathbf{v} + \mathbf{j}_A) + (\dot{\rho}_A)_{gen}$$

$$\frac{\partial \rho_B}{\partial t} = -\nabla \cdot (\rho_B \mathbf{v} + \mathbf{j}_B) + (\dot{\rho}_B)_{gen} \tag{30}$$

$$\vdots$$

Conservation of mass requires that the constituent mass generation rates sum to zero:

$$(\dot{\rho}_A)_{gen} + (\dot{\rho}_B)_{gen} + \ldots = 0 \quad (\text{conservation of mass}) \tag{31}$$

2.5 Differential Species Material Balance

The total molar flux of species A is the sum of the convective and diffusive molar fluxes:

$$\mathbf{N}_A = c_A \mathbf{v} + \mathbf{J}_A \tag{32}$$

where $c_A \mathbf{v}$ is the convective molar flux and $\mathbf{J}_A$ is the diffusive molar flux:

$$c_A \mathbf{v} = \text{convective (molar) flux of A}$$
$$\mathbf{J}_A = \text{diffusive (molar) flux of A}$$

The differential material balances for a solution consisting of species A, B, . . . are

$$\frac{\partial c_A}{\partial t} = -\nabla \cdot (c_A \mathbf{v} + \mathbf{J}_A) + (\dot{c}_A)_{gen}$$

$$\frac{\partial c_B}{\partial t} = -\nabla \cdot (c_B \mathbf{v} + \mathbf{J}_B) + (\dot{c}_B)_{gen} \tag{33}$$

$$\vdots$$

The sum of the constituent molar generation terms is the total molar volumetric generation rate:

$$(\dot{c}_A)_{gen} + (\dot{c}_B)_{gen} + \ldots = \dot{c}_{gen} \tag{34}$$

2.6 Differential Energy Balance

Energy may be transported by heat, by work, and by convection. Thus, the energy flux can be written as the sum of three terms:

$$\mathbf{E} = \mathbf{q} + \boldsymbol{\sigma} \cdot \mathbf{v} + \mathbf{m}\hat{E} \tag{35}$$

where $\mathbf{q}$ is the *heat flux*, $\boldsymbol{\sigma}$ is the *stress tensor* (defined as the force per unit area), and $\mathbf{m}\hat{E}$ is the convective energy flux.

As used in Eq. (35), the stress tensor σ accounts for the forces exerted on the surface of the differential control volume by the surrounding material. Multiplying the stress σ by the material velocity $\mathbf{v}$ gives the rate of work done (per unit area) on the surface of the control volume:

Rate of work by material stresses (per unit area) $= \sigma \cdot \mathbf{v}$

Conservation of energy requires that the energy generation rate be zero:

$$\dot{e}_{gen} = 0 \quad \text{(conservation of energy)} \tag{36}$$

The differential energy balance may be written as

$$\frac{\partial e}{\partial t} = -\nabla \cdot (\mathbf{q} + \sigma \cdot \mathbf{v} + \mathbf{m}\hat{E}) \tag{37}$$

It is common practice to express the energy concentration as the product of the mass density and the specific energy: $e = \rho\hat{E}$. Hence, the energy balance becomes

$$\frac{\partial \rho\hat{E}}{\partial t} = -\nabla \cdot (\mathbf{q} + \sigma \cdot \mathbf{v} + \mathbf{m}\hat{E}) \tag{38}$$

If the material flux is measured in moles, the energy balance may be written in the equivalent form

$$\frac{\partial \rho\hat{E}}{\partial t} = -\nabla \cdot (\mathbf{q} + \sigma \cdot \mathbf{v} + \mathbf{N}\hat{E}) \tag{39}$$

2.7 Differential Entropy Balance

Entropy may be transported by heat transfer and by convection. Thus, the entropy flux can be written as the sum of two terms:

$$\mathbf{S} = \frac{\mathbf{q}}{T} + \mathbf{m}\hat{S} \tag{40}$$

According to the second law of thermodynamics, the entropy generation rate must be non-negative:

$$\dot{s}_{gen} \geq 0 \quad \text{(second law of thermodynamics)} \tag{41}$$

The differential entropy balance is

$$\frac{\partial s}{\partial t} = -\nabla \cdot \left(\frac{\mathbf{q}}{T} + \mathbf{m}\hat{S}\right) + \dot{s}_{gen} \tag{42}$$

If the molar flux is used instead of the mass flux, the entropy balance becomes

$$\frac{\partial s}{\partial t} = -\nabla \cdot \left(\frac{\mathbf{q}}{T} + \mathbf{N}\tilde{S} \right) + \dot{s}_{\text{gen}} \tag{43}$$

2.8 Differential Momentum Balance

The momentum concentration is the product of density and velocity: $[\mathbf{p}] = \rho\mathbf{v}$. The momentum flux is the product of the momentum concentration and the velocity: $\mathbf{P} = \rho\mathbf{vv}$. The differential momentum balance is

$$\frac{\partial \rho\mathbf{v}}{\partial t} = -\nabla \cdot (\rho\mathbf{vv}) + \mathbf{f} \tag{44}$$

where $\mathbf{f}$ is the net force (per unit volume) acting on the control volume. In most fluid systems, $\mathbf{f}$ is the sum of a stress term and a body-force term:

$$\mathbf{f} = -\nabla \cdot \sigma + \mathbf{b} \tag{45}$$

The most important body force is gravity, for which $\mathbf{b} = \rho\mathbf{g}$.

The momentum balance becomes

$$\frac{\partial \rho\mathbf{v}}{\partial t} = -\nabla \cdot (\rho\mathbf{vv}) - \nabla \cdot \sigma + \mathbf{b} \tag{46}$$

3 FLUX AND TRANSPORT EQUATIONS

In most cases, balance equations are not sufficient by themselves; additional equations are needed to compute the fluxes and transport rates.

3.1 Diffusive Material Fluxes

As shown before, the overall material flux is the sum of the convective and diffusive fluxes:

$$\mathbf{m}_A = \rho_A\mathbf{v} + \mathbf{j}_A \tag{47}$$

$$\mathbf{N}_A = c_A\mathbf{v} + \mathbf{J}_A \tag{48}$$

The diffusive flux ($\mathbf{j}_A$ or $\mathbf{J}_A$) is driven by gradients of concentration, electrical potential, temperature, pressure, and so on. In the simplest situations, the rate of diffusion can be described by some variation of Fick's law:

$$\mathbf{j}_A = \rho D_A \nabla \omega_A \tag{49}$$

$$\mathbf{J}_A = -c D_A \nabla \chi_A \tag{50}$$

where ω_A is the mass fraction of A, χ_A is the mole fraction of A, and D_A is *the diffusion coefficient* or *diffusivity* of A. In some cases, other models may be required for describing diffusion (see Refs. 4 and 5).

3.2 Heat Conduction

Heat conduction (also called *heat diffusion*) is the transport of thermal energy by random molecular motion. Conduction is driven by a temperature gradient. In most cases, the heat flux can be described by Fourier's law:

$$\mathbf{q} = -k\nabla T \tag{51}$$

where k is the *conductivity* of the material. Conductivity values may be found in many standard handbooks and heat transfer textbooks (9,10).

3.3 Radiation Heat Transfer

Radiation is the transport of energy by electromagnetic waves. Radiative heat transfer tends to be especially important at high temperatures, such as occur in combustion. Consider an object having a surface temperature T_s. The magnitude of the radiation heat flux leaving the surface of the object is given by the Stefan-Boltzmann law:

$$|\mathbf{q}_{rad}| = \varepsilon\sigma T_s^4 \tag{52}$$

where σ is the *Stefan-Boltzmann constant,* a universal constant ($\sigma = 5.67 \times 10^{-8}$ W/m^2 K^4); and ε is the *emissivity*, a material property. Values of the emissivity for many different materials are tabulated in standard handbooks (9,10).

3.4 Fluid Stress

The *stress tensor* $\boldsymbol{\sigma}$ appears in both the differential energy balance and the differential momentum balance. In general, stress is defined as force per unit area. The force and the area can be represented as vectors having magnitude and direction:

$$\mathbf{F} = (F_x, F_y, F_z) \quad \text{and} \quad \mathbf{A} = (A_x, A_y, A_z) \tag{53}$$

Stress is a tensor having a magnitude and *two* directions. For example, the yx component of the stress tensor is*

$$\sigma_{yx} = \frac{F_x}{A_y} \tag{54}$$

The stress tensor $\boldsymbol{\sigma}$ is often represented as a 3×3 matrix having nine components:

---

*Some texts transpose the subscripts, so that $\sigma_{ij} = F_i/A_j$.

$$
\boldsymbol{\sigma} = \begin{pmatrix} \sigma_{xx} & \sigma_{xy} & \sigma_{xz} \\ \sigma_{yx} & \sigma_{yy} & \sigma_{yz} \\ \sigma_{zx} & \sigma_{zy} & \sigma_{zz} \end{pmatrix} = \begin{pmatrix} F_x/A_x & F_y/A_x & F_z/A_x \\ F_x/A_y & F_y/A_y & F_z/A_y \\ F_x/A_z & F_y/A_z & F_z/A_z \end{pmatrix} \tag{55}
$$

If the force is directed perpendicular to the area on which it acts, the result is a *normal stress*. If the force acts tangentially to the area, the result is a *shear stress*. Thus, the diagonal elements of the stress tensor are normal stresses, and the off-diagonal elements are shear stresses:

Normal stresses: $\sigma_{xx}, \sigma_{yy}, \sigma_{zz}$

Shear stresses: $\sigma_{xy}, \sigma_{xz}, \sigma_{yx}, \sigma_{yz}, \sigma_{zx}, \sigma_{zy}$

For a fluid, it is customary to resolve the stress tensor into two components:

$$
\boldsymbol{\sigma} = P\boldsymbol{\delta} + \boldsymbol{\tau} \tag{56}
$$

where P is the *fluid pressure,* $\boldsymbol{\delta}$ is the *identity tensor,* and $\boldsymbol{\tau}$ is the *viscous stress tensor.* In matrix notation,

$$
\begin{pmatrix} \sigma_{xx} & \sigma_{xy} & \sigma_{xz} \\ \sigma_{yx} & \sigma_{yy} & \sigma_{yz} \\ \sigma_{zx} & \sigma_{zy} & \sigma_{zz} \end{pmatrix} = -P \begin{pmatrix} 1 & 0 & 0 \\ 0 & 1 & 0 \\ 0 & 0 & 1 \end{pmatrix} + \begin{pmatrix} \tau_{xx} & \tau_{xy} & \tau_{xz} \\ \tau_{yx} & \tau_{yy} & \tau_{yz} \\ \tau_{zx} & \tau_{zy} & \tau_{zz} \end{pmatrix} \tag{57}
$$

The viscous stress tensor depends on the material. For an incompressible Newtonian fluid such as water, the diagonal elements of the viscous tensor are given by

$$
\tau_{xx} = 2\mu \frac{\partial v_x}{\partial x} \qquad \tau_{yy} = 2\mu \frac{\partial v_y}{\partial y} \qquad \tau_{zz} = 2\mu \frac{\partial v_z}{\partial z} \tag{58}
$$

where μ is the *viscosity.* For the same fluid, the off-diagonal elements of τ are given by

$$
\tau_{xy} = \tau_{yx} = \mu \left(\frac{\partial v_x}{\partial y} + \frac{\partial v_y}{\partial x} \right) \qquad \tau_{xz} = \tau_{zx} = \mu \left(\frac{\partial v_x}{\partial z} + \frac{\partial v_z}{\partial x} \right)
$$
$$
\tau_{yz} = \tau_{zy} = \mu \left(\frac{\partial v_z}{\partial y} + \frac{\partial v_y}{\partial z} \right) \tag{59}
$$

In vector notation, these Eqs. (58) and (59) become

$$
\boldsymbol{\tau} = \mu[\nabla \mathbf{v} + (\nabla \mathbf{v})^t] \tag{60}
$$

where the superscript t denotes a transpose. Other equations for the viscous stress are given in textbooks on fluid mechanics and transport phenomena (1–4).

3.5 Interfacial Transfer Coefficients

Many situations of practical engineering importance involve the transport of energy or material between two phases. It is common practice to compute the interfacial heat transfer rate using a relation of the form

$$\dot{Q} = \pm hA \, \Delta T \tag{61}$$

where h is the *heat transfer coefficient,* A is the interfacial surface area over which heat transfer occurs, and ΔT is the difference in temperature between the two phases. The sign is chosen so that the heat flows from high temperature to low temperature. Similarly, mass transfer between two phases can be described by a relation of the form

$$\dot{N}_A = \pm k_A A \, \Delta c_A \tag{62}$$

where k_A is the *mass transfer coefficient* for species A, A is the interfacial surface area over which mass transfer occurs, and Δc_A is the difference in the concentration of species A between the two phases.

Heat and mass transfer coefficients are determined experimentally; the results of such experiments are typically presented in terms of dimensionless parameters. Table 3 lists a number of important dimensionless parameters. Thus, heat transfer data are often presented in terms of the Nusselt number, Nu, a dimensionless heat transfer coefficient. In many cases, the Nusselt number is found to be a function of the Reynolds number Re and the Prandtl number Pr:

$$Nu = f(Re, Pr) \tag{63}$$

The equivalent relation for mass transfer is

$$Sh = f(Re, Sc) \tag{64}$$

where Sh is the Sherwood number and Sc is the Schmidt number (see Table 3). Many standard handbooks and heat transfer textbooks list dimensionless transport relationships for engineering systems (9,10).

4 STOICHIOMETRY AND CHEMICAL KINETICS

Stoichiometry is the study of the proportions in which elements combine to form compounds. *Chemical kinetics* is the science dealing with the rates of chemical reactions.

4.1 Stoichiometry

Consider a general chemical reaction in which reactants A, B, ... react to form products X, Z, ... according to the equation

TABLE 3   Some Important Dimensionless Groups

Group	Definition	Significance
Nusselt number, Nu	$Nu \equiv \dfrac{hL}{k}$	Dimensionless heat transfer coefficient
Péclet number, Pe (thermal)	$Pe \equiv Re \cdot Pr$	Ratio of heat conduction to convection
Péclet number, Pe_A (chemical)	$Pe_A \equiv Re \cdot Sc_A$	Ratio of species A diffusion to convection
Prandtl number, Pr	$Pr \equiv \dfrac{\mu/\rho}{k/(\rho \hat{C}_p)} = \dfrac{\mu \hat{C}_p}{k}$	Ratio of momentum transport to heat conduction
Reynolds number, Re	$Re \equiv \dfrac{\rho L v}{\mu}$	Ratio of inertial momentum transport to viscous momentum transport
Schmidt number, Sc_A	$Sc_k \equiv \dfrac{\mu/\rho}{D_k}$	Ratio of viscous momentum transport to species diffusion
Sherwood number, Sh_A	$Sh_A \equiv \dfrac{k_A L}{D_A}$	Dimensionless mass transfer coefficient for species A

Notes:
$\hat{C}_p \equiv$ specific heat
$D_i \equiv$ diffusivity of species i
$h \equiv$ heat transfer coefficient
$k_i \equiv$ mass transfer coefficient for species i
$k \equiv$ thermal conductivity
$L \equiv$ characteristic length
$\mu \equiv$ viscosity
$\rho \equiv$ density

$$aA + bB + \ldots = \ldots + xX + zZ$$

This can be written in the equivalent form

$$0 = -aA - bB - \ldots + xX + zZ$$

or, more compactly,

$$0 = \sum_{k=1}^{n} v_k I_k \tag{65}$$

where v_k is the *stoichiometric coefficient* of species I_k. Note that the stoichiometric coefficient is positive for a product and negative for a reactant.

4.2 Reaction Rate

The rate of a chemical reaction can be defined in terms of the stoichiometric coefficient:

$$\dot{\xi} = \frac{(\dot{N}_A)_{gen}}{v_A} = \frac{(\dot{N}_B)_{gen}}{v_B} = \ldots = \frac{(\dot{N}_k)_{gen}}{v_k} = \ldots = \frac{(\dot{N}_X)_{gen}}{v_X} = \frac{(\dot{N}_Z)_{gen}}{v_Z} \quad (66)$$

Here, ξ is the *extent of reaction variable* or *reaction coordinate*. The *volumetric reaction rate r* may be defined by:

$$r = \frac{\dot{\xi}}{V} \quad (67)$$

The volumetric reaction rate is usually a function of the temperature and the concentrations of the various reactants and products:

$$r = \pm k(T)f(c_A,c_B, \ldots ,c_X,c_Z) \quad (68)$$

where $k(T)$ is the so-called *rate constant* (which is not really a constant, but rather is a function of temperature).

In most cases, the rate constant is given by the *Arrhenius equation*:

$$k = k_0 \exp\left(\frac{-E_a}{RT}\right) \quad (69)$$

In this equation, k_0 is the *preexponential factor* and E_a is the *activation energy*, both of which must be determined experimentally.

The concentration-dependent part of the rate expression, $f(c_A,c_B, \ldots , c_X, c_Z$, must also be determined experimentally. In many cases, the reaction rate is found to depend on the concentrations of the reacting species as follows:

$$f(c_A,c_B, \ldots) = c_A^{\alpha}c_B^{\beta} \ldots \quad (70)$$

A reaction that obeys this equation is said to be of "α order" in species A, of "β order" in species B, and so on. The overall order of the reaction n is the sum of the individual reaction orders:

$$n = \alpha + \beta + \ldots \quad (71)$$

More information on the determination of rate constants and their applications can be found in standard texts on kinetics and reactor design (6–8).

5   EQUILIBRIUM THERMODYNAMICS

Thermodynamics is the study of the relationship between heat, work, and various forms of energy. Of particular interest to thermodynamics are the conditions for equilibrium.

5.1   Auxiliary Thermodynamic Functions

In addition to the energy and entropy, it is common in thermodynamics to define certain auxiliary functions. One of these is the *enthalpy* or *heat function H*:

$$H = U + PV \tag{72}$$

Two common free-energy functions are defined. The *Helmholtz free-energy function A* is defined by

$$A = U - TS \tag{73}$$

The *Gibbs free-energy function G* is defined by

$$G = H - TS \tag{74}$$

The Gibbs free-energy function is of special interest because it provides a convenient criterion of spontaneity and equilibrium.

5.2   Spontaneity and Equilibrium

A *spontaneous process* is one that occurs in a closed system without the assistance of some external agency. The Gibbs free energy decreases in a spontaneous process:

$$\Delta G \leq 0 \quad \text{(spontaneous process)} \tag{75}$$

At *equilibrium*, the system does not change with time. The Gibbs free energy reaches a minimum at equilibrium:

$$\frac{dG}{dt} = 0 \quad \text{(equilibrium)} \tag{76}$$

5.3   Fugacity

The change in Gibbs free energy for an ideal gas undergoing an isothermal compression or expansion from pressure P_1 to pressure P_2 is given by

$$\Delta G = RT \ln \frac{P_2}{P_1} \quad \text{(ideal gas)} \tag{77}$$

Although Eq. (77) does not apply for a real gas, it is common practice to retain the same functional form for real gases:

$$\Delta G = RT \ln \frac{f_2}{f_1} \quad \text{(real gas)} \tag{78}$$

Here, f_1 is the fugacity at pressure P_1. The fugacity may be considered as an adjusted pressure. It is defined so as to coincide with the pressure at low densities:

$$\lim_{P \to 0} \frac{f}{P} = 1 \tag{79}$$

The ratio of fugacity to pressure is called the fugacity coefficient, ϕ. Thus, the previous equation may be written

$$\phi \equiv \frac{f}{P} \implies \lim_{P \to 0} \phi = 1 \tag{80}$$

5.4 Chemical Potential

Consider a solution consisting of n species A, B, ... The Gibbs free energy of the solution is given by

$$G = \sum_{k=1}^{n} N_k \mu_k \tag{81}$$

where μ_k is the *chemical potential* of species k, defined by

$$\mu_k = \left(\frac{\partial G}{\partial N_k} \right)_{T,P,N_{j \neq k}} \tag{82}$$

5.5 Fugacity and Activity

The chemical potential is generally a function of temperature, pressure, and composition. It is common practice to write

$$\mu_A = \mu^{\circ}_A(T) + RT \ln a_A \tag{83}$$

where $\mu^{\circ}_A(T)$ is the standard chemical potential and a_A is the activity of species A. The activity is defined as the ratio of the fugacity f_A to a standard-state fugacity f°_A:

$$a_A = \frac{f_A}{f^{\circ}_A} \tag{84}$$

Activities and standard states are discussed in greater detail in many texts on chemical thermodynamics (11,12).

5.6 Phase Equilibrium

Consider a multicomponent system that separates into two or more phases (denoted I, II, . . .). The criteria for phase equilibrium are

$$T_I = T_{II} = \ldots \quad \text{(thermal equilibrium)} \tag{85}$$

$$P_I = P_{II} = \ldots \quad \text{(mechanical equilibrium)} \tag{86}$$

$$(\mu_A)_I = (\mu_A)_{II} = \ldots \quad \text{(equilibrium for species A)} \tag{87}$$

$$(\mu_B)_I = (\mu_B)_{II} = \ldots \quad \text{(equilibrium for species B)} \tag{88}$$

$$\vdots$$

5.7 Reaction Equilibrium

Consider a reaction

$$a\text{A} + b\text{B} + \ldots = \ldots + x\text{X} + z\text{Z}$$

which, as before, can be expressed in the shorthand notation

$$0 = \sum_{k=1}^{n} \nu_k I_k$$

The criterion for chemical reaction equilibrium is

$$\Delta G = \sum_{k=1}^{n} \nu_k \mu_k = 0 \tag{89}$$

Reaction equilibrium may also be expressed in terms of the standard Gibbs free-energy change:

$$\Delta G^0 = RT \ln K_a \tag{90}$$

Here, K_a is the equilibrium constant, defined by

$$K_a = \prod_{k=1}^{n} a_k^{\nu_k} \tag{91}$$

6 ENGINEERING FLUID MECHANICS

Fluid mechanics deals with the flow of liquids and gases. For most engineering applications, a macroscopic approach is usually taken.

6.1 Engineering Bernoulli Equation

Most engineering problems in fluid mechanics can be solved using the *engineering Bernoulli equation*, also called the *mechanical energy balance*. It can be derived from the macroscopic energy balance (see Ref. 13), subject to the following restrictions: (a) the system is at steady state; (b) the system has a single fluid intake and a single outlet; (c) gravity is the sole body force, with constant $|g|$; (d) the flow is incompressible; (e) the system may include one or more pumps or turbines. Under these conditions, the macroscopic energy balance becomes

$$\Delta\left(\frac{P}{\rho} + |g|z + \frac{\alpha}{2}\langle v\rangle^2\right) = \frac{\dot{W}_s}{\dot{m}} - \frac{\dot{F}}{\dot{m}} \tag{92}$$

where

P = the fluid pressure

$\langle v\rangle$ = the velocity averaged over the cross section of the pipe or conduit

α = average velocity correction factor (2.0 for laminar flow and 1.07 for turbulent flow)

$\dot{W}_2$ = rate of work done by pumps or turbines (positive for pumps, negative for turbines)

$\dot{F}$ = frictional loss rate

Dividing by the acceleration of gravity $|g|$ yields the so-called *head form* of the Bernoulli equation:

$$\Delta\left(\frac{P}{\rho|g|} + z + \frac{\alpha}{2|g|}\langle v\rangle^2\right) = \frac{\dot{W}_s}{\dot{m}|g|} - \frac{\dot{F}}{\dot{m}|g|} \tag{93}$$

Each of the terms in this equation has the dimensions of length.

6.2 Fluid Friction in Pipes and Conduits

The frictional loss rate $\dot{F}$ equals the rate at which useful mechanical energy is converted to thermal energy by friction. It is usually computed from an equation of the form

$$\frac{\dot{F}}{\dot{m}} = 4f\left(\frac{L}{D}\right)\frac{\langle v\rangle^2}{2} \tag{94}$$

where f is the Fanning friction factor.* In general, the friction factor is a function

---

*This is not the only friction factor in widespread use. Some authors prefer the Darcy-Wiessbach friction factor, $f_{DW} = 4f$.

of the pipe diameter D, the surface roughness ε, and the Reynolds number Re, the latter being defined as

$$\text{Re} = \frac{\rho D |v|}{\mu} \tag{95}$$

where μ is the fluid viscosity.

In the laminar-flow regime, $f = 16/\text{Re}$. For turbulent flows, a number of charts, graphs, and equations are available to compute the friction factor. The Colebrook equation has traditionally been used, although it requires a trial-and-error solution to find f:

$$\frac{1}{\sqrt{f}} = -4 \log \left(\frac{\varepsilon/D}{3.7} + \frac{1.255}{\text{Re}\sqrt{f}} \right) \tag{96}$$

Wood's approximation (Ref. 13) gives f directly, without a trial-and-error procedure:

$$f = a + b\,\text{Re}^{-c} \tag{97}$$

where

$$a = 0.0235 \left(\frac{\varepsilon}{D} \right)^{0.225} + 0.1325 \left(\frac{\varepsilon}{D} \right) \tag{98}$$

$$b = 22 \left(\frac{\varepsilon}{D} \right)^{0.44} \tag{99}$$

$$c = 1.62 \left(\frac{\varepsilon}{D} \right)^{0.134} \tag{100}$$

The relations presented in this section were developed for cylindrical pipes or tubes; however, the same equations may be used for noncylindrical ducts if the pipe diameter D is replaced in Eqs. (94–100) by the hydraulic diameter D_H:

$$D_H = 4 \frac{\text{volume of fluid}}{\text{area wetted by fluid}} \tag{101}$$

6.3 Minor Losses

The relations developed in the previous section apply only to straight pipes or conduits. Most pipelines, however, include bends, valves, and other fittings which create additional frictional losses. These additional losses are often called "minor losses," although they may actually exceed the friction caused by the pipe itself.

There are two common ways to account for minor losses. One is to define an *equivalent length* L_{eq} which equals the length of straight pipe that would give the same frictional loss as the valve or fitting in question. The total equivalent length L_{total} is the sum of the true length of the pipe and the individual equivalent lengths of the valves and fitting:

$$L_{total} = L_{pipe} + \sum_{\text{fitting } i} (L_{eq})_i \tag{102}$$

The total equivalent length L_{total} is used in Eq. (94) in place of L to compute the total frictional losses.

The second common approach to computing minor losses relies on the concept of a *loss coefficient* K_L, defined for each type of valve or fitting according to the equation

$$\frac{\dot{F}}{\dot{m}} = K_L \frac{\langle v \rangle^2}{2} \tag{103}$$

A comprehensive listing of typical equivalent lengths and loss coefficients is published by the Crane Company (14).

6.4 Fluid Friction in Porous Media

A porous medium is a solid material containing voids through which fluids may flow. The most important single parameter used to described porous media is the *porosity* or *void fraction* ε:

$$\varepsilon = \frac{\text{void volume}}{\text{bulk volume}} = \frac{V_{voids}}{V_{voids} + V_{solid}} \tag{104}$$

Fluid friction in a porous medium of thickness L is usually described by Darcy's law:

$$\frac{\dot{F}}{\dot{m}} = \frac{\mu}{\rho} \frac{L}{k} \langle v \rangle \tag{105}$$

where k is a material property called the *permeability*.

REFERENCES

1. R. B. Bird, W. E. Stewart, and E. N. Lightfoot, *Transport Phenomena*. New York: Wiley, 1960.
2. M. M. Denn, *Process Fluid Mechanics*. Englewood Cliffs, NJ: Prentice-Hall, 1980.
3. R. W. Fahien, *Fundamentals of Transport Phenomena*. New York: McGraw-Hill, 1983.
4. W. M. Deen, *Analysis of Transport Phenomena*. New York: Oxford University Press, 1998.
5. E. L. Cussler, *Diffusion Mass Transfer in Fluid Systems*, 2nd ed. Cambridge, U.K.: Cambridge University Press, 1997.

6. O. Levenspiel, *Chemical Reaction Engineering*, 2nd ed. New York: Wiley, 1972.

7. H. S. Fogler, *Elements of Chemical Reaction Engineering*. Englewood Cliffs, NJ: PTR Prentice-Hall, 1992.

8. L. D. Schmidt, *The Engineering of Chemical Reactions*. New York: Oxford University Press, 1998.

9. F. P. Incropera and D. P. DeWitt, *Fundamentals of Heat and Mass Transfer*, 3rd ed. New York: Wiley, 1990.

10. D. R. Lide (ed.), *CRC Handbook of Chemistry and Physics*, 80th ed. Cleveland, OH: CRC Press, 1999.

11. I. M. Klotz and R. M. Rosenberg. *Chemical Thermodynamics: Basic Theory and Methods*, 3rd ed. Menlo Park, CA: W. A. Benjamin, 1972.

12. K. Denbigh, *The Principles of Chemical Equilibrium*, 3rd ed. Cambridge, U.K.: Cambridge University Press, 1971.

13. N. De Nevers, *Fluid Mechanics for Chemical Engineers*, 2nd ed. New York: McGraw-Hill, 1991.

14. *Flow of Fluids Through Valves, Fittings, and Pipe,* Crane Technical Paper 410. Chicago: The Crane Company, 1988.

10

Biotechnology Principles

Teresa J. Cutright
The University of Akron, Akron, Ohio

1 INTRODUCTION

As mentioned throughout this text, waste minimization encompasses recycling/reuse, waste reduction (material substitution, process changes, good housekeeping, etc.), and waste treatment on-site (1,2). Biotechnology has a direct impact on, and applicability to, an engineer's ability to achieve waste minimization goals. It has had demonstrated success with recycling programs via the generation of biogas as an alternative fuel. Bioremediation approaches have also been used for: point source reduction via biopolishing (3) and individual stream treatment (4); by-product utilization (5); material substitution (6); facilitation of new enzymatic/metabolic pathways to produce "cleaner" organic substances (7); and end-of-pipe treatments (8–10). This chapter will highlight a few of the biotechnology approaches to waste minimization.

When utilizing any biotechnology, it is important to remember that the primary function of a microorganism is not to destroy man's unwanted contaminants. Instead, a microbe must reproduce itself and maintain its cellular functions. To that end, as shown in Figure 1, every microorganism must: (a) protect itself from the environment, (b) secure nutrients (catabolism), (c) produce energy in a usable form (catabolism), (d) convert nutrients/food into cellular material (anabolism); (e) discard unnecessary waste products, and (f) replication genetic infor-

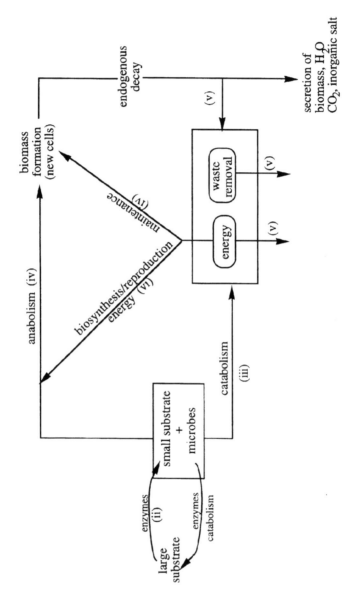

FIGURE 1 Overview of general cellular functions applicable to all living cells.

mation. It is an added benefit to mankind that the result from the microbial metabolism of substrates (i.e., step d), that the unwanted contaminants are degraded. Once it was realized that microorganisms could degrade unwanted contaminants, engineers started to manipulate the surrounding environment to ensure that the microbes would thrive and utilize the contaminant as the substrate.

Engineers currently use microorganisms to treat drinking water, municipal wastewater, and various industrial effluents. Usually the chemical and petrochemical industry is considered the only "real" contributor to industrial effluent (11). However, as shown in Table 1, more than just the chemical industry utilizes microorganisms for the treatment of waste on-site.

Regardless of whether the microorganisms are being used for cleaning drinking water, or municipal or industrial wastewaters; for end-of-pipe treatment at contaminated sites; or for waste minimization applications, certain key aspects apply (75). The following sections will outline the key aspects applicable to any biological treatment, provide a brief description and design criteria for the common waste minimization technologies, as well as highlight a few of the

TABLE 1 Some of the Industries that Utilize Biological Treatment for the Reduction of Waste

Industry	References
Coal processing	12–14
Cosmetics	15,16
Dyes	16–20
Fertilizer plants	21,22
Food	23,24
Citric acid	23,24
Dairy	25–30
Poultry	31–34
Slaughterhouses	22,35,36
Vegetables	37,38
Heavy metals processing	13,39–45
Oil processing and refineries	46–49
Paint	50–53
Paper	6,21,22,54–57
Pesticides	13,18,58–60
Pharmaceutical	61–66
Printing	19,20,67,68
Soap	46,69
Tannery	22,70,71
Textiles	3,18,72–74

innovative of bioprocesses that enable the selective removal of unwanted chemicals in product streams.

2 KEY ELEMENTS ESSENTIAL TO ALL BIOLOGICAL TREATMENT METHODS

Several basic biological requirements are essential for any biological treatment process to be successful. They are based on the principles required to support all ecosystems and include the presence of: appropriate microbes for degrading the contaminant(s), substrate for carbon and energy source, required terminal electron acceptor (TEA), inducer to facilitate enzyme synthesis, nutrients for supporting microbial growth, microbes to degrade metabolic byproducts, environmental conditions to minimize growth of competitive organisms (76–78). These factors will be discussed below.

2.1 Adequate Microbial Population

The primary requirement for any successful biological treatment or waste minimization strategy is the presence of an adequate microbial population. Luckily for environmental engineers, Mother Nature has supplied a wide variety of microbes to select from and to cultivate. The organisms are subdivided into different categories based on their metabolic capabilities and/or requirements. Table 2 contains the classifications based on the microbial carbon source, energy source, and respiration mode. If a contaminant can only be degraded in the presence of another organic material that serves as the primary electron source, then cometabolism is occurring. If the interaction of the two organisms is nonobligatory, then it is a synergistic relationship. Mutalism occurs if the interaction is beneficial yet obligatory. Since microbes are very versatile, it is important to remember that they may belong to more than category.

The microbe's versatility may also enable it to treat more than one particular type of contaminant. As shown in Table 3, different species of *Pseudomonas* have demonstrated success at reducing agricultural, heavy metal, food, and solvent wastes. In each instance, the primary requirement for successful treatment is the presence of an adequate population. Researchers have determined that a microbial count of 10^3–10^8 cfu/liter, 10^4–10^7 cfu/g would be adequate for groundwater and soil applications, respectively (77,79–81). Therefore, to ensure a successful biological treatment for waste minimization, a minimum of 10^8 cfu/liter would be recommended. If the contaminant concentration or toxicity increases, the microbial population will have to increase as well. If the increase in biomass concentration does not result in the desired treatment efficiency, the microbes being utilized may have to be changed to another source.

TABLE 2 Microbial Classification Based on Growth Requirements: Carbon and Energy Source, Respiration Mode

Term	Definition
Prototrophs	Most self-sufficient. Can synthesize all required growth compounds given CO_2 or a single organic compound.
Auxotrophs	Cannot synthesize all compounds required for growth.
Carbon source:	
Autotrophs	C source from CO_2. Ex: algae, photosynthetic bacteria.
Heterotrophs	C obtained from reduced form of organic compound.
Methanotrophs, methanogens	C and energy source from methane.
Energy source:	
Phototroph	Derive energy from light (i.e., algae).
Chemotroph	Derive energy from chemical oxidation.
Organotrophs	Energy from oxidation of organic chemicals.
Lithotrophs	Energy from oxidation of inorganic chemicals.
TEA:	
Aerobe	Require oxygen source for growth.
Anaerobe (oligate)	Cannot grow in the presence of oxygen. Obtain TEA from different source.
Facultative anaerobe	Can utilize O_2 if present; however preferable growth in absence of O_2 via different TEA.
Fermentation	TEA obtained from organic compound.

The species indigenous to a natural environment contain several different microbes living together (i.e., mixed population). Therefore, it is reasonable to presume that it is highly unlikely that one bacterial strain will be successful for a complete waste minimization scheme. Table 3 includes bacteria that have been successful at degrading the parent compound of the specified waste. Most of the references did not report the achievement of complete mineralization (conversion of contaminant into CO_2, biomass, H_2O, and salts). Complete mineralization would require the use of a microbial consortium. A consortium is more than a simple group of bacteria that can grow together. The overall net effect of the consortium is greater than what the individual microbes can accomplish on its own. Consortiums facilitate the degradation sequence where one microbe de-

TABLE 3 Select Microbial Genus with Demonstrated Success for Treating Industrial Contaminants Based on Waste Classification

Waste type	Microbial genus	
Aliphatics	Achromobacter	Micrococcus
	Acinetobacter	Mycobacterium
	Arthrobacter	Pseuodmonas
	Bacillus	Vibrio
	Flavobacterium	
Agricultural (pesticides, herbicides)	Achromobacter	Flavobacterium
	Alcaligenes	Methylomonas
	Arthrobacter	Penicillium*
	Athiorocaceae	Pseudomonas
	Corneybacterium*	Zylerion*
Dyes	Aeromonas	Phaneorchaete*
	Micrococcus	Shigella
	Klebsiella	Trametes*
	Pseudomonas	
Food, dairy, slaughter	Acinetobacter	Nitrosomonas
	Arthrobacter	Pseudomonas
	Bacillus	Rhodococcus
	Brevibacterium	Vibrio
Metals	Aeromonas	Pseudomonas
	Alteromonas	Saccharomyces†
	Bacillus	
	Enterobacter	
Pulp and paper	Arthrobacter	Talaromyces*
	Eisenia	Trichoderma*
	Chromobacter	Xanthomonas
	Sporotrichum*	
Solvents	Alcaligenes	Nitrosomonas
	Citrobacter	Nocardia
	Desulfomonite	Pseudomonas
	Enterobacter	Rhodococcus
	Morganeela	Xanthobacter
	Mycobacterium	

*Fungi; †yeast.

grades the metabolites of the first. The number and types of microbes required for a successful consortium depends on the contaminant classification, complexity, and concentration.

Natural mixed populations (i.e., consortiums) can be viewed as an interactive community that require each individual presence in order to thrive. When unmanipulated, the mixed population will contain one or two species that dominate the culture. These species are the most adaptable to the surrounding environment, have the most efficient energy utilization, and often facilitate the first step in the metabolic pathway. As time progress, the population may shift to one in which a different species dominates to continue the metabolic pathway or adjust for changes in substrate, nutrients, or terminal electron acceptor (82–84).

The microbes used for waste minimization applications have three basic modes of growth: attached (fixed), suspended, or free growth. Attached growth is similar to the biofilms used in wastewater treatment (trickling filters) and air emissions. Bioflims are surface aggregates composed of layers of bacteria that are embedded in a polysaccharide matrix. Biofilms differ from suspended growth in that they are fixed in a stationary place. Suspended growth systems (i.e., activated sludge) still have the bacteria attached to a surface, but the surface is freely moving within the reactor. Free growth systems are similar to slurry treatments where the bacteria can sorb and desorb from a surface.

2.2 Terminal Electron Acceptor

Without an adequate supply or specific type of TEA, biological treatment will fail. Table 4 includes the primary electron acceptors used. Aerobic microbes utilize O_2 as the TEA. For strict aerobes, the oxygen source is typically obtained from air or H_2O_2. Incorporation of alternative TEA sources provide another name to identify the respiration–microbe interaction. For instance, denitrifying bacteria utilize nitrate ($NO_3^- \rightarrow NO_2^- \rightarrow N_2$) as the TEA. *Nitrobacter sp.* is one of the microbes capable of nitrification, the conversion of NH_3-N to nitrates and nitrates. Sulfate reducers, such as *Desulfovibrio*, utilize $SO_4{2-}$ and generate

TABLE 4 Typical TEAs and Their Associated Respiration Modes

TEA	Form	Mode
Oxygen	O_2	Aerobic
Nitrate	NO_3	Anaerobic
Sulfate	SO_4	Anaerobic
Carbon dioxide	CO_2	Methane fermentation
Organic compound	Various	Aerobic, anaerobic, fermentation

S^{2-} (85). Methanogens require CO_2 as the TEA. Various bacteria can obtain the TEA from organic compounds.

The specific type of electron acceptor will dictate the metabolism mode, and thus the subsequent degradation reactions. Therefore the combination of microbes and TEAs utilized will enable a specific degradation pathway to be followed to facilitate cometabolism, prevent accumulation of toxic intermediates, etc. (86,87).

2.3 Nutrients

The nutrient requirements for microbes are approximately the same as the composition of their cells (76,88). The exception to this is carbon, which is sometimes needed at higher quantities and can be supplied by the contaminant for heterotrophic microorganisms. There are three categories of nutrients based on the quantity and essential need for them by the microorganism: macro, micro, and trace nutrients (89,90). For example, the macronutrients carbon, nitrogen, and phosphorus are known to comprise 50%, 14%, and 3% dry weight, respectively, of a characteristic microbial cell. Sulfur, calcium, and magnesium, which are micronutrients, comprise only 1%, 0.5%, and 0.5%, respectively of the cells dry weight (75,91,92). Trace nutrients, which are found in the least quantity, are not required by all organisms. The most common trace elements are iron, manganese, cobalt, copper, and zinc. Based on this approach, the optimal C:N:P mole ratio recommended for bioremediation applications is 100:10:1 (77,93,94). For example, 150 mg of nitrogen and 30 mg of phosphorous would be required to degrade 1 g of a theoretical hydrocarbon into cellular material. If the carbon source were easily and rapidly converted into carbon dioxide, then more carbon would be required in order to sustain microbes. For this reason, an additional carbon supplement may need to be supplied to facilitate the degradation of the contaminant depending on the microbe–contaminant interaction. The limits of the carbon (substrate) concentration will dictate the source and concentration of supplemental carbon and nutrients that are required (95).

Nitrogen is the nutrient most commonly added at bioremediation projects. It is primarily used for cellular growth (NH_4^+ or NO_3^-) for the synthesis of cellular proteins and cell-wall components. It can also be used as an alternative electron acceptor (NO_3^-). It is commonly added as urea or as ammonium chloride but may also be supplied as any ammonia salt or ammonium nitrate (92). All of these forms are readily assimilated in bacterial metabolism. However, if an ammonium ion is used to supply nitrogen, an increased oxygen demand on the surrounding system will be created facilitating the need for additional TEA supplements (96).

Phosphorus is the second most commonly added nutrient in bioremediation and is supplied to serve as a source for cellular growth. Often the ability of the microbes to secure the required phosphorous levels depends on the correct nitrogen level to be already in place (97). Phosphorus may be added as potassium phosphate, sodium phosphate, or orthophosphoric and polyphosphate salts. Potassium phosphate (mono- and di-basic) can also serve as a buffering agent to control pH. However, there are also cautions regarding the addition of a phosphorus source. The addition of potassium phosphate may accelerate the cleavage of the hydrogen peroxide added as an oxygen source. If the hydrogen peroxide cleaves too quickly, the oxygen source may be depleted before it reaches the contaminated zone. The maximum orthophosphate addition to provide microbial nutrients while avoiding significant precipitation, peroxide cleavage, or toxicity in most environments is 10 mg/liter (81).

Currently, there are no specific methods for predetermining the exact nutrient sources to utilize for a given situation or the role they will play. The specific ratio will depend on the chemicals to be treated, the microbes to be utilized, and the presence of inorganic nutrients already in the waste stream (98). The presence of a limiting nutrient will greatly affect the extent of remedial treatment. Therefore, researchers have conducted studies to determine some generalizations based on the type of microbe incorporated. For instance, nitrogen is required in smaller quantities when fungi are utilized. Magnesium is required in greater concentrations for photosynthetic bacteria. For aerobic cultures, iron is necessary in higher quantities than their anaerobic counterparts. Even with these generalizations, it is always a good idea to conduct a quick feasibility experiment to determine the exact nutrients required (77). This is particularly true for biological waste minimization processes with "exotic" chemicals, such as synthetic polymers, pesticides, and pharmaceuticals.

2.4 Environmental Conditions

There are several other environmental conditions that are critical to biological applications. The two major conditions are temperature and pH. The majority of microorganisms can grow only in a specific temperature range. If the microbes can grow only over a 10°C range, they are termed stenothermoal. Eurythermal organisms can grow over a 40°C range. Three cardinal temperatures are used to further describe microbial growth, the minimum, maximum, and optimum temperatures. Psychrophilic organisms have an optimum temperature of ~15°C with a minimum of <0°C (99). Most of the organisms used for degrading organic chemicals are mesophiles that have an optimum of 30°C. Thermophilic organisms have an optimal growth temperature above 45°C. Some researchers report the thermophilic optimum to be ~60°C, while some have listed a temperature closer to 90°C (100–102). When the surrounding temperature is beyond the optimum, microbial activity decreases (89). Typically, the loss of enzymatic activity is attributed to thermal denaturation that alters the membrane structure of the microorganism, thereby decreasing its normal cellular functions.

The pH level can affect the microbe's ability to conduct cellular functions, cell membrane transport, and the equilibrium of enzyme-catalyzed reactions. The minimum and maximum pH for microbial growth typically differs by 3 log units. Most bacteria can exist at a pH between 5 and 9, but have an optimal value near 7. Some bacteria are like fungi in that they prefer a more acidic environment (pH 1–3). For instance, *Thiobacillus thioxidans* has an optimum of 2.5 and is therefore classified as an acidophilic organism. Other bacteria, such as *Cyanobacteria* and *Bacillus sphaericus*, have a more alkaline optimum. These species are referred to as alkalophilic bacteria (103).

The pH and supplemental nutrients may effect the redox potential, which indicates the available TEA. Redox potential defines the electron availability as it affects the oxidation states of hydrogen, carbon, nitrogen, oxygen, sulfur, manganese, iron, etc. (104). As the surrounding environment is reduced, the electron density is increased and the redox potential becomes more negative. For an optimal aerobic environment, the redox potential must be greater than 50 mV (78,105). If nutrient additions or deviations in pH change the redox potential the necessary TEA/respiration mode may not be present for the desired degradation reaction.

3 COMMON BIOPROCESSES USED FOR WASTE CONTROL

3.1 Suspended Growth

The most widely studied and used suspended growth process is the activated sludge system used for the majority of industrial pretreatment and municipal wastewater treatment plants. The more common orientations for activated sludge include completely mixed, plug flow, oxidation ditch, contact stabilization, and sequencing batch reactors (106,107). Each configuration requires that the appropriate biomass, TEA level and source, pH, and contaminant loading be in place in order for the treatment to be successful. Since activated sludge processes are the most widely implemented suspended growth system, they have been the first used for waste minimization applications (10,92,108,109). The design equations and common operating parameters will be covered only briefly. More detailed information pertaining to activated sludge systems can be found in any wastewater or general civil engineering textbook.

Figure 2 represents the completely mixed activated sludge model typically found in industrial pretreatment applications. Here *i*, *e*, *r*, and *w* are used to denote the influent, effluent, recycle, and waste conditions, respectively. Other parameters include:

Q = volumetric flow rate, m^3/d

Va = volume of the aeration tank, m^3

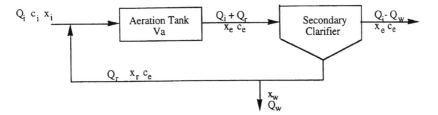

FIGURE 2 Process diagram for completely mixed activated sludge model.

c = contaminant concentration, g/m^3

x = biomass concentration, g/m^3

The substrate (contaminant) and biomass material balances are used to obtain the required operating design equations. In order to develop the substrate and biomass material balances, the following several assumptions must be made:

Complete mixing occurs.
Influent flows and substrate concentrations are constant, i.e., the system is at steady state.
Influent biomass (x_i) is zero.
All of the substrate, food, is soluble.
Biological activity occurs only in the aeration tank (no substrate is removed in the clarifier).
Sludge wasting occurs only in the clarifier.
Solids mass in clarifier and recycle line << mass solids in aeration tank (xV_a).

3.1.1 Biomass Material Balance

Using the assumptions stated above, a material balance initially neglecting the recycle stream is made:

Influent biomass + biomass production =
effluent biomass + wasted biomass

$$Q_i x_i + V \frac{dx}{dt} = (Q_i - Q_w)x_e + Q_w x_w \qquad (1)$$

In order to simplify Eq. (1), relationships pertaining to the sludge age, θ_c, and biomass generation must be known. Depending on the wastewater, the amount of biomass in the aeration tank is 800–6000 mg/liter. The cell age is a measure of the average amount of time the biological solids remain in the aeration tank and is used to help maintain the desired active biomass concentration. The sludge age and biomass generation terms are typically expressed as shown in Eqs. (2) and (3), respectively:

$$\theta_c = \frac{Vx}{(dx/dt)_R} = \frac{xV_a}{Q_w x_r + (Q_i - Q_w)x_e} \tag{2}$$

$$\left(\frac{dx}{dt}\right)_{gen} = Y\left(\frac{dc}{dt}\right)_u - k_d x \tag{3}$$

where the $k_d x$ term is used to describe the amount of energy to keep the biomass alive and Y is the true cell yield. Substituting Eqs. (2), (3), and the Monod relationship for (dc/dt) into Eq. (1) will yield the common design equation. After rearranging, Eq. (1) becomes

$$c_e = \frac{K(\theta_c^{-1} + k_d)}{Yk - [\theta_c^{-1} + k_d]} \tag{4}$$

where

K = Monod saturation constant, g/m^3

k_d = microbial decay coefficient, d^{-1}

k = maximum specific substrate utilization rate, d^{-1}

Y = true cell growth yield, g cell produced per g cell removed

c_e = effluent concentration, g/m^3

θ_c = mean cell residence time, d

Using Eq. (4), the only variable is the cell age, which can be used to control the effluent concentration. In other words, if the desired effluent is not achieved, the cell age is increased which increases the substrate residence time, thereby enabling the biomass to have a longer time to facilitate the substrate degradation.

3.1.2   Substrate Material Balance

One of the uses of the substrate material balance is to determine the volume of the recycle stream required for achieving the desired effluent concentration. In this situation, the system boundary is around the aeration tank and includes the recycle stream.

Using the basic material balance, accumulation = in − out + generation, and the assumptions stated previously, the substrate material balance is

$$Q_i c_i + Q_r c_e = (Q_i + Q_r)c_e + \left(\frac{dc}{dt}\right)_u V_a \tag{5}$$

Grouping like terms and incorporating the relationship for cell age yields

$$\frac{\theta_c^{-1} + k_d}{Y} = \frac{Q(c_i - c_e)}{xV_a} \tag{6}$$

Rearranging Eq. (6) to group process variables yields

$$\frac{xV_a}{Q} = \frac{Y(c_i - c_e)}{\theta_c^{-1} + k_d} \tag{7}$$

Equation (7) can be further simplified by introducing the hydraulic residence time, θ_H:

$$x\theta_H = \frac{Y(c_i - c_e)}{\theta_c^{-1} + k_d} \tag{8}$$

The form given in Eq. (8) is the operating substrate design equation used by most wastewater treatment facilities. This form allows, for a fixed effluent substrate concentration and cell age, the hydraulic residence time and required biomass concentration to be determined.

3.1.3 Typical Design Values and Waste Applications

As stated previously, the average initial biomass concentration for activated sludge systems is in the range of 800–6000 mg/liter. The specific value used depends on the food substrate (F) to microbe (M) ratio. Most F/M ratios are in the range 0.04–0.07 [107]. The higher ratio is implemented for elevated organic loading rates that require more microbial activity to achieve the desired effluent concentrations. For completely mixed activated sludge systems with a 5-day cell age, the optimal loading rate of 1 kg substrate/day/m^3 would require at least 25% of the water to be recycled through the aeration tank.

Activated sludge systems have had demonstrated success with conventional municipal wastewater. They have also been used for dairy waste, nonpesticide agricultural waste, pharmaceutical waste, and low-strength singular solvents in industrial pretreatment. Activated sludge is not effective for exotic organic chemicals such as pesticides, high-strength solvents, elevated heavy metal concentrations, alkaline waters, or mixed wastes. Copper, nickel, and zinc at concentrations as low as 1 mg/liter have been shown to inhibit microbial activity (110). As effluent constraints become more stringent, the applicability of basic activated sludge systems for treating most industrial pretreatment waste streams will decrease unless the waste streams are segregated and the biomass acclimated to the specific waste present (111,112).

3.2 Attached Growth

Attached growth systems encompasses processes from biofilms to slime layers. Biofilms are used for facilitating biooxidation reactions of contaminated gas emissions. Slime layers are utilized for biofiltration.

3.2.1 Biofiltration

The two most common processes employed for biofiltration are trickling filters (TFs) and rotating biological contactors (RBCs). For wastewater applications, the slime layers are comprised of a mixed population of bacteria, protozoa, and fungi. Depending on the waste constituents, the dominant species in the population will change. Also, the protozoa are employed as an auxiliary control on the slime thickness. If the slime layer is too large, the treatment efficiency will be decreased due to mass transfer limitations.

Trickling Filters. Trickling filters are not filters as their name implies, but a means of providing a large surface area where the microbes can attach as they "feed" on the organics. The media support is usually in the form of rocks, stones, or ceramics, depending on the surface area and throughput required. As depicted in Figure 3, water trickles over the top of the media to provide the contact between the microbial population and contaminant. TFs are classified as an aerobic treatment since aerobic cultures are responsible for over 75% of the degradation. However, aerobic cultures are located only to a film depth of 0.1–0.2 mm; toward the center of the media, faculative anaerobes dominate the microbial population, as dissolved oxygen concentrations at this point are minimal (84,113,114).

TFs have had demonstrated success at both the pilot scale (for exotic chemicals) and full scale for municipal-industrial pretreatment. At the pilot scale, TFs in different orientations have been effective for both inorganics and organic constituents. One study demonstrated a 94% reduction of manganese, while another study degraded 97% of a 300-ppm styrene waste stream (115). TFs have been used for over 30 years as a secondary or tertiary treatment for municipal wastewater (116). Full-scale effectiveness has also be proven for some recalcitrant organics as well. For instance, one plant was used to treat a waste stream containing 460 mg/liter of alkyl ethoxylate sulfates and alkylbenzene sulfonates to between 46 and 130 mg/liter (115).

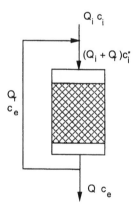

FIGURE 3 General representation of a trickling filter.

The three primary design variables for TFs are the hydraulic loading rate (HLR), organic loading rate (OLR), and recycle ratio (α). The parameters are determined from material balances around the system shown in Figure 4 as

$$\text{HLR} = \frac{Q_i + Q_r}{A} \tag{9}$$

$$\text{OLR} = \frac{Q_i c_i + Q_r c_e}{V} \tag{10}$$

$$\alpha = \frac{Q_r}{Q} \tag{11}$$

Both the HLR and OLR will directly affect and the depth of media required for treating the contaminant. As shown in Eqs. (9) and (10), OLR and HLR are interconnected, thus providing engineers some room to manipulate actual values and still achieve the desired effluent constraints. For instance, changing the medium depth will increase the residence time, thereby enabling the microbes to have greater contact with the contaminant. Varying will assist with maintaining the biofilm thickness and minimum flow rate, dampen the OLR variation, and increase the contact with the organics. Recirculation ratios will typically vary from 0.4 to 4, depending on the effluent constraints. Values greater than 4 are used to treat higher-strength waste streams. With this approach, a system with high-strength wastewater may have a low HLR and an elevated OLR. However, if the OLR is sustained beyond the design capacity, the filter will become clogged and ultimately fail (106).

The first design equations for TFs were empirical. Over the past 20 years, these equations were modified to incorporate biochemical kinetics. The most frequently used design equation was developed by Eckenfelder (116,117) as

$$\frac{c_e}{c_i^*} = \exp\left\{-k\left[\frac{A_s}{(1 + \alpha)Q}\right]^n z a_v^m\right\} \tag{12}$$

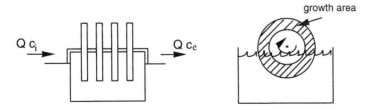

growth area

Q c_i Q c_e

FIGURE 4 General representation of a rotating biological contactor.

where

$$c_e = \text{effluent concentration, g/m}^3$$
$$c_i^* = \text{influent concentration, including recycle, g/m}^3$$
$$k = \text{reaction rate constant, d}^{-1}$$
$$A_s = \text{cross-sectional filter area, m}^2$$
$$\alpha = \text{recycle ratio}$$
$$z = \text{filter depth, m}$$
$$a_v = \text{specific surface area per packing piece}$$
$$m, n = \text{emperical constants determined by lab testing}$$

Biotowers. Biotowers are the common biological treatment implemented for minimizing waste in the dairy, milk, and food processing industries (107). They are similar to TFs in design, operation, and use. HLR, OLR, and are still the key design and operating parameters. The primary difference is that biotowers use plastic media instead of the stones or rocks used in TFs. By changing the media material to plastic, higher hydraulic and organic loading rates are obtainable, thus enabling biotowers to treat high-strength industrial wastes. Some pilot-scale trials have also investigated the use of soil, granular activated carbon, and perlite as the support material. However, polyethylene is still the primary support/packing for biotower applications.

Biotowers have been used to treat waste streams containing more than 200 ppm of the more recalcitrant compounds. For instance, methyl ethyl ketone, styrene, trichloromethane, and kerosene were each degraded by over 90% (82). These efficiencies were achieved when each compound was present as the sole contaminant. If the contaminants were present as a composite waste stream, the biomass would have to be altered. Even with a different microbial consortium, the remediation efficiency may not be as high, due to the degradation by products that would be formed. Researchers are currently addressing this situation. Based on preliminary successes, researchers are investigating the applicability of biotowers for treating composite streams in the pharmaceutical, textile, surface coating, and polymer processing industries (82).

Rotating Biological Contactors. Rotating biological contactors (RBCs) employ aerobic fixed film treatment to degrade either organic or ammonia-based constituents in wastewater. As shown in Figure 4, a RBC is comprised of a series of corrugated disks 3–4 mm in diameter mounted on a horizontal shaft. Spacing between the disks is in the range of 30–40 mm. Microbial slime layers forms on the disks' surface as they slowly rotate through the water (~1–2 rpm). The biofilm formed on each disk is approximately 0.3–4 mm thick. A minimum of 40% of the disk surface area is submerged in the waste stream as the RBC is rotated, to provide the required moisture as well as to bring the microbes in contact with the contaminants. The remaining surface area is exposed to the atmosphere, to supply the required oxygen as well as to facilitate the biooxidation. One RBC unit can provide up to 10,000 m^2 of surface area for biological growth.

Often, several RBCs are connected in parallel to treat the wastewater in a timely manner. Placement of modular units in series enables the attainment of treatment levels exceeding that of conventional secondary treatment (118).

As with TFs, Eckenfelder et al. (119) developed a relationship to determine the removal efficiency based on contaminant concentration:

$$\frac{c_e}{c_i} = 14.2 \left[\frac{(Q/A)^{0.558}}{\exp(0.32N_s)c_i^{0.684}T^{0.248}} \right] \qquad (13)$$

where

Q = mass rate of pollutant, g/d

A = disk area, ft^2

N_s = number of RBC units used

T = water temperature, °C

It is important to note that Eq. (13) provides a conservative estimate for the biodegradation efficiency, since it does not incorporate the reaction kinetics. Eckenfelder and his associates did modify the equation to include a reaction rate term. However, pilot-plant studies would be required to obtain the actual k value for the specific biomass–waste combination used. Currently, the kinetic data for RBCs are extremely limited, even for municipal wastewater applications (120).

RBCs are advantageous because they have a higher biomass concentration than most treatment systems and are easier to operate under varying load conditions. Unfortunately, they are prone to operational problems attributed to biomass buildup (114). Although proven effective for low-strength organic (<1% organic contaminants) and ammonia streams, RBCs are inefficient for high heavy metal concentrations and complex organics such as pesticides and chlorinated solvents (106,121). RBCs also tend to be very sensitive to temperature fluctuations. For these reasons, RBCs are currently used only as secondary or tertiary treatments.

3.2.2 Bioxidation—Scrubbing

With the newest amendment to the Clean Air Act, contaminants present in the gas/air phase are beginning to come under closer scrutiny due to the adverse affects they pose to human health and the environment. Part of the stricter regulation is attributed to the acute toxicity that the contaminants pose, as well as the fact that they are an unseen threat and disperse quickly (122). For instance, the 1984 accidental spill in Bhopal resulted in a death toll of 2500 people in less than 2 h. Although incidents such as that at Bhopal are not frequent, air emissions normally released from processing facilities still pose a serious threat. These gaseous contaminants can be efficiently and cost-effectively treated biologically via biofilters, TFs, and bioscrubbers.

Biofilters and TFs for gaseous systems operate on the same principles as their liquid counterparts, the only difference being the medium to be treated. Biofilters are essentially packed beds through which the contaminated gas stream is ventilated (14,50,123). The beds have depths of 3–4 ft and can be packed with ceramic material, soils, plastic, or inovative materials such as volcanic rocks (124,125). The gas enters the bottom of the bed with an upward gas velocity ranging from 0.0005 to 0.5 ft/s and residence times of 15–60 s, depending on the bed volume (126,127). TFs are also attached growth systems. Instead of water being trickled through the support system, liquid nutrients and contaminated gas are passed through (128).

Bioscrubbers are comprised of biological oxidation and an adsorption step in which the pollutants are adsorbed to water and a bioreactor for treating the water (129). The incorporation of thermophilic bacteria or coolants for the air stream enables bioscrubbers to be modified and extends their applicability. With such modifications, the adsorption column serves as both the contaminant sorption mechanisms as well as the biological support media, thereby combining the two steps in conventional bioscrubbers (130).

As with the liquid-based approach, gas treatments employ a consortia of bacteria instead of a pure culture. Consortiums are used since they are more effective and often occur naturally as the introduction of new, unidentified cultures that result from nonsterile operations (131). The combination of seeded and unidentified cultures has resulted in 75–90% remediation of the waste constituents, depending on the initial contaminant type and concentration (132,133).

For instance, a biotrickling reactor developed in Germany has demonstrated full-scale success at treating solvent-laden air. Over a 3-mo period, reactors removed 200 g/m^3 of solvent per hour at an OLR of 600 m^3/m^3 filter/h (134). Other researchers have proven the ability of mixed consortia to remove over 90% of a BTEX-contaminated off-gas (131,135). A different continuous biofilter achieved a maximum elimination capacity of 195 g/m^3/h for ethanol vapors (132). Biological treatment of off-gases has also been effective for methyl *tert*-butyl ether, toluene, and styrene (52,115,136). In most bioscrubbing applications, the removal performance based on contaminant classification follows a sequence similar to most biodegradation trends. In other words, removal rates follow the order alcohols > esters > ketones > aromatics > alkanes (137).

4 INNOVATIVE BIOLOGICAL APPROACHES TO WASTE MINIMIZATION

4.1 Microorganisms for Material Substitution or By-product Utilization

Two of the key approaches to waste minimization are material substitution and by-product utilization (138,139). If the raw material that produces the waste can be substituted with one that either minimizes the waste or yields a viable product, then the waste minimization goal has been realized. Furthermore, waste minimization is achievable if the intermediates or by-products of the process can also be reused. Microorganisms have the capability to be effective vehicles for both of these approaches, as outlined by the brief examples provided below.

4.1.1 Material Substitution

One example of a biological approach to material substitution in waste minimization is used in the formation of phenylacetylcarbinol (PAC). PAC is the key precursor in the pharmaceutical industry for nasal decongestants. During PAC formation, too much of a benzaldehyde intermediate is also made and disposed of as a waste stream (140,141). Recent studies have shown that by changing the yeast used to produce the enzyme for PAC formation, the benzaldehyde concentration is lowered (142,143). Some of these studies simultaneously increased the product yield by 140 mmol/liter (144).

p-Hydroxybenzoate (HBA) is another key raw material used in the pharmaceutical industry. Typically, HBA is produced by the Kolbe-Schmidt carboxylation of phenol, but this approach produces hazardous wastes (145,146). Bioconversion of toluene via *Pseudomonas putida* can also produce p-HBA. In fact, substitution with a biological approach increased the product formation while simultaneously eliminating the hazardous intermediate(s) (147).

The two aforementioned material substitution examples are slightly different from the nonbiological methods in that a raw material was not substituted. Instead, the seed culture or culture for processing an intermediate was substituted for another process. In both scenarios the bacterial substitution altered the metabolic pathway to remove an unwanted by-product.

4.1.2 Biodegradable Polymers

Perhaps the biggest area of material substitution involves the development of biodegradable polymers (148–150). Engineers and environmentalists know that synthetic polymers are essentially nonbiodegradable within an acceptable timeframe. Natural polymers, such as cellulose, proteins, starch, and polydhroxyalkanoates (PHAs), can be successfully constructed from both bacteria and plant matter (151–153). These new, natural polymers are particularly advantageous to waste minimization processes since not only do they eliminate the toxic waste streams generated via synthetic polymers, they are completely degradable by bacterial and plant enzymes. Furthermore, the end product is often "cleaner" than the synthetic counterpart (154). For these reasons, biopolymers are showing great promise for use in food packaging, medical delivery devices, and cosmetics (155–158).

Cellulose-based biopolymers have been extensively studied and therefore will not be presented here (159–162). It is important to note that the information gained by studying cellulose-based biopolymers led to the identification of several new approaches to natural polymers. One of the key developments was the substitution of PHAs for synthetic materials. PHAs are attracting attention as the most viable biopolymers due to their similarity to synthetic polymers while maintaining their inherent biodegradability (163–165). PHAs are produced via both plant and bacterial metabolic pathways (166–168).

Lee and Choi (169) documented the four primary metabolic pathways for the biosynthesis of PHAs. Two pathways involve *Ralstonia eutrophia* and *Rhodospirillium rubrum*. The other two pathways are found in most *Pseudomonads*. Due to the prevalence of the *Pseudomonad* species, these two particular biosynthetic pathways can yield an almost endless supply of PHAs. Other bacteria and their associated pathways have also been studied for PHA production [170]. Since the bacterial source and pathway used for generating the PHA is known, it is easy to identify and employ the enzyme required for degrading the polymer (171). For instance, cellulose-based and PHA polymers are easily degraded by *Hyphomicrobium sp.*, *Rhodococcus rhidochrous*, and *Trichoderma reesei* (172). This approach has also been used for the production and eventual degradation of other relatively new biopolymers such as polyhydroxybutyrates (173,174).

4.1.3 Conversion of a Waste Stream to a Viable Product

For every 9 lb of cheese produced, 1 lb of whey is generated as a waste stream. Currently the worldwide production of whey is 260,000 lb per year (175). Due to the high percentage of crude protein found in the whey, it is highly desirable as a substrate for producing various enzymes and oligosaccharides. For instance, lactase, the enzyme used to hydrolyze lactose into the simple sugars of glucose and galactose, is found naturally in the digestive tract of most humans and in several plants. Lactase can also be generated by a number of different yeast (*Saccharomyces fragilis*, *Trouls utilis*, *Kluveromyces marcianus*, and *Kluveromyces fragilis*) cultivated on the whey (176).

Other viable materials can be generated from the whey by changing the yeast, microbial, or fungi source. Ten different fungal strains are capable of producing proteinase enzymes from whey (177). Gellan gum is generated when *Sphingomonas paucimobilis* is utilized (178). A natural surfactanct, sophorolipid, is the end yield for processes utilizing species of *Cryptococcus curvaatus*, *Candida bombicola*, and *Mucor miehei* (179). The aforementioned are just of few of the innovations that have demonstrated the capability of biotechnology to convert a useless waste stream into a viable product.

4.2 Isolated Enzymes and Biocatalysts

While microbial processing may offer effective decomposition of organic compounds, the practice of microbial degradation on a large scale can be complicated by a few inherent limitations (pH sensitivity, nutrient amendments, population dynamics, etc.), depending on the waste stream to be treated. In addition, compounds such as organic solvents often present great toxicity toward microorganisms. Above all of these, most of the recalcitrant organic contaminants usually possess minimal solubility in the aqueous phase, where the microorganisms are hosted and considered the most active. Recent advances in biocatalysis have demonstrated that it is feasible to carry out biotransformations in nearly pure organic media (180–182).

Enzymes are biocatalysts, or proteins secreted by microorganisms to accelerate the rate of a specific biochemical reaction without being consumed in the reaction. The key environmental parameters critical to microorganisms must be in place for producing the enzymes from the intact microbial cells. The difference between normal microbial degradation and biocatalysis occurs once the enzymes have been harvested. After harvesting, the enzymes do not require subsequent nutrient, TEA, or energy source amendments. As with intact cells, the enzymes can either be utilized in a free suspension or immobilized on a support. Immobilized enzymes often show improved stability during variations in environmental conditions over free enzymes. For example, when laccase was immobilized for treating elevated concentrations of phenol, it maintained over 80% of its activity when subjected to changes in pH, temperature, and storage conditions (183).

Isolated enzymes can also afford much faster reactions, which are usually several orders of magnitude higher than those resulting from traditional microbial process (184). The increase in reaction time has been attributed to the absence of competing processes that are present with the natural metabolism of intact organisms. Since the competing processes are absent, enzymes possess a greater specificity toward a particular compound. An example of an enzyme's ability to treat recalcitrant compounds has been demonstrated by the textile industry.

Lacasse is the primary extracellular enzyme produced by *Trametes versicolor*. Lignin peroxidase is one of the many enzymes secreted by *P. chrysosporium*. Both of these enzymes have been used for the decolorization of the synthetic dyes contained in textile effluents (20). Utilization of these isolated enzymes yielded greater remediation efficiencies in less time and was independent of microbial growth rates. Decolorizations of 80% and over 90% were achieved for anthraquinone dyes with lignin peroxidase and lacasse (20,185). Lacasse was also capable of degrading the reaction by-products present in the textile effluent. Lacasse and lignin peroxidase were not the only enzymes that have been used for decolorizing textile effluents. Horseradish peroxidases and manganese peroxidases have also shown success for decolorizing effluents from the textile and paper industries (186–189).

As stated earlier, enzymes can also possess greater affinity toward a particular compound. Therefore, utilization of enzymes as biocatalysts is one of the processes that enable the selective removal of unwanted chemicals in product streams (190,191). For example, researchers have recently utilized enzymes to minimize waste at a cattle dipping facility. The cattle dipping liquid must be discarded when potasan, a dechlorination by-product, accumulates to a specified level. Due to the high degree of chlorination, the waste cannot be reused. Grice et al. (192) employed parathion hydrolase to selectively hydrolyze the potasan. Since the potasan concentration was reduced, the use of the dipping liquid was extended, thereby decreasing the amount of waste generated.

Phenol and substituted phenolics are among the constituents present in most industrial wastewaters. Due to their prominence in the waste streams and associated toxicity, phenolics have been identified for selective removal via enzymes. One approach is to use tyrosinase to convert phenol to o-quinones, which are then easily adsorbed (193). This two-step process was effective even in the presence of microbial inhibitors. When the tyrosine was not immobilized, the remediation efficiency was decreased. Other illustrations of treating phenolic wastewaters include the treatment of 4-chlorophenol with horseradish peroxidase (194), pentachlorophenol with horseradish peroxidase (195), and various phenol substitutions with immobilized peroxidases (196,197).

4.3 Biological Recovery and/or Treatment of Heavy Metals

4.3.1 Background

Heavy metals are among the contaminant classifications receiving the greatest scrutiny in waste minimization programs. These compounds are present in normal municipal wastewater and in various industrial effluents such as those of electroplating and metal finishing. Not all heavy metals pose a threat to the microorganisms used in treatment operations. For instance, Fe, Mo, and Mn are important trace elements with very low toxicity, while Zn, Ni, Cu, V, Co, W, and Cr are classified as toxic elements. Although toxic, Zn, Ni, and Cu have moderate importance as trace elements. As, Ag, Cd, Hg, and Pb have a limited beneficial function to most microorganisms (198).

Activated sludge consortiums and other microbial processes can tolerate most heavy metals at very low concentrations. As heavy metal concentration increases, the metals can become harmful to the bacteria (nontoxic), achieve toxicity status, or pass through the system unaltered. All three of these scenarios pose a serious threat to microbial and human health. When metal toxicity becomes too great, the microbes that are required for degrading the organic compounds perish. If metals such as Cu, Cr, and Zn pass through the system and are ingested, chronic human health disorders will arise (199). Recently, scientists have found that certain microorganisms either possess an inherent resistance to heavy metals or can develop the resistance through changing their internal metabolism.

Metal-resistant microbes have potential applications to facilitate a biotechnology process that occurs in the presence of, but does not require, heavy metals. Bacteria that can resist toxic metal concentrations could be utilized to convert the organic constituent present into a readily usable or disposable form. Microorganisms that can do more than simply tolerate elevated metal concentration can be used in the recovery of metals or bioremediation of contaminated media (198,200–202). The microbial processes that remediate or recover heavy metals include leaching (biological solubilization), precipitation, sequestration, and biosorption (203–206). Biosorption has been the most widely studied aspect for waste minimization activities.

4.3.2 Biosorption

Donmez et al. (207) defined biosorption as the "accumulation and concentration of contaminants from aqueous solutions via biological materials to facilitate recovery and/or acceptable disposal of the target contaminant." This definition can be expanded to depict the difference between active and passive biosorption. Active biosorption entails passage across the cell membrane and participation in the metabolic cycle. Passive sorption is the entrapment of heavy metal ions in the cellular structure and subsequent sorption onto the cell binding sites (208). Live biomass involves both active and passive sorption, whereas inactive cells entail only the passive mode.

Bioadsorbent materials encompass a broad range of biomass sources, including cyanobacteria, algae, fungi, bacteria, yeasts, and filamentous microbes. The biomass can be cultivated specifically for metal sorption, or a waste biomass can be utilized. Table 5 includes a brief compilation of the microorganisms studied for the biosorption of heavy metals via passive sorption with dead biomass. Although the research contained in Table 5 focused on the sorption of dead biomass, live cells can also be used.

Both live and inactivated (dead) biomass possess interesting metal-binding capacities due to the high content of functional groups contained in their cell walls (39,222). A few differences have been exhibited between the use of live and dead biomass. In general, live biomass can accumulate more metal ions per unit cell weight. Live cells, if processed correctly, can be reused almost indefinitely. For instance, live *Candida sp.* can sorb (per gram of live biomass) 17.23 mg, 10.37 mg, and 3.2 mg of Cd, Cu, and Ni, respectively (212). The sorption capacity was

TABLE 5 Examples of a Few of the Yeast, Bacteria, Algae, and Fungi Used for the Biosorption of Heavy Metals from Waste Effluents (metals listed in order of selectivity)

Species	Heavy metal	References
Yeast:		
Saccharomyces cerevisiae	Cu, Cr(VI), Cd, Ni, Zn	209–211
Candida sp.	Cd, Cu, Ni	212
Bacteria:		
Thiobacillus ferrooxidans	Cd, Zn, Cu	213
Bacillus cereus	Pb, Mn, Ni, Zn	214
Pseudomonas aeuriginosa	Cd, Hg, Cu	200
	Cr, Cu, Pb	215
	Cu	216
Rhisopus arrhisus	Cr(VI)	199
	Cr, Cu, Pb	211,217–219
Synechocystis sp.	Cu, Ni, Cr	207
Algae:		
Chlorella vulgaris	Cu, Ni, Cr	207
Scenedesmus obliquus	Cu, Ni, Cr	207
Fungi:		
Aspergillus niger	Pb, Cd, Cu, Ni	208
Bacillus thuringiensis	Cd, Hg, Cu	200
Fusarium oxysporum	Ni, Mn, Pb, Co, Cu	214
Phanerochaete chrysosporium	Cd, Zn, Cu	220,221

reduced by over 35% of each metal when inactive cells were used. Unfortunately, active sorption with live cells usually takes longer and poses stricter environmental controls than the use of inactivated biomass (223). Moreover, the use of dead biomass from pharmaceutical or other industrial operations is a waste minimization technique in itself (224). It converts a previously useless waste stream into a viable process step.

Also, as indicated in Table 5, sorption selectivity and specificity will depend on the source of biomass used. Research conducted by Kanosh and El-Shafei (214) showed that fungi had a metal selectivity of Ni >> Mn > Pb > Co > Cu >> Zn >> Al, whereas bacterial strain selectivity was Pb > Mn > Ni > Zn >> Al > Li > Cu > Co. Isotherm studies with brewer's yeast (*Saccharomyces cerevisiae*) resulted in a sorption order of Pb > Cu > Cd ≈ Zn > Ni (225). When *Candida sp.* is used as the source of live biomass, selectivity order changes for three of the metals, to Cd > Cu > Ni. The differences in degree of sorption and specificity are attributed to the inherent differences in cellular metabolism and functional groups adhered to the cell walls.

It is important to note that the success of metal biosorption has led researchers to investigate this technique for other waste compounds. For example, biosorption has also received a cursory investigation for the decolorization of dye effluents. The yeast strain *Kluyveromyces marxianus* IMB3 was used to treat effluents containing azo and diazo dyes. The biomass was able to adsorb over 68 mg dye/mg biomass during the 3-day study (226).

4.3.3 Heavy Metal Recovery

Microbial immobilization of heavy metals is used more for the recovery of specific metal ions. The ability of the microbes to immobilize the metals is an extension of its normal survival mechanism. Initially the microbes capable of immobilization were studied for their ability to transform the compound into a less toxic form (227). The first studies documented the ability of bacteria to change the valence state of soluble chromium from 6 to 3. When the experiment was conducted over a longer period of time, other bacteria were able to facilitate the precipitation of the metals (228). Precipitation occurred as a result of the microbial excretion of organic acids, enzymes, and polymeric substance. When the metals have been precipitated, they no longer pose a threat to the microorganisms and enable easy recovery.

Biosorption can also be used for the recovery of metals. However, the procedure is often more involved and costly than microbial immobilization. Recovery from dead biomass entails the digestion of the cellular material, followed by the separation of the different metal ions by chemical or electrolytic methods. Washing live cells with different electrolytic and buffer solutions can cause the elution of the bound metal. Again, separation and purification of the different metals in the eluted solution can be achieved by traditional chemical or electrolytic methods.

4.4 Minimization of Biomass

As stated in the activated sludge section, part of the biomass (i.e., sludge) is recycled and the remainder is wasted. In the past, the wasted biomass was either used as a nutrient amendment for farm lands, compost supplement, or disposed of in a landfill. However, depending on the waste being treated, the degradation by-products and inorganics that have sorbed to the suspended flocs may cause the wasted biomass to be classified as a hazardous waste. Even when not classified as hazardous, changing legislation and rising processing and disposal costs have led to studies that focus on biomass growth to decrease the amount of excess microbial matter (229,230).

The U.S. Air Force developed a two-step process to decrease the volume of biomass disposed of. First, the wasted biomass and suspended solids were solublized via a hot acid hydrolysis step. After cooling, the solublized material was recycled through the normal activated sludge treatment area. Utilization of this process has resulted in a 90% reduction of the hazardous biological solid waste disposed of in controlled landfills (231).

A second innovation was to eliminate excess sludge production by treating it in the aeration tank. The simultaneous activated sludge treatment and sludge reduction was brought about by the addition of ozone. In some facilities, ozone is used to supply the necessary oxygen to the microbial population. For this application, care has to be taken to ensure that too much ozone is not added, so that normal chemical oxidation reactions do not occur. The phenomenon led engineers to investigate ozonation for the reduction of excess biomass. Controlling the ozone dosage to 0.05 g per gram of suspended solids with a recycle ration of 0.3 reduced the excess sludge production to essentially zero (232). Although successful, this technique is not fully utilized due to the extensive monitoring and controls needed to ensure that the required population is kept viable and not seared.

Other approaches have encompassed increases in process temperatures or extending the aeration step. Unfortunately, neither avenue is economical or particularly effective (233). Perhaps a more efficient and controlled approach would be the direct manipulation of the microbial population. As stated earlier, bacterial populations will shift and alter their growth based on the available carbon, energy, and TEA sources. If the metabolic pathway is successfully uncoupled, substrate (i.e., contaminant) catabolism will continue, while biomass anabolism is restricted. Most microorganisms must satisfy their maintenance energy requirements before reproduction, so excess biomass formation will be reduced (234).

Any easy way to uncouple metabolic pathways and cause a population shift is to either switch the respiration mode from aerobic to anaerobic or add chemicals to deter growth rate (235,236). For most activated sludge processes, switching the respiration mode would be detrimental to the entire process. This approach is recommended only for the facultative anaerobes found in anaerobic digestors (237,238). Chemical additions, however, have proven to be quite successful at minimizing excess biomass formation without jeopardizing the overall process efficiency.

One study utilized nitrate as the nitrogen source instead of ammonia. This was a very cost-effective, easily adaptable method that decreased excess biomass generation by over 70% (239). A more unique means of manipulating biomass formation was the introduction of nickel to the raw influent. When nickel was added at a concentration of 0.5 mg/liter, the observed cell yield was reduced by approximately 65% (240). Unfortunately, a better biomass reduction could not be achieved. If too high a nickel concentration was implemented, steady state could not be attained.

4.5 Innovative Bioreactors

Manipulation of reactor design can decrease excess biomass formation, increase oxygen transfer, decrease shear stress, enhance contact time, enable treatment of recalcitrant compounds, and improve overall treatment efficiencies (127,229). The incorporation of baffles to bioreactors is known to enhance mixing and decrease shear (241). Although effective, the use of baffles is not considered a new technique. One of the most innovative and versatile developments is the use of membranes. Membranes can facilitate a multitude of different scenarios, one of which employs extractive membranes.

In general, extractive membranes are permeable to organics and virtually impermeable to water and heavy metals, thereby facilitating the biological treatment of the organics contained in mixed waste (198,242,243). Extractive membranes can be modified to allow the selective transfer of specific pollutants into an area where a specialized strain of bacteria will utilize the pollutant as the sole carbon source. Brookes and Livingston (112) used this approach to demonstrate the selective degradation of 3,4-dichloroanaline in wastewater. A similar methodology was used by Inguva et al. (244) to treat TCE and 1,2-dichloroethane contaminated wastewater. A sequence of such membranes can be employed in conjunction with the appropriate bio-areas containing different microorganisms for facilitating the treatment of high-molecular-weight compounds and/or mixed wastes.

Traditional activated sludge system efficiencies have been compared to those obtained by membrane bioreactors for high-molecular-weight compounds. The membrane bioreactor removed 99% of the chemicals, while the activated sludge system treated only 94% (245). Although this may not appear to be a significant difference, the membrane bioreactor was also highly effective in degrading the degradation by-products, whereas the activated sludge system was not. This was due largely to the enhancement of the biomass' viability by increasing the oxygen transfer when the membrane was in place (115,246). Membrane versatility also allows for high-strength wastes to be treated in both plug flow and completely mixed conditions (247,248).

The successful implementation of membranes is not limited to aerobic systems. Previously, ultrastrength wastewater was considered treatable only by a two-stage anaerobic process. Membrane anaerobic reactors enable an effective, single-phase treatment to be employed (38). Developments with hollow-fiber membranes have resulted in the simultaneous aerobic-anaerobic treatment of chlorinated compounds (249,250).

Hollow-fiber membranes have also been used as the support system for immobilized enzymes. The membranes provide a very large surface area for the immobilized enzymes. As with other enzyme applications, the reaction times and subsequent remediation efficiencies are greater than those typically achieved with intact cells. This approach has been used for wastewater denitrification, heavy metal recovery, and treatment of chlorinated compounds (251–255).

5 CONCLUSIONS

As shown by the brief examples presented here, biotechnology has demonstrated applicability in all areas of waste minimization—material substitution, selective removal, recycle/reuse of intermediates, end-of-pipe treatment, and source reduction. These demonstrations have included traditional processes such as activated sludge and trickling filters to innovative techniques involving heavy metal biosorption and immobilized enzymes. With continued advances in analytical and technological expertise, effluent constraints will promulgate to become more stringent. Biotechnology will continue to rise to the challenge. As scientists and engineers strive to learn more about microorganisms and their metabolic pathways, they will be able to convince Mother Nature to do as they wish. Natural approaches (i.e., green chemistry) are more effective and acceptable. Furthermore, if complete mineralization of waste effluents is achieved, or successful biosubstitutions are made, a secondary stream will not have to undergo subsequent treatment as with other technologies, thereby eliminating the need for waste minimization.

6 NOMENCLATURE

a_v	specific surface area per packing piece
A	RBC disk area, ft^2
A_s	cross-sectional filter area, m2
c	contaminant concentration, g/m^3
cfu	colony-forming unit
e	effluent
HBA	hydroxybenzoate
HLR	hydraulic loading rate
i	influent condition
k	maximum specific substrate utilization rate, d^{-1}
k_d	microbial decay coeffient, d^{-1}
K	Monod saturation constant, g/m^3
m,n	emperical constants in Eckenfelder equation
N_s	number of stages (RBC units) used
OLR	organic loading rate
PAC	phenylacetylcarbinol
Q	volumetric flow rate, m^3/d
r	recycle
RBC	rotating biological contactor

T	water temperature, °C
TEA	terminal electron acceptor
TF	trickling filter
Va	volume of the aeration tank, m^3
w	waste volume
x	biomass concentration, g/m^3
Y	true cell growth yield, g cell produced per g cell removed
z	filter depth, m
α	recycle ratio
θ_c	mean cell residence time, d

REFERENCES

1. S. Al-Muzaini, Waste Minimization Program in Shuaiba Industrial Area. *Water. Sci. Technol.*, vol. 39, no. 10–11, pp. 289–295, 1999.

2. J. A. Elterman, R. S. Reimers, L. M. Young, and C. G. Simms, Waste Minimization Study at Louisiana Oil & Gas Exploration and Production Facilities. Industrial Wastes Technical Conference, New Orleans, LA, 12/1–12, March 2–5, 1997.

3. T. Basu and M. Chakrabarty. Recent Achievements in the Field of Eco-processing of Textiles. *Colourage*, vol. 44, no. 2, pp. 17–23, 1997.

4. Y. Cohen and F. Giralt, Strategies in Pollution Prevention: Waste Minimization and Source Reduction. *AFINIDAD*, vol 53, no. 462, pp. 80–92, 1996.

5. G. E. Tong, Integration of Biotechnology to Waste Minimization Programs. In G. S. Sayler (ed.), *Environmental Biotechnology for Waste Treatment*, vol. 41, pp. 127–136, New York: Plenum Press, 1991.

6. D. J. Hardman, M. Huxley, A. T. Bull, H. Slater, and R. Bates, Generation of Environmentally Enhanced Products: Clean Technology for Paper Chemicals. *J. Chem. Tech. Biotechnol.*, vol. 70, pp. 60–66, 1997.

7. R. Fringuelli, P. Pellegrino, and F. Pizzo, Synthetic Pathways for a Cleaner Production of Organic Substances. *Life Chem. Rep.*, vol. 10, pp. 171–179, 1994.

8. E. D. Battalones and A. Castro. A Feasibility Study on the Biotreatability of Industrial Wastewaters. *SE Asian J. Trop. Med. Public Health*, vol. 5, no. 2, pp. 290–298, 1974.

9. I. Bhattacharya, Role of Biotechnology in Pollution Control. In H. S. Sohal and A. K. Srivastava (eds.), *Environment and Biotechnology*, pp. 163–170. New Dehli: Ashish Publishing, 1994.

10. B. J. D'Arcy, R. B. Todd, and A. W. Wither, Industrial Effluent Control and Waste Minimization: Case Studies by UK Regulators. *Water Sci. Technol.*, vol. 39, no. 10–11, pp. 281–287, 1999.

11. A. J. Englande, Jr., and C. F. Guarino, Toxics Management in the Chemical and Petrochemical Industries, *Water Sci. Technol.*, vol. 26, no. 1–2, pp. 263–274, 1992.

12. A. F. Gorovoi and N. A. Gorovaya, Assessment of the Toxicity of Mining Products and Wastes from the Processing of Donbass Anthracites. *Ugol. Ukr.*, vol. 12, pp. 38–39, 1997.

13. C. D. Litchfield, Practices, Potential, and Pitfalls in the Application of Biotechnology to Environmental Problems. In G. S. Sayler (ed.), *Environmental Biotechnology for Waste Treatment*, vol. 41, pp. 147–157. New York: Plenum Press, 1991.

14. F. M. Naqvi, Biological Techniques in Environmental Cleanups. *Indian J. Environ. Protect.*, vol. 13, no. 4, pp. 256–260, 1993.

15. M. Dias and J. D'Souza, Utilization of Photosynthetic Bacteria in Recycling Domestic Waste. *Reg. J. Energy, Heat Mass Transfer*, vol. 2, no. 3, pp. 187–191, 1980.

16. J. S. Knapp and P. S. Newby, The Microbiological Decholorization of an Industrial Effluent Containing a Diazo-Linked Chromophore. *Water Res.*, vol. 29, no. 7, pp. 1807–1809, 1995.

17. T. L. Hu, Degradation of Azo Dye RP$_2$B by *Pseudomonas luteola*. *Water Sci. Technol.*, 38(4–5), pp. 299–306, 1998.

18. P. P. Kanekar, Bioremediation of Chemopollutants from Industrial Effluents. *Proc. Acad. Environ. Biol.*, vol. 6, no. 1, pp. 1–6, 1997.

19. B. Wang, G. Li, L. Wang, J. Shi, and J. Li, Treatment of Waste Water from a Dye-Printing Factory in China by a Physicochemical-Biological System. *Eur. Water Manage.*, vol. 1, no. 1, pp. 25–30, 1998.

20. Y. Wong and J. Yu, Laccase-Catalyzed Decolorization of Synthetic Dyes. *Water Res.*, vol. 33, no. 16, pp. 3512–3520, 1999.

21. O. Ezeronye and N. Amogy, Microbiological Studies of Effluents from the Nigerian Fertilizer and Paper Recycling Mill Plants. *Int. J. Environ. Studies*, vol. 54, no. 3/4, pp. 213–221, 1998.

22. N. L. Nemerow and F. J. Agardy, *Strategies of Industrial & Hazardous Waste Management*. New York: Van Nostrand Reinhold, 1998.

23. C. L. Traviesa and E. F. Benitez, Microalgae Continuous Culture on Pretreated Cane Sugar Mill Wastes. *Leuna-Merseburg*, vol. 29, no. 2, pp. 217–221, 1987.

24. D. R. Wilson, I. C. Page, A. A. Cocci, and R. C. Landine, Case History—Two Stage, Low-Rate Anaerobic Treatment Facility for South American Alcochemical/Citric Acid Wastewater. *Water Sci. Technol.*, vol. 38, no. 4–5, pp. 45–52, 1998.

25. T. J. Cutright, Y. S. Hun, and S. Lee, Effects of Operating Parameters on the Growth of *Kluveromyces fragilis* on Cheese Whey. *Chem. Eng. Commun.*, 137, pp. 47–51, 1995.

26. J. R. Danalewich, T. G. Papagiannis, R. L. Gelyea, M. E. Tumbleson, and L. Raskin, Characterization of Dairy Waste Streams, Current Treatment Processes and Potential for Biological Nutrient Removal. *Water Res.*, vol. 32, no. 12, pp. 3555–3568, 1998.

27. W. Dzwolak and S. Ziajka, Enzymatic Hydrolysis of Milk Proteins under Alkaline and Acidic Conditions. *J. Food Sci.*, vol. 64, no. 3, pp. 393–395, 1999.

28. J. Gonzalez, P. Valdes, G. Nieves, and B. Guerrero, Application of Anaerobic Digestion to Dairy Industry Wastes. *Rev. Int. Contam. Ambiental.*, vol. 10, no. 1, pp. 37–41, 1994.

29. E. Papachristou and C. T. Lafazanis, Application of Membrane Technology in the Pretreatment of Cheese Dairy Waste and Co-treatment in a Municipal Conventional Biological Unit. *Water Sci. Technol.*, vol. 36, no. 2–3, pp. 361–367, 1997.

30. C. C. Ross, G. E. Valentine, Jr., and J. L. Walsh, Jr., Food-Processing Wastes. *Water Environ. Res.*, vol. 71, no. 5, pp. 812–815, 1999.

31. T. H. Chen and W. H. Shyu, Chemical Characterization of Anaerobic Digestion Treatment of Poultry Mortalities. *Bioresources Technol.*, vol. 63, no. 1, pp. 37–48, 1998.

32. B. Rusten, J. G. Siljudalen, A. Wien, and D. Eidem, Biological Pretreatment of Poultry Processing Wastewater. *Water Sci. Technol.*, vol. 38, no. 4–5, pp. 19–28, 1998.

33. E. A. Salminen and J. A. Rintala, Anaerobic Digestion of Poultry Slaughtering Wastes. *Environ. Technol.*, vol. 20, no. 1, pp. 21–28, 1999.

34. M. Takemasa, Nutritional Strategies to Reduce Nutrient Waste in Livestock and Poultry Production. *Kagaku*, vol. 36, no. 11, pp. 720–726, 1998.

35. B. Kherrati, M. Faid, M. Elyachioui, and A. Wahmane, Process for Recycling Slaughterhouse Wastes and Byproducts of Fermentation. *Bioresources Technol.*, vol. 63, no. 1, pp. 75–79, 1998.

36. C. Rose, T. Sastry, V. Madhavan, and R. Ranganathan, Chemical Modification of Serum Globulins Recovered from Slaughterghouse Waste and Its Partial Characterization. *Iran Polymer J.*, vol. 7, no. 1, pp. 47–52, 1998.

37. N. Garg and Y. Hang, Microbial Production of Organic Acids from Carrot Processing Waste, *J. Food Sci. Technol.*, vol. 32, no. 2, pp. 119–121, 1995.

38. P. M. Sutton, Innovative Biological Systems for Anaerobic Treatment of Grain and Food Processing Wastewaters. *Starch/Starke*, vol. 38, no. 9, pp. 314–318, 1986.

39. M. N. Akthar and P. M. Mohan, Bioremediation of Toxic Metals Ions from Polluted Lake Waters and Industrial Effluents by Fungal Biosorbent. *Chem. Sci.*, vol. 69, no. 12, pp. 1028–1030, 1995.

40. R. Bhagat and S. Srivastava, Biorecovery of Zinc by *Pseudomonas stutzeri* RS34. *Biohydrometallics Technol.*, vol. 2, pp. 209–217, 1993.

41. M. Chen and H. J. Oliver, Microbial Chromium (VI) Reduction. *Environ. Sci. Technol.*, vol. 28, no. 3, pp. 219–251, 1998.

42. D. Couillard, M. Chartier, and G. Mercier, Major Factors Influencing Bacterial Leaching of Heavy Metals (Cu and Zn) from Anaerobic Sludge. *Environ. Pollut.*, vol. 85, no. 2, pp. 175–184, 1994.

43. L. Fude, B. Harris, M. Urrutia, and T. Beveridge, Reduction of Cr(VI) by a Consortium of Sulfate-Reducing Bacteria (SRB III). *Appl. Environ. Microbiol.*, vol. 60, no. 5, pp. 1525–1531, 1994.

44. R. Glombitza, U. Iske, M. Bullmann, and J. Ondruschka, Biotechnology Based Opportunities for Environmental Protection in the Uranium Mining Industry. *Acta Biotechnol.*, vol. 12, no. 2, pp. 79–85, 1992.

45. D. P. Smith and R. Kalch, Minerals and Mine Drainage. *Water Environ. Res.*, vol. 71, no. 5, pp. 822–827, 1999.

46. S. Abou-Elela and R. Zaher, Pollution Prevention in the Oil & Soap Industry: A Case Study. *Water Sci. Technol.*, vol. 38, no. 4–5, pp. 139–144, 1998.

47. R. Bentham, N. McClure, and D. Catcheside, Biotreatment of an Industrial Waste Oil Condensate. *Water Sci. Technol.*, vol. 36, no. 10, pp. 125–1259, 1997.

48. K. E. Richardson, Refinery Waste Minimization. Hazardous Material Management Conference 12, pp. 505–513, 1995.

49. J. M. Wong, Petrochemicals. *Water Environ. Res.*, vol. 71, no. 5, pp. 828–832, 1999.

50. M. Mascarenhas, Novel Uses of Microorganisms in the Paint Industry: A Review. *Paintindia*, vol. 49, no. 5, pp. 35–40, 1999.

51. N. V. Men'shutina, A. I. Shamber, V. Men'shikov, and R. Salakhetdinov, Installations for Treatment of Wastewaters and Waste Gases from Paint and Varnish Industries. *Lakokras Mater. Ikh. Primen.*, vol. 5, pp. 24–27, 1998.

52. T. S. Webster, A. P. Togna, W. J. Guarinik, and L. McKnight, Application of a Biological Trickling Filter Reactor to Treat Volatile Organic Compound Emissions from a Spray Paint Booth Operation. *Metal Finishing*, vol. 97, no. 3, pp. 24–26, 1999.

53. A. Yasmin, S. Afrasayab, and S. Hasnain, Mercury Resistant Bacteria from Effluents of Paint Factory: Characterization and Mercury Uptake Ability. *Sci. Int.*, vol. 9, no. 3, pp. 315–320, 1997.

54. C. Elvira, L. Sampedro, E. Genitez, and R. Nogales, Vermicomposting of Sludges from Paper Mill and Dairy Industries with *Eisenia andrei*: A Pilot Scale Study. *Bioresources Technol.*, vol. 63, no. 3, pp. 205–211, 1998.

55. K. A. Kahmark and J. P. Unwin, Pulp & Paper Management. *Water Environ. Res.*, vol. 71, no. 5, pp. 836–852, 1999.

56. U. Kaluza, P. Klingelhofer, and K. Taeger, Microbial Degradation of EDTA in an Industrial Wastewater Treatment Plant. *Water Res.*, vol. 32, no. 9, pp. 2843–2845, 1998.

57. G. Rozmarin and V. Gazdaru, Connections between Biotechnology—Pulp and Paper Industry: Recent Applications, Possible Options. *Celul. Hirtie*, vol. 43, no. 1, pp. 29–37, 1994.

58. D. E. Mullins, S. E. Gabbert, J. E. Leland, R. W. Young, G. H. Hetzel, and D. R. Berry, Organic Sorption/Biodegradation of Pesticides. *Rev. Toxicol.*, vol. 2, no. 1–4, pp. 195–201, 1998.

59. H. Poggi-Varaldo, Agricultural Wastes. *Water Environ. Res.*, vol. 71, no. 5, pp. 737–785, 1999.

60. X. Shichong, Research on Treating Method of Pesticide Wastewater. *Water Treatment*, vol. 6, no. 3, pp. 343–350, 1991.

61. P. Babu, M. S. Kumar, D. G. Reddy, P. M. Kumar, and A. K. Sanhukhan, Biodegradation of Bulk Drug Industrial Effluents by Microbial Isolates from Soil. *J. Sci. Ind. Res.*, vol. 58, no. 6, pp. 431–435, 1999.

62. B. Gulmez, I. Ozturk, K. Alp, and O. Arikan, Common Anaerobic Treatability of Pharmaceutical and Yeast Industry Wastewater. *Water Sci. Technol.*, vol. 38, no. 4–5, pp. 37–44, 1998.

63. I. Kabdasli, M. Gurel, and O. Tunay, Pollution Prevention and Waste Treatment in Chemical Synthesis Processes for Pharmaceutical Industry. *Water Sci. Technol.*, vol. 39, no. 10–11, pp. 265–271, 1999.

64. B. Ruggerri and V. Specchia, Waste Control in Food and Pharmaceutical Industries. *Kem. Ind.*, vol. 39, no. 12, pp. 579–597, 1990.

65. H. F. Schroder, Substance Specific Detection and Pursuit of Non-eliminable Compounds During Biological Treatment of Wastewater from the Pharmaceutical Industry. *Waste Manage.*, vol. 19, no. 2, pp. 111–123, 1999.

66. W. Yin, Use of Clean Processes in Industry. *Huanjing Baohu*, vol. 12, pp. 15–17, 1993.

67. H. Herlitzius, Solvent Preparation with Biological Trickling Filters in a Printing Ink Factory. *VDI-Ber.*, vol. 1034, pp. 564–576, 1993.
68. C. Kellner and E. Vitzthum. Biological Waste Gas Cleaning in a Biological Trickling Filter. Conversion from a Pilot-Plant Scale to Industrial Scale with Toluene-Loaded Waste Gas at Printing Ink Producer. In W. Prins and J. Van Ham. (eds.), *Biological Waste Gas Clean*, pp. 6–66. Stuttgart: VDI Verlag, 1997.
69. K. Lerche, W. Kneist, H. Rohbeck, B. Hillemann, and R. Schwarz, Biological Treatment of Surfactant Loaded Wastewaters, Such as Laundry Effluents. German Patent 950420, 1994.
70. E. Pfleiderer and R. Reiner, Microorganisms in Processing of Leather. *Biotechnology*, vol. 6, pp. 729–743, 1988.
71. G. Ummarino, R. Mora, and A. Russo, Tannery Emissions: Possibilities of Biological Treatment. *Mater. Concianti*, vol. 68, no. 2, pp. 207–215, 1992.
72. D. P. Bakshi, K. G. Gupta, and P. Sharma, Enchanced Biodecolorization of Synthetic Textile Dye Effluent by *Phaneorchaete chrysosporium* Under Improved Culture Conditions. *World J. Microbiol. Biotechnol.*, vol. 15, pp. 507–509, 1999.
73. P. Nigam, I. M. Banat, D. Oxspring, R. Marchant, D. Singh, and W. Smyth, A New Facultative Anaerobic Filamentous Fungi Capable of Growth on Recalcitrant Textile Dyes as Sole Carbon Source. *Microbios*, vol. 84, no. 340, pp. 171–185, 1995.
74. R. Ul Haq and A. R. Shakoori, Short Communication: Microbiological Treatment of Industrial Wastes Containing Toxic Chromium Involving Successive Use of Bacteria, Yeast & Algae. *World J. Microbiol. Biotechnol.*, vol. 14, no. 4, pp. 583–585, 1998.
75. M. G. Roig, M. J. Martin-Rodrogiuez, J. M. Cachaza, S. L. Mendoza, and J. F. Kennedy, Principles of Biotechnological Treatment of Industrial Wastes. *Crit. Rev. in Biotechnol.*, vol. 13, no. 2, pp. 99–116, 1993.
76. K. H. Baker and D. S. Herson, *Bioremediation*. New York: McGraw-Hill, 1994.
77. J. R. Cookson, Jr., *Bioremediation Engineering: Design & Application*. New York: McGraw-Hill, 1995.
78. R. D. Norris, R. E. Hinchee, R. Brown, P. L. McCarty, J. T. Wilson, M. Reinhard, E. J. Bouwer, R. C. Borden, and T. M. Vogel, *Handbook Bioremediation*. Boca Raton, Florida: Lewis, 1994.
79. A. C. Palmisano, D. A. Maruscki, G. J. Ritchie, B. S. Schwab, S. R. Harper, and R. A. Rapapoprt, A Novel Bioreactor Simulating Composting of Municipal Solid Waste. *J. Microbiol. Meth.* vol. 18, no. 2, pp. 99–112, 193.
80. R. Bellandi (ed.), *Innovative Engineering Technologies for Hazardous Waste Remediation* (O'Brian & Gere Engineers, Inc.). Boston: Van Nostrand Reinhold, 1995.
81. E. Riser-Roberts, *Remediation of Petroleum Contaminated Soils*. Boca Raton, Florida: Lewis, 1998.
82. J. B. Eweis, S. J. Ergas, D. P. Y. Chang, and E. D. Schroeder, *Bioremediation Principles*. New York: McGraw-Hill, 1998.
83. V. Lazarova, D. Bellahcen, D. Rybacki, B. Rittmann, and J. Manem, Population Dynamics and Biofilm Composition in a New Three-Phase Circulating Bed Reactor. *Water Sci. Technol.*, vol. 37, no. 4–5, pp. 149–158, 1998.
84. M. Van Loosdrecht, L. Tijhuis, A. Wijdieks, and J. Heijnen, Population Distribution

in Aerobic Biofilms on Small Suspended Particles. *Water Sci. Technol.*, vol. 31, no. 1, pp. 163–171, 1995.

85. R. Hedderich, O. Klimmek, A. Kroger, R. Dirmeier, M. Keller, and K. Stetter, Anaerobic Respiration with Elemental Sulfur and with Disulfides. *FEMS Microbiol. Rev.*, vol. 22, no. 5, pp. 353–381, 1998.

86. J. F. Stolz and R. S. Oremland, Bacterial Respiration of Arsenic and Selenium. *FEMS Microbiol. Rev.*, vol. 23, no. 5, pp. 615–627, 1999.

87. S. Zala, A. Nerurkar, A. Desai, J. Ayyer, and V. Akolkar, Biotreatment of Nitrate Rich Industrial Effluent by Suspended Bacterial Growth. *Biotechnol. Lett.*, vol. 21, no. 6, pp. 481–485, 1999.

88. M. T. Madigan, J. M. Martinko, and J. Parker, *Brock's Biology of Microorganisms*, 8th ed., Upper Saddle River, New Jersey: Prentice Hall, 1997.

89. S. S. Sutherson, *Remediation Engineering: Design Concepts.* Boca Raton, Florida: Lewis, 1997.

90. E. Ward-Liebig and T. J. Cutright, The Investigation of Enhanced Bioremediation Through the Addition of Macro and Micro Nutrients in a PAH Contaminated Soil. *Int. Biodeter. Biodegrad.*, vol. 44, no. 1, pp. 55–64, 1999.

91. J. E. Bailey and D. F. Ollis, *Biochemical Engineering Fundamentals*, 2nd ed. New York: McGraw-Hill, 1986.

92. J. E. Burgess, J. Quarmby, and T. Stephenson, Role of Micronutrients in Activated Sludge-Based Biotreatment of Industrial Effluents. *Biotechnol. Adv.*, vol. 17, pp. 49–70, 1999.

93. J. R. Boulding (ed.), *EPA Environmental Engineering Source Book.* Ann Arbor, MI: Ann Arbor Press, 1996.

94. D. R. Schneider and R. J. Billingsley, *Bioremediation: A Desk Manual for the Environmental Professional.* Des Plaines, Iowa: Cahners, 1990.

95. O. M. El-Tayeb, S. Megahed, and M. El-Azizi, Microbial Degradation of Aromatic Substances by Local Bacterial Isolates. III. Factors Affecting Degradation of 2,4-Dichlorophenoxyacetic Acid by *Pseudomonas stutzeri*. *Egypt J. Biotechnol.*, vol. 4, pp. 84–90, 1998.

96. B. R. Patel, Nutrient Control in Pulp & Paper Wastewater Treatment Plants Using Online Measurements. *Environ. Conf. Exhib. 1*, pp. 79–81. Atlanta, GA: TAPPI Press, 1997.

97. K. Kouno, H. P. Lukito, and T. Ando, Minimum Available N Requirement for Microbial Biomass Formation in a Regosol. *Soil Biol. Biochem.*, vol. 31, no. 6, pp. 797–802, 1999.

98. C. Chien, E. Leadbetter, and W. Godchaux, III, *Rhodococcus spp.* Utilize Taurine (2-Aminoethanesulfonate) as Sole Source of Carbon, Energy, Nitrogen, and Sulfur for Aerobic Respiratory Growth. *FEMS Microbiol. Lett.*, vol. 176, no. 2, pp. 333–337, 1999.

99. C. Knoblauch, K. Sahm, and B. Jorgensen, Psychrophilic Sulfate-Reducing Bacteria Isolated from Permanently Cold Arctic Marine Sediments: Description of *Desulfofrigus oceanense gen nov. sp.*, nobv., *Desulfofrigus fragile sp.* nov., *D. gelida gen. nov.*, *D. psychrophilia gen. nov.* and *D. arctica sp.* nov. *Int. J. Syst. Bacteriol.*, vol. 49, no. 4, pp. 1631–1643, 1999.

100. G. Bharadwaj and R. Maheshwari, A Comparison of Thermal Characteristics and

Kinetic Parameters of Trehalases from a Thermophilic and a Mesophilic Fungus. *FEMS Microbiol. Lett.*, vol. 181, no. 1, pp. 187–193, 1999.

101. L. McKane and J. Kandel, *Microbiology: Essentials & Applications*, 2nd ed. New York: McGraw-Hill, 1996.

102. J. Simpa, P. Lens, A. Vieira, Y. Miron, J. B. van Lier, L. W. Hulshoff-Pol, and G. Lettinga, Thermophilic Sulfate Reduction in Upflow Anaerobic Sludge Bed Reactors Under Acidifying Conditions. *Process Biochem.*, vol. 35, pp. 509–522, 1999.

103. J. Singh, R. M. Vohra, and D. K. Sahoo, Alkaline Protease from a New Obligate Alkalophilic Isolate of *Bacillus sphaericus*. *Biotechnol. Lett.*, vol. 21, no. 10, pp. 921–924, 1999.

104. P. L. Bishop and Y. Tong, A Microelectric Study of Redox Potential Change in Biofilms. *Water Sci. Technol.*, vol. 39, no. 7, pp. 179–185, 1999.

105. M. Nay, M. Snozzi, and J. Zehnder, Fate and Behavior of Organic Compounds in an Artificial Saturated Subsoil Under Controlled Redox Conditions: The Sequential Soil Column System. *Biodegradation*, vol. 10, no. 1, pp. 75–82, 1999.

106. R. L. Droste, *Theory and Practice of Water and Wastewater Treatment*. New York: Wiley, 1997.

107. G. Kiely, *Environmental Engineering*. New York: McGraw-Hill, 1997.

108. P. N. Cheremisinoff, Biotechnology: Treating Industrial/Municipal Wastes and Wastewater. *Pollut. Eng.*, vol. 19, no. 9, pp. 74–87, 1987.

109. J. M. Wyatt, Biotechnological Treatment of Industrial Wastewater. *Microbiol. Sci.*, vol. 5, no. 6, pp. 186–190, 1988.

110. H. Chua, P. H. F. Yu, S. N. Sin, and M. W. L. Cheung, Sub-lethal Effects of Heavy Metals on Activated Sludge Microorganisms. *Chemosphere*, vol. 39, no. 15, pp. 2681–2692, 1999.

111. S. Baccella, G. Cerichelli, M. Chiarini, C. Ercole, E. Fantauzzi, A. Lepidi, L. Toro, and F. Veglio, Biological Treatment of Alkaline Industrial Wastes. *Process Biochem.*, vol. 35, no. 2, pp. 595–602, 2000.

112. P. R. Brookes and A. G. Livingston, Biotreatment of a Point-Source Industrial Wastewater Arising in 3,4-Dichloroaniline Manufacture Using an Extractive Membrane Bioreactor. *Biotechnol. Prog.*, vol. 10, no. 1, pp. 65–73, 1994.

113. W. Viessman, Jr., and M. J. Hammer, *Water Supply and Pollution Control*, 5th ed. New York: Harper Collins, 1993.

114. T. F. Yen, *Environmental Chemistry: Essentials of Chemistry for Engineering Processes*. New Jersey: Prentice Hall, 1999.

115. M. W. Fitch, J. B. Murphy, and S. S. Sowell, Biological Fixed-Film Systems. *Water Environ. Res.*, vol. 71, no. 5, pp. 638–655, 1999.

116. T. D. Reynolds and P. A. Richards, *Unit Operations & Processes in Environmental Engineering*, 2nd ed. New York: PWS, 1996.

117. W. W. Eckenfelder, Jr., *Water Quality Engineering for Practicing Engineers*. New York: Barnes & Noble, 1970.

118. G. M. Masters, *Introduction to Environmental Engineering & Science*, 2nd ed. Upper Saddle River, New Jersey: Prentice Hall, 1998.

119. W. W. Eckenfelder, Jr., Y. Argaman, and E. Miller, Process Selection Criteria for the Biological Treatment of Industrial Wastewaters. *Environ. Prog.*, vol. 8, no. 1, pp. 40–45, 1989.

120. W. W. Eckenfelder, Jr., and A. J. Englande, Jr., Innovative Biological Treatment for Sustainable Development in the Chemical Industries. *Water Sci. Technol.*, vol. 38, no. 4–5, pp. 111–120, 1998.

121. P. A. Vesilind, *Introduction to Environmental Engineering*. Boston: PWS, 1997.

122. C. D. Cooper and F. C. Alley, *Air Pollution Control: A Design Approach*, 2nd ed. Boston: Waveland Press, 1994.

123. M. S. McGrath, J. Nieuwland, and C. van Lith, Case Study: Biofiltration of Styrene and Butylacetate at a Dashboard Manufacturer. *Environ. Prog.*, vol. 18, no. 3:197–204, 1999

124. C. Colella, M. Pansini, F. Alfani, M. Cantarella, and A. Gallifuoco, Selective Water Adsorption from Aqueous Ethanol-Containing Vapors by Phillipsite-Rich Volcanic Tufts. *Microporous Mater.*, vol. 3, no. 3, pp. 219–226, 1994.

125. N. De Nevers, *Air Pollution Control Engineering*, 2nd ed. New York: McGraw-Hill, 2000.

126. J. Joyce and H. Sorensen, Bioscrubber Design. *Water Environ. Technol.*, vol. 11, no. 2, pp. 37–41, 1999.

127. D. Johnson, M. J. Semmens, J. S. Gulliver, A Rotating Membrane Contactor: Application to Biologically Active Systems. *Water Environ. Res.*, vol. 71, no. 2, pp. 163–168, 1999.

128. T. S. Webster, H. Cox, and M. A. Deshusses, Resolving Operational Problems Encountered in the Use of a Pilot/Full-Scale Trickling Filter Reactor. *Environ. Prog.*, vol. 18, no. 3, pp. 162–172, 1999.

129. J. W. Van Groenestijn and M. E. Lake, Elimination of Alkanes from Off-gases Using Biotrickling Filters Containing Two Liquid Phases. *Environ. Prog.*, vol. 18, no. 3:151–155, 1999.

130. F. G. Edwards and N. Nirmalakhandan, Modeling an Airlift Bioscrubber for Removal of Air Phase BTEX. *J. Environ. Eng.*, vol. 125, no. 11, pp. 1062–1070, 1999.

131. M. C. Veiga, M. Fraga, L. Amor, and C. Kennes, Biofilter Performance and Characterization of a Biocatalyst Degrading Alkylbenzene Gases. *Biodegradation*, vol. 19, no. 3, pp. 169–176, 1999.

132. D. Arulneyam and T. Swaminathan, Biodegradation of Ethanol Vapor in a Biofilter. *Bioprocess Eng.*, vol. 22, no. 1, pp. 63–67, 2000.

133. B. Hodkinson, J. B. Williams, and J. E. Butler, Development of Biological Aerated Filters: A Review. *Water Environ. Manage.*, vol. 13, no. 4, pp. 250–254, 1999.

134. F. Wittorf, S. Knauf, and H. E. Windberg, Biotrickling-Reactor: A New Design for the Efficient Purification of Waste Gases. *Proc. 4th Int. Biological Waste Gas Cleaning*, vol. 4, pp. 329–335. Stuttgart, VDI Verlag, 1997.

135. W. Schiettecatte, W. Leys, R. Gerards, J. Koning, H. Verachtert, and J. F. Van Impe, Evaluation of a Biofiltration Process: Purification of a BTEX Contaminated Airflow. *Toegepaste Biol. Wet.*, vol. 64, no. 5a, 197–200, 1999.

136. N. Fortin and M. A. Deshusses, Treatment of Methyl *tert*-Butyl Ether Vapors in Biotrickling filters 2. Analysis of Rate-Limiting Step and Behavior Under Transient Condtions. *Environ. Sci. Technol.*, vol. 33, no. 17, pp. 2987–2991, 1999.

137. H. Cox and M. A. Deshusses, Chemical Removal of Biomass from Waste Air

Biotrickling Filters: Screening of Chemicals of Potential Interest. *Water Res.*, vol. 33, no. 10, pp. 2383–2391, 1999.

138. P. T. Anastas, L. B. Bartlett, M. M. Kirchhoff, and T. C. Williamson, The Role of Catalysts in the Design, Development, and Implementation of Green Chemistry. *Catalysis Today*, vol. 55, no. 1, pp. 11–22, 2000.

139. Y. H. Huang and G. S. Hao, Green Chemistry and New Technology for Zero Pollutant Discharge. *Ziran Kexueban*, vol. 22, no. 3, pp. 348–350, 1999.

140. H. S. Shin and P. L. Rogers, Production of L-Phenylacetylcarbinol from Benzaldehyde Using Partially Purified Pyruvate Decarboxylase (PDC). *Biotechnol. Bioeng.*, vol. 49, no. 1, pp. 52–62, 1996.

141. C. M. Tripathi, S. C. Agarwal, and S. K. Basu, Production of L-Phenylacetylcarbinol by Fermentation. *J. Ferment. Bioeng.*, vol. 84, no. 6, pp. 487–492, 1997.

142. A. L. Oliver, F. A. Roddick, and B. N. Anderson, Cleaner Production of Phenylacetylcarbinol by Yeast Through Productivity Improvements and Waste Minimization. *Pure Appl. Chem.*, vol. 69, no. 11, pp. 2371–2385, 1997.

143. O. Popa, A. Ionita, C. Popa, I. Paraschiv, and A. Vamanu, Studies Regarding L-PAC Formation by Yeast Bioconversion. *Biotechnol. Lett.*, vol. 3, no. 2, pp. 115–122, 1998.

144. Y. Yao and J. Li, Production of L-Phenylacetylcarbinol by Pyruvate Decarboxylase. *Zhongguo Yaoke Daxue Xuebao*, vol. 30, no. 2, pp. 115–118, 1999.

145. J. Gao, L. Cheng, and H. He, Study of Kinetics of Solvent Carboxylation of Phenol. *Dalian Ligong Daxue Xuebao*, vol. 36, no. 3, pp. 288–292, 1996.

146. Y. Kharma, G. Sergeev, Y. Gordash, V. Bludilin, and L. Lekhter, Modeling of Equilibria in the Kolbe-Schmitt Carboxylation of Higher Alkylphenols. *Neftepererab. Neftekhim.*, vol. 38, pp. 47–49, 1990.

147. E. S. Miller, Jr., and S. W. Peretti, Bioconversion of Toluene to *p*-Hydroxybenzoate. *Green Chem.*, vol. 1, no. 3, pp. 143–152, 1999.

148. P. L. Bishop, *Pollution Prevention: Fundamentals and Practice*. New York: McGraw-Hill, 2000.

149. F. Huang, Present Status and Development of Biodegradable Polymers. *Huaxue Shijie*, vol. 40, no. 11, pp. 570–574, 1999.

150. Y. Poirier, Green Chemistry Yields a Better Plastic. *Nat. Biotechnol.*, vol. 17, no. 10, pp. 960–961, 1999.

151. C. Nawrath, Y. Poirier, and C. Somerville, Targeting of the Polyhydroxybutyrate Biosynthetic Pathway to the Plastids of *Arabidppsis thaliana* Results in High Levels of Polymer Accumulation. *Proc. Natl. Acad. of Sci. USA*, vol. 91, no. 26, pp. 12760–12764, 1994.

152. M. Matavulj, D. Radnovic, S. Gajin, O. Petrovic, Z. Svircev, I. Tamas, M. Bokorov, V. Divjakovic, F. Gassner, and H. P. Molitoris, Microorganisms, Producers and Degraders of Biosynthetic Plastic Materials. *Mikrobiologija*, vol. 32, no. 1, pp. 169–178, 1995.

153. J. Woodard and B. R. Evans, Utilization of Biocatalysts in Cellulose Waste Minimization. *Biotechnol. Res.*, vol. 7, pp. 157–179, 1998.

154. U. J. Haenggi, Requirements on Bacterial Polyesters as Future Substitute for Conventional Plastics for Consumer Goods. *FEMS Microbiol. Rev.*, vol. 16, no. 2–3, pp. 213–220, 1995.

155. C. Chu, A. Lu, M. Liszkowski, and R. Sipehia, Enhanced Growth of Animal and Human Endothelial Cells on Biodegradable Polymers. *Biochim. Biophys. Acta.*, vol. 1472, no. 3, pp. 479–485, 1999.

156. A. De la Maza, L. Codech, O. Lopez, J. L. Parra, M. Sabes, and J. Guinea, Biopolymer Excreted by *Psuedoalteromonas antartica* NF3, as a Coating and Protective Agent of Liposomes Against Dodecyl Maltoside. *Biopolymers*, vol. 59, no. 6, pp. 579–588, 1999.

157. J. Fu and S. Li, Biodegradable Polymers Used in the Biomedical Field (2). *Wuhan Gongye Daxue Xuebao*, vol. 21, no. 5, pp. 19–22, 1999.

158. M. Schulz, T. Blunk, and A. Gopfercih, Carrier for Drug Formulations and Artificial Living Tissues. *Pharm. Z.*, vol. 144, no. 45, pp. 3661–3668, 1999.

159. P. Biely and L. Kremnicky, Yeasts and Their Enzyme Systems Degrading Cellulose, Hemicelluloses, and Pectin. *Food Technol. Biotechnol.*, vol. 36, no. 4, pp. 305–312, 1998.

160. M. Itavaara, M. Siika-Aho, and L. Viikari, Enzymatic Degradation of Cellulose Based Materials. *J. Envrion. Polymer Degradation*, vol. 7, no. 2, pp. 67–73, 1999.

161. J. M. Krochta and L. C. De Mulder-Johnston, Biodegradable Polymers from Agricultural Products. *ACS Symp. Ser.*, vol. 64, no. 7, pp. 120–140, 1996.

162. D. B. Wilson and D. C. Irwin, Genetics and Properties of Cellulases. *Adv. Biochem. Eng. Biotechnol.*, vol. 65, pp. 1–21, 1999.

163. C. Nawrath, Y. Poirier, and C. Somerville. Plant Polymers for Biodegradable Plastics: Cellulose, Starch, and Polyhdroxyalkanoates. *Mol. Breed.*, vol. 1, no. 2, pp. 105–122, 1995.

164. Y. Poirier, C. Nawrath, and C. Somerville, Production of Polyhdroxyalkanoates, a Family of Biodegradable Plastics and Elastomers in Bacteria and Plants. *Bio/Technology*, vol. 13, no. 2, pp. 142–150, 1995.

165. T. Wang, L. Ye, and Y. Song, Progress of PHA Production in Transgenic Plants. *Chin. Sci. Bull.*, vol. 44, no. 19, pp. 1729–1736, 1999.

166. K. M. Elborough, A. J. White, S. Z. Hanley, and A. R. Slabas, Production of the Biodegradable Plastic Polyhydroxyalkanoates in Plants. *Portland Press Proc.*, vol. 14, pp. 125–131, 1998.

167. H. Miyasaka, H. Nakano, H. Akiyama, S. Kanai, and M. Hirano, Production of PHA by Genetically Engineered Marine Cyanobacterium. *Studies Surface Sci. Catalysis*, vol. 114, pp. 237–242, 1998.

168. R. J. Van Wegen, Y. Ling, and APJ Middelberg, Industrial Production of Poly-hydroxyalkanoates Using *Escherichia coli*: An Economic Analysis. *Chem. Eng. Res. Des.*, vol. 76, no. A3, pp. 417–426, 1998.

169. S. Y. Lee and J. Choi, Production and Degradation of PHAs in Waste Environment. *Waste Manage.*, vol. 19, pp. 133–139, 1999.

170. G. Braunegg, G. Lefebvre, and K. Genser, Polyhydroxyalkanoates Biopolyesters from Renewable Resources: Physiological and Engineering Aspects. *J. Biotechnol.*, vol. 65, no. 2–3, pp. 127–161, 1998.

171. K. Sakai, T. Yamauchi, F. Nakasu, and T. Ohe, Biodegradation of Cellulose Acetate by *Neisseria sicca*. *Biosci. Biotechnol. Biochem.*, vol. 60, no. 10, pp. 1617–1622, 1996.

172. L. A. Vanderberg, T. M. Foreman, M. Attrep, Jr., J. R. Brainard, and N. N. Sauer,

Treatment of Heterogenous Mixed Wastes: Enzyme Degradation of Cellulosic Materials Contaminated with Hazardous Organics, Toxic and Radioactive Metals. *Environ. Sci. Technol.*, vol. 33, no. 8, pp. 1256–1262, 1999.

173. R. Mutitakul, S. Kulpreecha, and S. Shioya, Production of PHB, a Biodegradable Polymer form *Bacillus sp.* BA-019. *Biotechnol. Sustainable Util. Biol. Resource Trop.*, vol. 12, pp. 407–413, 1998.

174. G. D. Reyes, R. S. So, and M. M. Ulep, Isolation, Screening and Identification of Bacteria for poly-β-Hydroxybutyrate (PHB) Production. *Study Environ. Sci.*, vol. 66, pp. 737–748, 1997.

175. M. I. Foda and M. Lopez-Leiva, Continuous Production of Oligosaccharides from Whey Using a Membrane Reactor. *Process Biochem.*, vol. 35, pp. 581–587, 2000.

176. T. J. Cutright, The Production of α-Lactase. M. S. thesis, The University of Akron, Akron, OH, 1992.

177. M. M. Saad, A. N. Saad, M. Abdel-Hadi, I. Ghany, and H. Ismail, Production and Some Properties of Proteinase by Fungi Utilizing Whey. *Egypt J. Microbiol.*, vol. 32, no. 3, pp. 411–421, 1997.

178. A. M. Fialho, L. Martins, M. Donval, J. Leitao, M. Ridout, A. Jay, V. Morris, and I. Sa-Correia, Structures and Properties of Gellan Polymers Produced by *Sphingomonas paucimobilis* ATCC 31461 from Lactose Compared with Those Produced from Glucose and from Cheese Whey. *Appl. Environ. Microbiol.*, vol. 65, no. 6, pp. 2485–2491, 1999

179. R. T. Otto, H. J. Daniel, G. Pekin, K. Muller-Decker, G. Furstenberger, M. Reuss, and C. Syldatk. Production of Sophorolipids from Whey. II Whey Composition, Surface Active Properties, Cytotoxicity, and Stability Against Hydrolases by Enzymatic Treatment. *Appl. Microbiol. Biotechnol.*, vol. 52, no. 4, pp. 495–501, 1999.

180. J. S. Dordick, Enzymatic Catalysis in Monophasic Organic Solvents. *Enzyme Microb. Technol.*, vol. 11, pp. 194–211, 1989.

181. P. Wang, M. S. Sergueeva, L. Lim, and J. S. Dordick, Biocatalytic Plastics as Active and Stable Materials for Biotransformations. *Nature Biotechnol.*, vol. 15, pp. 789–793, 1997.

182. P. Wang, C. A. Woodard, and E. N. Kaufman, Poly(ethylene glycol)-Modified Ligninase Enhances Pentachlorophenol Biodegradation in Water Solvent Mixtures. *Biotechnol. Bioeng.*, vol. 64, no. 3, pp. 290–297, 1999.

183. A. D' Annibale, S. R. Stazi, V. Vinciguerra, E. Di Mattia, and G. G. Sermanni, Characterization of Immobilized Lacasse from *Lentinula edodes* and Its Use in Olive-Mill Wastewater Treatment. *Process Biochem.*, vol. 34, pp. 697–706, 1999.

184. F. Alfani, M. Cantarella, A. Gallifuoco, and V. Romano, On the Effectiveness of Immobilized Enzymes with Linear Mixed-Type Product Inhibition Kinetics. *Chem. Eng. J.*, vol. 57, no. 1, pp. B23–B29, 1995.

185. T. Nishida, Y. Tsutsumi, M. Kemi, T. Haneda, and H. Okamura, Decolorization of Anthraquinone Dyes by White Rot Fungi and Its Related Enzymes. *Mizu Kankyo Gakkaishi*, vol. 22, no. 6, pp. 465–4714, 1999.

186. S. K. Garg and D. R. Modi, Decolorization of Pulp-Paper Mill Effluents by White-Rot Fungi. *Crit. Rev. Biotechnol.*, vol. 19, no. 2, pp. 85–112, 1999.

187. C. J. Jaspers, G. Jimenez, and M. J. Penninck, Evidence for a Role of Manganese Peroxidase in the Decolorization of Kraft Pulp Bleach Plant Effluent by *P. chryso-*

sporium: Effects of Initial Culture Conditions on Enzyme Production. *J. Biotechnol.*, vol. 37, no. 3, pp. 229–234, 1994.

188. P. Peralta-Zambor, S. Gomes de Moraes, E. Esposito, R. Antunes, J. Reyes, and N. Duran, Decolorization of Pulp Mill Effluents with Immobilized Lignin and Manganese Peroxidases from *Phanerochaete chrysosporium*. *Environ. Technol.*, vol. 19, no. 5, pp. 521–528, 1998.

189. T. Zhang, S. Wada, T. Yamagishi, I. Hiroyasu, K. Tatsumi, and Q. X. Zhao, Treatment of Bleaching Wastewater from Pulp-Paper Plants in China Using Enzymes and Coagulants. *J. Environ. Sci. (China)*, vol. 11, no. 4, pp. 480–484, 1999.

190. N. Caza, J. K. Bewtra, N. Biswas, and K. E. Taylor, Removal of Phenolic Compounds from Synthetic Wastewater Using Soybean Peroxidase. *Water Res.*, vol. 33, no. 13, pp. 3012–3018, 1999.

191. J. C. Rozzell, Commercial Scale Biocatalysis: Myths and Realities. *Bioorg Med Chem.*, vol. 7, no. 10, pp. 2253–2261, 1999.

192. K. J. Grice, G. F. Payne, and J. S. Karns, Enzymatic Approach to Waste Minimization in a Cattle Dipping Operation: Economic Analysis. *J. Agric. Food Chem.*, vol. 44, no. 1, pp. 351–357, 1996.

193. W. Q. Sun and G. F. Payne, Tyrosine-Containing Chitosan Gels: A Combined Catalyst and Sorbent for Selective Phenol Removal. *Biotechnol. Bioeng.*, vol. 51, no. 1, pp. 79–86, 1996.

194. T. Zhang, Q. X. Zhao, S. Wilson, and M. Qi, Study of the Removal of 4-Chlorophenol from Wastewater Using Horseradish Peroxidase. *Toxicol. Environ. Chem.*, vol. 71, no. 1–2, pp. 115–123, 1999.

195. G. Zhang and J. A. Nicell, The Characteristics of Ezymatic Treatment of Pentachlorophenol in Wastewater. *Chin. Sci. Bull.*, vol. 44, no. 2, pp. 178–180, 1999.

196. S. Davis and R. G. Burns, Decolorization of Phenolic Effluents by Soluble and Immobilized Phenol Oxidases. *Appl. Microbiol. Biotechnol.*, vol. 32, no. 6, pp. 721–726, 1990.

197. J. P. Solyanikova and L. A. Golovleva, Phenol Hydroxylases: An Update. *Biochemistry*, vol. 64, no. 4, pp. 365–372, 1999.

198. D.H. Nies, Microbial Heavy-Metal Resistance. *Appl. Microbiol. Biotechnol.*, vol. 51, pp. 730–750, 1999.

199. R. S. Prakasham, J. S. Merrie, R. Sheela, N. Saswathi, and S. V. Ramakrishna, Biosorption of Chromium VI by Free and Immobilized *Rhizopus arrhizus*. *Environ. Pollut.*, vol. 104, pp. 421–427, 1999.

200. A. Hassen, N. Saidi, M. Cherif, and A. Boudabous, Effects of Heavy Metals on *Pseudomonas aeruginosa* and *Bacillus thuringiensis*. *Bioresource Technol.*, vol. 65, pp. 73–82, 1998.

201. V. M. Ibeanusi and E. A. Archibold, Mechanisms for Heavy Metal Uptake in a Mixed Microbial Ecosystem. *ASTM Spec. Technol. Pub.*, pp. 191–203, 1995.

202. M. Mergaey, Microbial Resources for Bioremediation of Sites Polluted by Heavy Metals. *NATO ASI Ser. 3*, vol. 19, pp. 65–73, 1997.

203. J. Adams, T. Pickett, and J. Montgomery, Biotechnologies for Metal and Toxic Inorganic Mining Processes and Waste Solutions. In N. McPherson and R. Nora, (eds.), *Randol Gold Forum Conf. Proc.*, pp. 143–146. Golden CO: Randol International, 1996.

204. G. Bunke, P. Gotz, and R. Bucholz, Metal Removal by Biomass: Physio-Chemical Elimination Methods. *J. Biotechnol.*, vol. 11a, pp. 431–452, 1999.

205. K. M. Paknikar, P. R. Puranik, and A. V. Pethkar, Development of Microbial Biosorbents—A Need for Standardization of Experimental Protocols. *Process Metall.*, vol. 9B, pp. 363–372, 1999.

206. B. Volesky, Biosorption for the Next Century. *Process Metall.*, vol. 9B, pp. 161–170, 1999.

207. G. C. Donmez, Z. Aksu, A. Ozturk, and T. Kutsal, A Comparative Study on Heavy Metal Biosorption Characteristics of Some Algae. *Process Biochem.*, vol. 34, pp. 885–892, 1999.

208. A. Kapoor, T. Viraraghavan, and D. R. Cullimore, Removal of Heavy Metals Using the Fungus *Aspergillus niger*. *Bioresource Technol.*, vol. 70, pp. 95–104, 1999.

209. I. Bakkaloglu, T. J. Buter, L. M. Evison, F. S. Holland, and I. C. Hancock, Screening of Various Biomass for the Removal of Heavy Metals (Zn, Cu, Ni) by Biosorption, Sedimentation, and Desorption. *Water Sci. Technol.*, vol. 38, no. 6, pp. 269–277, 1998.

210. A. Stoll and J. R. Duncan, Enhanced Heavy Metal Removal from Waste Water by Viable, Glucose Pretreated *Saccharomyces cerevisiae* Cells. *Biotechnol. Lett.*, vol. 18, no. 10, pp. 1209–1212, 1996.

211. K. Tsekova, A. Kaimaktchiev, and A. Tsekova, Bioaccumulation of Heavy Metals by Microorgansims. *Biotechnol. Biotechnol. Equip.*, vol. 2, pp. 94–96, 1998.

212. M. Li, H. Jiang, W. Hou, and L. Xing, Heavy Metal Biosorption of Yeasts. *Junwu Xitong*, vol. 17, no. 4, pp. 367–373, 1998.

213. A. Ruiz-Manirquez, J. A. Noriega, J. H. Yeomans, L. J. Orgega, and P. I. Magana, Biosorption of Heavy Metals by *Thiobacillus ferrooxidans*. *Rev. Soc. Quim. Mex.*, vol. 42, no. 5, pp. 228–233, 1998.

214. A. L. Kansoh and H. A. El-Shafei, Multiple Heavy Metal Tolerance in Some Fungal and Bacerial Strains. *Afr. J. Mycol. Biotechnol.*, vol. 6, no. 3, pp. 31–40, 1998.

215. J. S. Chang and J. C. Huang, Selective Adsorption/Recovery of Pb, Cu, and Cd with Multiple Fixed Beds Containing Immobilized Bacterial Biomass. *Biotechnol. Prog.*, vol. 14, no. 5, pp. 735–741, 1998.

216. L. Philip, L. Iyengar, and C. Venkobacher, Immobilized Microbial Reactor for Heavy Metal Pollution Control. *Int. J. Environ. Pollut.*, vol. 6, no. 2/3, pp. 277–284, 1996.

217. R. Ileri and A. Akkoyunlu, Biosorption of Immobilized Dead Biomass in a Continuous Sheet Bioreactor. *Fresenius Environ. Bull.*, vol. 8, no. 5/6, pp. 397–404, 1999.

218. Y. Sag, U. Acikel, Z. Zumriye, and T. Kutsal, Competitive Biosorption of Chromium (VI), Iron (III), and Copper (II) Ions from Binary Metal Mixtures by *R. arrhizus* and *C. vulgaris*. *Turkish J. Eng. Environ Sci.*, vol. 22, no. 2, pp. 145–154, 1998.

219. Y. Sag and T. Kutsal, An Overview of the Studies About Heavy Metal Adsorption Process by Microorganisms In: A. Hassen, N. Saidi, M. Cherif, and A. Boudabous, (eds.) Effects of Heavy Metals on *Pseudomonas aeruginosa* and *Bacillus thuringiensis*. *Bioresource Technol.*, vol. 65, pp. 73–82, 1998.

220. A. Tomasini-Campocosio, S. Escarcega-Cruz, E. Gonzalez-Iribarren, and M. Meraz-Rodriguez, Biosorption of Heavy Metals. *Inf. Technol.*, vol. 9, no. 6, pp. 73–77, 1998.

221. U. Yetis, G. Ozcengiz, F. B. Dilek, N. Ergen, A. Erbay, and A. Dolek, Heavy Metal Biosorption by White-Rot Fungi. *Water Sci. Technol.*, vol. 38, no. 4–5, pp. 323–330, 1998.

222. H. Seki and A. Suzuki, Biosorption of Heavy Metal Ions to Brown Algae, *Macrocystis pyrifera, Kjellmaniella crassifolia,* and *Undaria pinnatifida. J. Colloid Interface Sci.*, vol. 206, no. 1, pp. 297–301, 1998.

223. T. J. Butter, L. M. Evison, I. C. Hancock, and F. S. Holland, The Kinetics of Metal Uptake by Microbial Biomass: Implications for the Design of a Biosorption Reactor. *Water Sci. Technol.*, vol. 38, no. 6, pp. 279–286, 1998.

224. A. I. Ferra and J. A. Teixeira, The Use of Flocculating Brewer's Yeast for Cr(III) and Pb(II) Removal from Residual Wastewaters. *Bioprocess Eng.*, vol. 21, pp. 431–437, 1999.

225. A. Hammaini, A. Ballester, F. Gonzalez, M. L. Blazquez, and J. A. Munoz, Activated Sludge as a Biosorbent of Heavy Metals. *Process Metall.*, vol. 9B, pp. 185–192, 1999.

226. M. Bustard, G. McMullan, and A. P. McHale, Biosorption of Textile Dyes by Biomass Derived from *Kluyveromyces marxianum* IMB3. *Bioprocess Eng.*, vol. 19, pp. 427–430, 1998.

227. R. F. Unz and K. L. Shuttleworth, Microbial Mobilization and Immobilization of Heavy Metals. *Curr. Opin. Biotechnol.*, vol. 7, no. 3, pp. 307–310, 1996.

228. G. M. Gadd, Microbial Control of Heavy Metal Pollution. In J. C. Fry, G. M. Gadd, R. A. Herbert, C. W. Jones, and I. A. Watson-Clark, (eds.), *Microbial Pollution Control*, pp. 59–88. Cambridge, U.K.: Cambridge University Press, 1992.

229. L. F. Stratchan, D. J. Leak, and A. G. Livingston, Minimization of Excess Biomass Production in an Extractive Membrane Bioreactor. Eur. Conf. Young Researchers in Chemical Engineering, vol. 2. pp. 1043–1045. New York: IChE Publishing, 1995.

230. D. P. Smith, Submerged Filter Biotreatment of Hazardous Leachate in Aerobic, Anaerobic, and Anaerobic/Aerobic Systems. *Haz. Waste Haz. Mater.*, vol. 12, no. 2, pp. 167–187, 1995.

231. F. E. Hall, Jr., C-ALC Hazardous Waste Minimization Strategy: Reduction of Industrial Biological Sludge from Industrial Wastewater Treatment Facilities. Air and Waste Management Report WP82BO1/1, 1997.

232. H. Yasui and M. Shibata, An Innnovative Approach to Reduce Excess Sludge Production in the Activated Sludge Process. *Water Sci. Technol.*, vol. 30, no. 9, pp. 11–20, 1994.

233. M. Mayhew and T. Stephenson, Low Biomass Yield Activated Sludge: A Review. *Environ. Technol.*, vol. 18, no. 9, pp. 883–892, 1997.

234. E. Low and H. A. Chase, The Effect of Maintenance Energy Requirements on Biomass Production During Wastewater Treatment. *Water Res.*, vol. 33, no. 3, pp. 847–853, 1998.

235. M. Mayhew and T. Stephenson, Biomass Yield Reduction: Is Biochemical Manipulation Possible Without Affecting Sludge Process Efficiency? *Water Sci. Technol.*, vol. 38, no. 8–9, pp. 137–144, 1998.

236. I. Purtschert and W. Gujer, Population Dynamics by Methanol Addition in Denitrifying Wastewater Treatment Plants. *Water Sci. Technol.*, vol. 39, no. 1, pp. 43–50, 1999.

237. E. Low and H. A. Chase, Reducing Production of Excess Biomass During Wastewater Treatment. *Water Res.*, vol. 33, no. 5, pp. 1119–1132, 1999.

238. W. Verstrate and P. Vandevivere, New and Broader Applications of Anaerobic Digestion. *Crit. Rev. Environ. Sci. Technol.*, vol. 29, no. 2, pp. 151–173, 1999.

239. F. L. Smith, G. A. Sorial, M. T. Suidan, A. W. Breen, P. Biswas, R. C. Brenner, Development of Two Biomass Control Strategies for Extended, Stable Operation of Highly Efficient Biofilters with High Toluene Loadings. *Environ. Sci. Technol.*, vol. 30, no. 5, pp. 1744–1751, 1996.

240. C. F. Gokcay and U. Yetis, Effect of Nickel (II) on Biomass Yield of Activated Sludge. *Water Sci. Technol.*, vol. 34, no. 5–6, pp. 163–171, 1996.

241. R. Grover, S. S. Marwaha, and J. F. Kennedy, Studies on the Use of an Anaerobic Baffled Reactor for the Continuous Anaerobic Digestion of Pulp and Paper Mill Black Liquors. *Process Biochem.*, vol. 34, pp. 653–657, 1999.

242. D. Brady, P. D. Rose, and J. R. Duncan, The Use of Hollow Fiber Cross-Flow Microfiltration in Bioaccumulation and Continuous Removal of Heavy Metals from Solution by *Saccharomyces cerevisie*. *Biotechnol. Bioeng.*, vol. 44, no. 11, pp. 1362–1366, 1994.

243. M. Reiser, K. Fischer, and D. Bardtke, Biomembrane Reactor—From Laboratory to Industrial Use. In W. L. Prins, J. Van Ham (eds.), *Proc. 4th Int. Biological Waste Gas Clean*, pp. 181–188. Duesseldorf: VDI Verlag, 1997.

244. S. Inguva, M. Boensch, and G. Shreve, Microbial Enhancement of TCE and 1,2-DCA Solute Flux in UF-Membrane Bioreactors. *AIChE J.*, vol. 44, no. 9, pp. 2112–2123, 1998.

245. N. Cicek, J. P. Franco, M. T. Suidaon, V. Urbain, and J. Manem, Characterization & Comparison of a Membrane Bioreactor and a Conventional Activated Sludge System in the Treatment of Wastewater Containing High Molecular Weight Compounds. *Water Environ. Res.*, vol. 71, no. 1, pp. 64–70, 1999.

246. J. A. Scott, D. A. Neilson, and P. N. Boon, A Dual Function Membrane Bioreactor System for Enhanced Aerobic Remediation of High-Strength Industrial Wastewater. *Water Sci. Technol.*, vol. 38, no. 4–5, pp. 413–420, 1998.

247. J. S. Almeida, M. Maria, and J. G. Crespo, Development of Extractive Membrane Bioreactors for Environmental Applications. *Environ. Protect. Eng.*, vol. 25, no. 1–2, pp. 111–121, 1999.

248. K. Brindle, T. Stephenson, and M. J. Semmens, Pilot-Plant Treatment of a High-Strength Brewery Wastewater Using a Membrane-Aeration Bioreactor. *Water Environ. Res.*, vol. 71, no. 6, pp. 1197–1203.

249. M. A. Deshusses, W. Chen, A. Mulchandani, and I. J Dunn, Innovative Bioreactors. *Curr. Opin. Biotechnol.*, vol. 8, no. 2, pp. 165–168, 1997.

250. C. Wen, Q. Xia, and Y. Qian, Domestic Wastewater Treatment Using an Anaerobic Bioreactor Coupled with Membrane Filtration. *Process Biochem.*, vol. 35, no. 3,4, pp. 335–340, 1999.

251. M. Bodzek, Membrane Techniques in Wastewater Treatment. *Environ. Protect. Eng.*, vol. 25, no. 1–2, pp. 153–192, 1999.

252. H. L. Bohn and K. H. Bohn, Moisture in Biofilters. *Environ. Prog.*, vol. 18, no. 3, pp. 156–161, 1999

253. C. Hatanaka, Development of High Performance Bioreactor. *Chiaki Bio Ind.*, vol. 16, no. 9, pp. 40–50, 1999.
254. L. Diels, R. S. Van, K. Somers, I. Willems, W. Doyen, M. Mergeay, D. Springael, and R. Leysen, The Use of Bacteria Immobilized in Tubular Membrane Reactors for Heavy Metal Recovery and Degradation of Chlorinated Aromatics. *J. Membrane Sci.*, vol. 100, no. 3, pp. 249–258, 1995.
255. D. M. F. Prazeres and J. M. S Cabral, Enzymatic Membrane Bioreactors: Current State of the Art and Future Prospects. *Recent Adv. Mar. Biotechnol.*, vol. 2, pp. 181–223, 1998.

11

Novel Materials and Processes for Pollution Control in the Mining Industry

Alan Fuchs and Shuo Peng
University of Nevada, Reno, Reno, Nevada

Tremendous opportunities exist in the area of novel materials for environmental engineering applications. These opportunities have traditionally been in the areas of membranes, ion exchange, and adsorbents, but new areas relating to technological advances in "nanomaterials" and "bio-applications" have spawned new generations of designed materials for many pollution control applications. The emphasis in this chapter will be on new technologies which have been or will be useful for pollution control in the mining industry. This will require a review of developments in the general areas of membranes, ion exchange, and adsorption, and discussion of how these materials are useful in mining applications.

1 MEMBRANES MATERIALS AND PROCESSES

A great deal of work has been done on the use of membrane processes for treatment of mine waters. Some of the typical membrane configurations of membranes separators are shown in varied textbooks (e.g., Ref. 1). This text also has examples of hollow fiber membranes and typical flow arrangements of these systems. Recent examples of this include separation of rare earths using liquid membranes (2) and copper recovery from Chilean mine waters, also using liquid membranes (3). Earlier studies in this area related to the removal of ammonium and nitrate ions from mine effluents using nanofiltration and reverse osmosis (4,5).

A major focus in the area of membrane development during the past few years has been in the use of liquid membranes. Valenzuela et al. (3) describe the use of a hollow fiber-supported liquid membrane for the recovery and concentration of copper from Chilean mine water. This system works by impregnating the porous structure of the membrane with an organic film which acts as a selective extraction medium. The film is salicylaldoximic extractant. Utilizing a low concentration of the extractant, a high degree of copper recovery is possible using this technique. An organic solution, containing 5-dodecylsalicylaldoxime was dissolved in a solvent containing 91% aliphatics and 9% aromatics. A feedstream containing 0.8–1.4 g/liter Cu(II) was used for testing, with a pH of 2.8–3.2 and a density of 1.05 g/ml at 20°C. Concentrated sulfuric acid solutions were used as metal-acceptor stripping agents. The hollow fiber membranes were microporous PTFE fibers with a membrane pore size of 2.0 μm. Because of the hydrophobic nature of the membrane used, the pores are rapidly and easily filled with the solvent-containing "carrier" extractant. A similar system was also studied by Valenzuela et al. (3). After impregnation with the the solvent, the feed solution and acid strip solution was recirculated through the fiber bores and outside the fibers, respectively.

Copper extraction in a membrane extractor occurs by diffusion of copper ions from the bulk feed into the solvent. The reaction which takes place in the membrane is as follows:

$$Cu^{+2}_{(aq)} + 2HR_{(org)} = CuR_{2(org)} + 2H^+_{(aq)}$$

where HR is the acidic extractant and CuR_2 is the metal complex extracted into the organic phase. The extractant diffuses through the membrane into the acid, where the stripping reaction takes place. In this way the carrier is regenerated and copper ions are free to be collected in the stripping liquor.

The results of this work indicate that copper concentrations can be reduced from 1 g/liter to below 0.2 g/liter in 8 h. The effectiveness of copper removal is dependent on the sulfuric acid concentration in the stripping solution.

Yang et al. (2) describe the use of a combined extraction/electrostatic psuedo liquid membrane (ESPLIM) for extraction and separation of rare earths in a simulated mine water. In this process, a continuous organic phase, consisting of 20% di-(2-ethylhexyl) phosphoric acid and 80% kerosene serves as the bulk liquid membrane. This stream contains a specific extractant for the metal ions to be extracted and separated. A discontinuous, aqueous stream, <0.2 mm in diameter, is used for phase settlement. The system consists of a grounded electrode coated with polyethylene film mounted at the sides of extraction and stripping cells in a rectangular reaction tank. A high-voltage electrode coated with polyethylene is wound on the perforated baffle plate separating the extraction and stripping cell. When a high-voltage electrostatic field is applied to the reaction tank, the aqueous drops in the organic continuous phase disintegrate into numerous smaller droplets under the action of the electrostatic field. This provides a great deal of surface area for separation. The extractant dissolved in the continuous organic phase acts as a shuttle to transport metal ions from the extraction cell to the stripping cell.

A summary of opportunities for membrane technologies in the treatment of mining and mineral process streams was presented by Awadalla and Kumar (4). This study indicated a variety of applications including acid mine drainage (AMD), treatment of flotation water, copper smelting and refining wastewater, mill wastewater, removal of ammonium and nitrate ions, membranes in the aluminum industry, treatment of groundwater, treatment of uranium wastewater, treatment of dilute gold cyanide solutions, recovery of zinc from pond water, rare earth (RE) concentration, and separation of selenium from barren solution.

AMD contains pollutants such as iron, manganese, calcium, magnesium, and sulfate ions. Although lime neutralization is considered the "best available technology economically achievable," it is no longer considered environmentally acceptable because of the low-level contamination of heavy metals which cannot be removed. Alternatively, almost complete removal of dissolved solids can be achieved by the use of ion exchange, distillation, and reverse osmosis (RO) to produce high-quality water which can be used by municipalities or industry. The use of RO is best implemented as a supplement to neutralization processes. The RO concentrate stream is neutralized and clarified prior to discharge or recycled. Coupled RO/ion exchange can be used when high concentration of calcium sulfate and/or iron fouling is a problem. For the case of water reuse in which completely demineralized water is not essential, a charged ultrafiltration process using negatively charged noncellulosic membranes was utilized. For the case of AMD for coal conversion processes, high-ultrafiltration recovery with high removal of calcium sulfate and iron and good flux are required. Recovery of up to 97% is achievable by introducing an interstage settling step. Commercially available charged ultrafiltration membranes by PSAL (millipore type of noncellulosic skin on cellulosic backing) were used in this study. Cost for treatment using UF with interstage settling are $1.33/1000 gal of AMD, including membrane replacement cost, pumping cost, and lime cost.

In order to avoid problems with recycling wastewater from flotation mills which contain the breakdown products of collector-frother reagents, the water must be purified before recycling to the mining operation. The traditional method for treatment of flotation water involves lime precipitation, ozonation, adsorption on activated carbon, and biological treatment (4). Biological treatment requires excessive holdup and is dependent on the climate, the presence of toxic heavy metals, and sensitive control of the microorganisms. Reverse osmosis has been used for the recovery of flotation reagents. Commercial RO membranes have been used to remove 95% of organic carbon, calcium, and magnesium from the flotation feed stream.

Scrubber blowdown from a primary copper smelting plant and acid processing water from a selenium-tellurium plant have been treated using negatively charged noncellulosic ultrafiltration membranes (4) . Removal of over 85% of As and Se from the acid processing water was made possible when the pH was adjusted to 10 and the solids were settled prior to ultrafiltration. Scrubber blowdown was effectively treated without pH adjustment to a pH of 4.5. Arsenic-containing wastewater was also pretreated with UF and polished using RO. This method produced a permeate stream containing less than 50 ppb arsenic.

Alkaline solutions of NaCN are used to leach gold-containing ores, producing dilute gold cyanide solutions (4). The two conventional methods of recovering gold from these solutions include the Merill-Crowc process of cementation using zinc powder and adsorption using activated carbon. Concentrated gold solutions are formed by elution. Reverse osmosis has been investigated as a means to concentrate the dilute gold solutions. In the case of metal finishing operations using gold and cyanide solutions, FilmTec FT-30 membranes have been used to provide rejections in the range of 91–99% for free and combined cyanides (with copper and zinc). Membrane performance was strongly pH dependent. Reverse osmosis has also been used for silver and copper cyanide concentration (Osmonics, Inc). This study utilized a nitrogen-containing aromatic condensation polymer. Experiments indicated that the feed could be concentrated three times with 70% removal of permeate, resulting in low gold content in the permeate.

Nanofiltration (NF) and RO have been used for removal of ammonium and nitrate ions from synthetic and actual mine effluents (5). In mine and mill water, ammonium and nitrate ions are generated from the degradation of cyanide from gold mill effluents and ammonium nitrate-fuel oil (ANFO) blasting agents in mines. Nitrogen-containing reagents are also used in ore processing and extractive metallurgy. The results of experiments using NF and RO membranes were reported for testing and actual mill effluent. The results of the testing were that good removal of ammonium (>99%) and nitrate ions (>97%) were achieved using RO, while NF was less effective. Lower effectiveness of the NF membrane was believed to be caused by ammonium being present in the sulfate form and not the larger ammonium iron sulfate complex which does not form because there is no iron in the mining effluent. No scaling or fouling problems were observed in these studies.

Cross-flow membrane technolgies have also been applied to mineral suspensions (6). In this study, using microporous filtration (0.1-μm membranes) suspensions of $CaCO_3$ were investigated using an intermittent cleaning approach in order to increase the permeate flux.

A thorough review of membrane technology for applications to industrial wastewater treatment has been made by Caetano (7). In this review, E. Drioli provides a broad overview in the areas of desalination, gas separation, pervaporation, membrane bioreactors, enzyme membrane reactors, and hybrid systems based on pervaporation and distillation.

In the more general area of environmental applications, significant work has been done on the treatment of streams containing metals. There has been a great deal of interest in the use of ion-exchange membranes in this area. Sengupta (8) has investigated electromembrane partitioning as a means for heavy metal decontamination. This is a unique and rather interesting new approach for the in-situ removal of metals from contaminated soils.

A low-level direct current (DC), less than 1 V/cm, is applied to the soil while a composite ion-exchange membrane is wrapped around the cathode. Upon imposition of the DC potential, the cations move toward the cathode, where they are captured by the composite membrane. By the design of the ion-exchange membrane, the nonselective ions should pass freely through the membrane. The membrane utilized for this work is a thin sheet prepared by grinding a cross-linked polymer ion exchanger and suspending the ion exchanger in a PTFE porous matrix. These membranes are 90% ion exchanger, 10% PTFE, and are microporous with >40% voids with a pore size distribution below 0.5 μm. One potential problem with this process is that periodically these membranes must be removed and chemically regenerated with strong (3–5%) mineral acid solution.

Electrodialytic decontamination of soil polluted with heavy metals has been investigated using ion-exchange membranes by Hansen et al. (9). The process for removal of metal ions from soils using electric current and passive membranes is known as electrokinetic soil remediation. This method involves the use of passive membranes to separate the polluted soil from the electrodes. There are several shortcomings to this approach, including addition of acid counterions into the soil, return of heavy metals back into the soil, and heavy-metal precipitation at the H^+ and OH^- front. By introduction of ion-exchange membranes into the electrokinetic soil remediation process, an electrodialytic soil remediation process results. The ion-exchange membranes are oriented in certain directions. This orientation, with pairs of anion- and cation-exchange membranes placed on both sides of the polluted soil, eliminates all three of the problems mentioned above. This configuration also provides two compartments containing liquid solutions and the heavy metals, which can be withdrawn as needed. In this situation, heavy-metal ions pass through the cation-exchange membrane in the direction of the cathode and are prevented from passing through the anion-exchange membrane and never reach the cathode. They end up in the compartment between the two ion-exchange membranes.

Li et al. (10) have investigated the use of a cation-selective membrane for removal of heavy metals from soils. An improvement in the traditional electroremediation approach is described. In this work a cation membrane is placed around the cathode to prevent hydroxyl species moving toward the anode. This prevents precipitation of metals in the soil, and the metals precipitate in the column of water around the cathode.

Membranes are also used for removal of metals for industrial applications (11). Bulk liquid membranes are used for facilitated transport of silver using a rotating film pertraction device. In this process two aqueous solutions are separated by an organic liquid. The membrane liquid is in contact with the donor and acceptor liquids adhering to the surfaces of the rotating disks. Transport of the solvent involves extraction from one solution and stripping in the other. This paper describes the recovery of silver from nitrate solutions using the rotating film pertraction method using tri-isobutylphosphine sulfide (TIBPS) in n-octane as the liquid membrane. Aqueous silver nitrate was the donor phase and the acceptor phase was aqueous ammonia. The results of the study indicated that because of low rates of transport of other metals, including copper, zinc, and nickel, rotating film pertraction can be used effectively to separate silver from solution.

Yang et al. (2) describe a unique metal extraction method using two sets of hydrophobic microporous hollow fiber membranes for separation of metals in solution. One set of hollow fibers carries an acidic organic extractant (LIX 84, anti-2-hydroxy-5-nonylacetophenone oxime) in a diluent. The other set of hollow fibers carries a basic organic extractant (TOA, tri-n-octylamine). The aqueous, metal-containing stream is carried on the shell side of the membrane system. Cations, including copper, zinc, and nickel, are transported into the acidic extractant. Anions, including chromium(VI), mercury, and cadmium, are extracted into the basic stream.

Palladium has also been separated from silver in a nitric acid solution using liquid surfactant membranes (12). The organic carrier used in these studies is LIX 860, which is a β-hydroxyoxime. The liquid surfactant membrane is Span 80, a commercially available surfactant, and the solvent is n-heptane. The aqueous donor phase contains silver and palladium and is acidified using nitric acid. The receiving phase contains thiourea and is tested in hydrochloric, perchloric, nitric, and sulfuric acids. Under optimal conditions, palladium was separated from silver recovered in entirety.

Another liquid membrane, investigated by Fu et al. (13), is trioctylamine (TOA) as a mobile carrier in kerosene. Precious metals, including gold, palladium, platinum, iridium, and ruthenium in hydrochloric acid, were extracted using this membrane system. The metals were extracted into perchlorate and nitric acid solutions. An inert PTFE polymer 80 μm thick, 74% porous, and 0.45 μm in average pore diameter was used as a support for the liquid membrane.

Low-pressure reverse osmosis (RO) was used by Ujang and Anderson (14) for separation of mono- and divalent ions. Sulfonated polysulfone membranes are used as a low-pressure reverse osmosis process for separation of mono- and divalent zinc ions. It was observed that the higher the operating pressure, the greater was the permeate flux for both species. At lower operating pressure, higher permeate fluxes were observed using divalent ions. Metal removal of divalent ions was greater for divalent ions than for monovalent ions for all concentrations.

2 ADSORPTION MATERIALS AND PROCESSES

Recovery of gold from cyanide has been evaluated using many different adsorbent materials. Petersen and Van Deventer (15) investigated the competitive role of gold and organics on adsorption by a variety of adsorbents, including activated carbon, ion-exchange resin, ion-exchange fibers, and membranes. A variety of adsorbents were investigated, including coconut shell activated carbon, macroporous ion-exchange resin, ion-exchange membrane, and ion-exchange fibers (polypropylene-based strong-base and weak-base fibers). Adsorbents were evaluated after being exposed to the organic compound, sodium ethyl xanthate, for 6 h. The absorbents were challenged with a variety of organic compounds, including ethanol, sodium ethyl xanthate, potassium amyl xanthate, and phenol. The two mechanisms investigated to explain the reduced adsorption of gold in the presence of the organics were (a) blockage of the carbon pores by the organic, and (b) competition between gold cyanide and organics for the active sites on the carbon surface. The results of the study indicated that both the rate of adsorption and the equilibrium loading were affected by the organic on the adsorption of gold cyanide onto activated carbon. The resin particles were only effected by the rate of adsorption, while the membranes and fibers experienced both kinetic and equilibrium changes. The results of this study indicated that the long-chain organics (xanthates) have a higher degree of inhibition of mass transfer of gold cyanide compared to the low-molecular-weight substance (ethanol). The aromatic substances did not affect the performance of the fibers or membrane. This is because the small pore diameters did not permit the large aromatics to penetrate. The results indicated that the second mechanism, a competitive effect between gold cyanide and the organic compounds, was responsible for the results observed for the gold-equilibrated absorbents.

Klein et al. (16) have investigated polymeric resins as adsorbents for industrial applications. The motivation for investigation of polymeric resins versus activated carbon is their ease of regeneration. Activated carbon systems are typically regenerated using steam or thermal methods, while polymeric resins can be regenerated using simple solvents such as aliphatic alcohols. The resins used were methylene-bridged styrene divinylbenzene-based co-polymer (Dow Chemical, Midland, MI). Some of the characteristics of these polymeric resins which may be controlled are hydrophobicity, pore size, and surface area. These resins were challenged with benzoic acid and chlorobenzene, and adsorption isotherms and bed regeneration curves were generated. The results of this study indicated that with only few bed volumes (15–25), using methanol as regenerant, 90–95% of the adsorbed solute could be recovered. The polymeric resins maintained good adsorptive capacity after repeated cycling.

3 ION-EXCHANGE MATERIALS AND PROCESSES

Applications of ion exchange to leaching solutions of an Algerian gold ore have been investigated by Akretche et al. (17). In the cyanide medium, the gold and other metals such as silver, copper, and iron attach to the anion-exchange resin. These metals are later eluted with acid thiourea to yield a concentrated solution which is treated by cementation or an electrolytic method. This work describes the use of electrodialysis of copper(I), which is normally not feasible due to the presence of formamidine disulfide. This is accomplished when the solutions are obtained by elution of cuprocyanides by thiourea.

4 CONCLUSIONS

There have been great strides in the development of new technologies for pollution control in the mining industry during the past five years, many in the development of new materials and processes. Many of these developments are in the areas of membranes, adsorbents, and ion exchange. In the area of membranes, a great of work has been done using liquid membranes. These are generally supported synthetic membrane systems with a variety of liquids to facilitate transport. Electroremediation and electrodialytic membrane approaches have also seen a great deal of attention. Activated carbon-based and other organic absorbents have been used for treatment of mining wastes. Polymeric resins have also been used as adsorbents for industrial applications. Anionic ion-exchange resins have also been used for treatment of leaching solutions.

REFERENCES

1. W. L. McCabe, J. C. Smith, and P. Harriott, *Unit Operations of Chemical Engineering, 5th Edition*, New York: McGraw-Hill, 1993.
2. Z. F. Yang, A. K. Guha, and K. Sirkar, *Ind. Eng. Chem. Res.*, vol. 35, pp. 1383–1394, 1996.
3. F. Valenzuela, C. Basualto, C. Tapia, and J. Sapag, *J. Membrane Sci.*, vol. 155, pp. 163–168, 1999.
4. F. T. Awadalla and A. Kumar, *Separation Sci. Technol.*, vol. 29, no. 10, pp. 1231–1249, 1994.
5. F. T. Awadalla, C. Striez, and K. Lamb, *Separation Sci. Technol.*, vol. 29, no. 4, pp. 483–495, 1994.

6. D. Si-Hassen, A. Ould-Dris, M. Y. Jaffrin, and Y. K. Benkahla, *J. Membrane Sci.*,vol. 118, pp. 185–188, 1996.

7. A. Caetano (ed.), *Membrane Technology: Applications to Industrial Wastewater Treatment.* Dordrecht, The Netherlands: Kluwer, 1995.

8. S. Sengupta and A. K. Sengupta, *Hazardous and Industrial Wastes—Proceedings of the Mid-Atlantic Industrial Waste Conference Proceedings of the 1997 29th Mid-Atlantic Industrial and Hazardous Waste Conference*, July 13–16, 1997, Blacksburg, VA, Lancaster, PA: Technomic Publishing Co. Inc., pp. 174–182.

9. H. K. Hansen, L. M. Ottosen, S. Laursen, and A. Villumsen, *Separation Sci. Technol.*, vol. 32, no. 15, pp. 2425–2444, 1997.

10. Z. Li, J. Yu, and I. Neretnieks, *Environ. Sci. Technol.*, vol. 32, pp. 394–397, 1998.

11. L. Boyadzhiev and K. Dimitrov, *J. Membrane Sci.*, vol. 68, p. 137–143, 1994.

12. T. Kakoi, M. Goto, and F. Nakashioo, *Separation Sci. Technol.*, vol. 32, no. 8, pp. 1415–1432, 1997.

13. J. Fu, S. Nakamura, and K. Akiba. *Separation Sci. Technol.*, vol. 32, no. 8, pp. 1433–1445, 1997.

14. Z. Ujang and G. K. Anderson, *Water Sci. Technol.*, vol. 38, no. 4–5, pp. 521–528, 1998.

15. F. W. Petersen and J. S. J. Van Deventer, *Separation Sci. Technol.*, vol. 32, no. 13, pp. 2087–2103, 1997.

16. J. Klein, G. M. Gusler, and Y. Cohen, Removal of Organics from Aqueous Systems: Dynamic Sorption/Regeneration Studies with Polymeric Resins. In *Novel Absorbents and Their Environmental Applications*, Y. Cohen and R. W. Peters (eds.), AIChE Symp. Ser., vol. 91, pp. 72–78.

17. D. E. Akretche, A. Gherrou, and H. Kerdjoudj, *Hydrometallurgy,* vol. 46, pp. 287–301, 1997.

12

Monitoring In-Situ Electrochemical Sensors

Joseph Wang
New Mexico State University, Las Cruces, New Mexico

1   INTRODUCTION

Electroanalytical methods are concerned with the interplay between electricity and chemistry, namely, the measurements of electrical quantities such as current or potential, and their relationship to chemical parameters. Electroanalytical chemistry can play a major role in pollution control and prevention. In particular, electrochemical sensors and detectors are very attractive for on-site and in-situ monitoring of priority pollutants. Such devices are highly sensitive, selective toward electroactive species, fast, accurate, compact, portable, and inexpensive. Several electrochemical devices, such as oxygen or pH electrodes, have been widely used for years for environmental analysis. Recent advances in electro-chemical sensor technology have expanded the scope of electrochemical devices toward a wide range of organic and inorganic contaminants.

The present chapter reviews recent efforts at the author's laboratory, aimed at in-situ monitoring of priority pollutants. Continuous monitoring, effected in the natural environment, offers a rapid return of the chemical information (with a proper alarm in case of a sudden discharge), avoids costs and errors associated with the collection of discrete samples, while maintaining the sample integrity. The use of remote sensors thus has significant technical and cost benefits over traditional sampling and analysis. Our latest developments of remote electro-chemical probes will be covered in the following sections.

2 REMOTELY DEPLOYED ELECTROCHEMICAL SENSORS

Remotely deployable submersible sensors capable of monitoring contaminants in both time and location are advantageous in a variety of applications. These range from shipboard marine surveys, downhole monitoring of groundwater contami-nation, to real-time analysis of industrial streams. The development of sub-mersible electrochemical probes requires proper attention to various challenges, including the effect of sample pH, ionic strength, dissolved oxygen, or natural convection, specificity and sensitivity, surface fouling, in-situ calibration, and miniaturization. By addressing these and other obstacles, we were able to develop remote sensors for a wide range of inorganic and organic contaminants.

2.1 Remote Monitoring of Metal Contaminants

Metal pollution has received enormous attention due to its detrimental impact on the environment. The need for continuous monitoring of trace metals in a variety of matrices has led to the development of submersible sensors based on electro-chemical stripping analysis (1,2). Stripping analysis has been established as a powerful technique for determining toxic metals in environmental samples (3,4). The remarkable sensitivity of stripping analysis is attributed to its unique "built-in" preconcentration step, during which the target metals are electroplated onto the surface. Both electrolytic and nonelectrolytic (adsorptive) accumulation schemes have thus been employed to achieve sub-parts-per-billion detection limits. The analytical current signal (i), obtained during the subsequent stripping (potential scanning) step is proportional to the metal concentration (C) and accumulation time (t_{acc}):

$$i = KC\, t_{acc} \tag{1}$$

Remote metal monitoring has been realized by eliminating the needs for mercury surfaces, oxygen removal, forced convection, or supporting electrolyte (which previously prevented the direct immersion of stripping electrodes into sample streams). This was accomplished through the development of nonmercury electrodes, judicious coupling of potentiometric stripping operation, and the use of advanced ultramicroelectrode technology (1). Compatibility with field opera-tions was achieved by connecting the three-electrode housing [including a gold fiber working electrode, in the polyvinyl chloride (PVC) tube], via environmen-tally sealed three-pin connectors, to a 25-m-long shielded cable. Convenient and simultaneous quantitation of several trace metal levels (e.g., Pb, Cu, Ag, Hg) has thus been realized in connection to measurement frequencies of 20–30/h (based on deposition periods of 1–2 min).

The in-situ monitoring capability of the remote metal sensor was demonstrated in studies of the distribution of labile copper in San Diego Bay (CA) (5). For this purpose, the probe was floating on the side of a small U.S. Navy vessel. The resulting map of copper distribution reflected the metal discharge and circulation pattern in the bay. We are currently collaborating with Prof. Daniele's group in using the remote probes for assessing the distribution of metal contaminants in the canals and lagoon of Venice, Italy (6).

The extension of remote stripping electrodes to additional metals that cannot be electroplated relies on the adaptation of adsorptive stripping protocol for a submersible operation (7). Such procedures rely on the formation and adsorptive accumulation of complexes of the target metals. Accordingly, remote adsorptive stripping sensors require a new probe design based on an internal solution chemistry. Such a renewable-reagent adsorptive stripping sensor relies on the continuous delivery of the ligand, its complexation reaction with the metal "collected" in a semipermeable microdialysis sampling tube, and transport of the complex to the working electrode compartment. Such dialysis sampling also offers extension of the linear range and protection against surface fouling (due to its dilution and filtration actions).

The new flow-probe format was employed for monitoring trace metals such as nickel, uranium, or chromium. As desired for effective in-situ monitoring, such adsorptive stripping probes have the capability to detect rapidly fluctuations in the analyte concentration continuously. Such ability is indicated from Figure 1, which displays the response of a chromium probe (8) upon switching from the 5- to 25 µg/l chromium solutions. Such behavior is attributed to the reversibility of the accumulation/stripping cycle, with the stripping and subsequent 10-s "cleaning" steps completely removing the accumulated complex. In addition, the

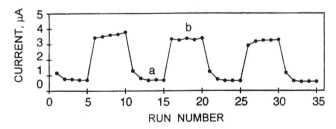

FIGURE 1 Response of the remote chromium(VI) probe to alternate exposures to (a) "low" (5 µg/l) and (b) "high" (25 µg/l) chromium(VI) levels. Accumulation for 30 s at –0.9 V; square-wave voltammetric stripping scan.

reagent flow continuously replenishes the solution, to "erase" an internal buildup of chromium.

Other groups have also been involved in the development of remote metal sensors. For example, Kounaves's team reported on probes based on mercury-plated iridium-based microelectrode arrays and square-wave voltammetric stripping detection (9). A solid-state reference electrode that eliminates leakage of electrolyte to the surrounding low-ionic-strength aquatic environment was employed. The device developed by Buffle's group (4,10) has been coupled to a thick agarose-gel antifouling membrane that facilitates measurements in complex media. There is no doubt that these and similar developments of submersible stripping sensors will have a major impact on the surveillance of our water resources.

2.2   Remote Modified Electrodes and Biosensors

Chemically and biologically modified electrodes (CMEs) have greatly enhanced the power of electrochemical detectors and devices (11). The ability to deliberately control and manipulate surface properties can lead to a variety of attractive effects. Electrochemical sensors based on modified electrodes combine the remarkable sensitivity of amperometry with new chemistries and biochemistries. Such manipulation of the molecular architecture of the detector surface offers new levels of reactivity that greatly expand the scope of electrochemical devices, and enhance the power of in-situ electrochemical probes.

2.3   Biosensors

Biosensors are small devices employing biochemical molecular recognition properties as the basis for a selective analysis. The major processes involved in any biosensor system are analyte recognition, signal transduction, and readout. The remarkable specificity of biological recognition processes has led to the development of highly selective electrochemical biosensors. In particular, enzyme electrodes, based on amperometric or potentiometric monitoring of changes occurring as a result of the biocatalytic process, have the longest tradition in the field of biosensors. Such devices are usually prepared by immobilizing an enzyme onto the electrode surface. The integration of these devices with remotely deployed probes should add new dimensions of specificity to in-situ electrochemical monitoring of pollutants. In the adaptation of enzyme electrodes to a submersible operation, one must consider the influence of actual field conditions (pH, salinity, temperature) on the biocatalytic activity.

The first remotely deployed biosensor targeted phenolic contaminants in connection to a submersible tyrosinase enzyme electrode (12). The enzyme, immobilized within a stabilizing carbon paste matrix, converted its phenolic substrates to easily reducible quinone products. The sensor responded rapidly to micromolar levels of various phenol contaminants, with no carryover (memory) effects.

We also developed a remote biosensor for field monitoring of organophosphate nerve agents (13). The device relied on the coupling the enzymatic activity of organophosphate hydrolase (OPH) with the submersible amperometric probe configuration. Low (micromolar) levels of paraxon or parathion have thus been measured directly in untreated natural water matrices. The OPH enzyme obviates the need for lengthy and irreversible enzyme inhibition protocols common to inhibition-based biosensors.

Finally, hydrogen peroxide and organic peroxides have been monitored at large instrument–sample distances by incorporating a reagentless peroxidase bioelectrode into the remote probe assembly (14). A low detection potential (~0.0 V) accrued from the use (co-immobilization) of a ferrocene co-substrate allowed convenient monitoring of micromolar peroxide concentrations in untreated samples.

2.4 Modified Electrodes

Chemical layers can also be used to enhance the performance of electrochemical devices. The use of electrocatalytic surfaces can expand the scope of remote electrodes to pollutants possessing slow electron-transfer kinetics. One example of the adaptation of modified electrodes for a submersible operation is a remote sensor for toxic hydrazine compounds, based an electropolymerized films of 3,4-dihydroxybenzadehyde (15). The low-potential detection accrued from this catalytic action offers convenient measurements of micromolar hydrazine concentrations in untreated groundwater or river water samples.

We also developed a submersible probe based on a carbon-fiber working electrode assembly, connected to a 50 ft-long shielded cable, for the continuous monitoring of the 2,4,6-trinitortoluene (TNT) explosive in environmental matrices (16). The facile reduction of the nitro moiety allowed convenient and fast (1–2 s) square-wave voltammetric measurements of parts-per-million levels of TNT.

3 SUBMERSIBLE ELECTROCHEMICAL ANALYZERS

The ability to perform metal–ligand complexation reactions on a cable platform, in connection to adsorptive stripping measurements, has led to the development of submersible electrochemical analyzers (17). As opposed to current in-situ sensors (which lack sample preparatory steps, essential for optimal analytical performance), the new on-cable automated microanalyzer will eventually incorporate all the steps of the analytical protocol into the submersible device. The new "lab-on-cable" concept thus involves the combination of sampling, sample pretreatment, separation of components, and detection step (along with self-calibration) into a single sealed submersible package. The first generation of this submersible microlaboratory integrates microdialysis sampling, with reservoirs for the reagent, waste, and calibration/standard solution, along with the micromump and necessary fluidic network on a cable platform (Figure 2). The sample and reagent are thus brought together, mixed, and allowed to react in a reproducible manner. Future generations will accommodate additional functions (e.g., preconcentration, filtration, extraction) for addressing the needs of complex environmental samples. Micromachining technology is being explored for further miniaturization and for facilitating these in-situ sample manipulations. Proper attention is also being given to the design of compact, low-powered, automated instrumentation for unattended operation, "smart" data processing, and signal transmission (via satellite links). Such a standalone "microlaboratory" can be submersed directly in the environmental sample, to provide real-time continuous information on a wide range of priority pollutants. The ability to perform in-situ all the necessary

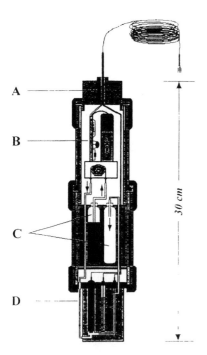

FIGURE 2 Schematic diagram of the electrochemical "lab-on-cable" system: (A) cable connection; (B) micropump; (C) reservoirs for reagent and waste solutions; (D) microdialysis sampling tube and an electrochemical flow detector.

steps of the analytical protocol should have an enormous impact on pollution control and prevention. Preliminary examples based on stripping monitoring of trace metals or in-situ biosensing of phenols and various enzyme inhibitors are presented below.

For example, we demonstrated the utility of the "lab-on-cable" probe for circumventing in-situ problems common to electrochemical stripping analysis (18). In particular, an internal delivery of an appropriate solution, containing a ligand, third element, or a conducting salt, was used to minimize errors due to overlapping peaks, intermetallic compounds, or ohmic distortions, respectively. Similarly, internal delivery of a strong acid was used for on-cable release of the metals from "collected" metal complexes, as desired for in-situ monitoring of the total metal content (19).

We also developed a submersible phenol analyzer, based on an enzymatic (tyrosinase) bioassay (17). The assay involved microdialysis sampling of the phenolic compounds, their mixing with the internally delivered tyrosinase solution, and amperometric monitoring of the quinone product. The internal buffer solution assured independence of sample conditions such as pH or ionic strength [which commonly influence the performance of remote biosensors (12)]. Another enzymatic assay was developed for the in-situ monitoring of cyanide (20). Such an enzyme inhibition assay relied on the internal delivery of tyrosinase and its catechol substrate using a flow-injection manifold. The flow probe thus addressed the challenges to in-situ enzyme-inhibition devices (e.g., the replacement of the inhibited enzyme and of the consumed substrate).

We envision the integration of multiple techniques and assays onto a single cable platform, i.e., a complete submersible laboratory. Eventually, we expect to eliminate the cable platform, and to use microlaboratories on miniaturized boats or submarines, which would travel across the water stream and provide the desired spatial and temporal information on target contaminants.

4 CONCLUSIONS AND FUTURE PROSPECTS

Electrochemical sensor technology is still limited in scope and cannot address all environmental monitoring needs, yet a vast array of electrochemical devices has been developed in recent years for in-situ monitoring numerous organic and inorganic pollutants. By providing a fast return of the analytical information in a timely, safe, and cost-effective fashion, the new, remotely deployed probes would offer direct and reliable assessment of the fate and gradient of contaminants sites, while greatly reducing the huge analytical costs. While the concept of "lab-on-cable" is still at infancy, such a strategy should revolutionize the way of monitoring priority pollutants, and would have a major impact on field analytical chemistry. Ongoing commercialization efforts, coupled with regulatory acceptance, should lead to the translation of these research efforts into large-scale environmental applications.

ACKNOWLEDGMENT

This work was supported by the U.S. Department of Energy Environmental Managenemt Science Program (grant DE-FG07-96ER62306) and by the DOE-WERC program.

REFERENCES

1. J. Wang, D. Larson, N. Foster, S. Armalis, J. Lu, X. Rongrong, K. Olsen, and A. Zirino, *Anal. Chem.*, vol. 67, pp. 1481–XXXX, 1995.
2. J. Wang, B. Tian, J. Lu, J. Wang, D. Luo, and D. MacDonald, *Electroanalysis*, vol. 10, pp. 399–402, 1998.
3. J. Wang, *Stripping Analysis*. VCH Publishers, New York, 1985.
4. M. Tercier and J. Buffle, *Electroanalysis*, vol. 5, pp. 187–200, 1993.
5. J. Wang, N. Foster, S. Armalis, D. Larson, A. Zirino, and K. Olsen, *Anal. Chim. Acta*, vol. 310, pp. 223–231, 1995.
6. S. Daniele, J. Wang, and J. Lu, *Analyst*, in press.
7. J. Wang, J. Lu, D. Luo, J. Wang, M. Jian, B. Tian, and K. Olsen, *Anal. Chem.*, vol. 69, pp. 2640–2645, 1997.
8. J. Wang, J. Wang, J. Tian, B., D. MacDonald, and K. Olsen, *Analyst*, vol. 124, pp. 349–352, 1999.
9. J. Herdan, R. Feeney, S. Kounaves, A. Flannery, C. Storment, and C. Kovacs. *Environ. Sci. Technol.*, vol. 32, pp. 131–137, 1998.
10. M. L. Tercier and J. Buffle, *Anal. Chem.*, vol. 68, pp. 3670–3678, 1996.
11. J. Wang, *Electroanalysis*, vol. 3, pp. 255–259, 1991.
12. J. Wang and Q. Chen, *Anal. Chim. Acta*, vol. 312, pp. 39–45, 1995.
13. J. Wang, L. Chen, A. Mulchandani, P. Mulchandani, and W. Chen, *Electroanalysis*, vol. 11, pp. 866–869, 1999.
14. J. Wang, G. Cepria, and Q. Chen, *Electroanalysis*, vol. 8, pp. 124–127, 1996.
15. J. Wang, Q. Chen, and G. Cerpia, *Talanta*, vol. 43, pp. 1387–1391, 1996.
16. J. Wang, R. K. Bhada, J. Lu, and D. MacDonald, *Anal. Chim. Acta*, vol. 361, pp. 85–91, 1998.
17. J. Wang , J. Lu, B. Tian, S. Ly, M. Vuki, W. Adeniyi, and R. Armennderiz, *Anal. Chem.*, in press.
18. J. Wang, J. Lu, D. MacDonald, and M. Augelli, *Fresenius Z. Anal. Chem.*, vol. 364, pp. 28–31, 1999.
19. J. Wang, J. Wang, J. Lu, B. Tian, D. MacDonald, and K. Olsen, *Analyst*, vol. 124, pp. 349–352, 1999.
20. J. Wang, B. Tian, D. MacDonald, J. Wang, and D. Luo, *Electroanalysis*, vol. 10, pp. 1034–1037, 1998.

13

Using Roadmaps in Pollution Prevention: The Los Alamos Model

Thomas P. Starke
Los Alamos National Laboratory, Los Alamos, New Mexico

James H. Scott
Abaxial Technologies, Los Alamos, New Mexico

1   INTRODUCTION

Roadmapping is a powerful technique for displaying the structural relationships among science, technology, applications, and results of applications. Because they can incorporate complex, multiple relationships, they are used to display the possible paths from the present state to a desired end state. Well-constructed, comprehensive roadmaps are used for science and technology management, including strategic planning, evaluating cost/risk, and program execution; for enhancing communications among researchers, technologists, managers, and stakeholders; for identifying deficiencies and opportunities in science and technology programs; and for identifying obstacles to achieving a desired end state. There are several roadmap methodologies in use today, including forecast roadmaps, retrospective roadmaps, and process evaluation roadmaps. Because roadmapping methodology is so flexible, it can be used in many applications; it is frequently used for process evaluation, technology forecasting, and for defining investment strategies. Roadmaps have been used successfully by the U.S. Department of Defense (DoD), the semiconductor industry, and various manufacturing concerns.

The starting point for all roadmapping methodologies consist of a defined current state and a very well-defined desired end state; in general, one cannot have a high-quality map without a carefully and comprehensively defined end state. A complex project or process will have a number of intermediate states or goals between the current state and the desired end state. The roadmap itself consists of a network of nodes representing activities, events, or processes. Nodes can contain a variety of information, depending on the purpose of the node. Nodes are linked by actions. The network of nodes and links ideally represents *all* pathways from the current state to the desired end state in such a way that schedule, cost, and technical risk can be evaluated along each pathway. Analysis of high-quality maps can help evaluate options relative to risk, cost, and schedule; define deficiencies in current programs; and identify opportunities.

2 ROADMAP METHODOLOGY

In 1997 the Environmental Stewardship Office (ESO) at Los Alamos National Laboratory decided to prepare a roadmap for reaching the laboratory pollution prevention goal of substantially eliminating waste generation and pollutant release by the year 2010. The purpose of the roadmap was to identify:

Areas in which waste minimization and pollution prevention would have the greatest impact

Options for preventing pollution or minimizing waste in those areas

Costs, technical risk, time, and return on investment associated with implementing those options

The most cost-effective strategies for reaching the goal of substantially eliminating pollution and waste resulting from laboratory operations

In order to prepare this roadmap, ESO chose a methodology that is based on technology roadmap principles developed by the Office of Naval Research and widely used in the DoD community (1). This methodology was modified by Los Alamos to incorporate the principals of process mapping developed by Robert Pojasek (2). The resulting methodology produces a roadmap with very broad scope but sufficient detail to allow identification of specific sources of pollution and waste and, consequently, specific remedial action options.

3 ROADMAP CONSTRUCTION

The DoD roadmap methodology is hierarchical and proceeds through a series of submaps or map elements from general to specific. Thus, the roadmap is made up of several levels, with the higher levels being more general and less detailed.

The highest level contains only the definition of the desired goal or end state, the overall strategy for achieving that goal, and the definition of the waste types to be considered in the roadmap. Normally, the highest-level map element is called the zero level, or the mission-level map element. This element is comprehensive in that it identifies the current condition and lays the foundation for the succeeding map elements. In the zero level map element the waste types from any particular set of operations are defined.

Level one maps take the waste types defined at level zero and develop process flow diagrams for each waste type. A process flow diagram is an overview of the process that generates the waste. Process flow diagrams provide a summary of the processes and activities that result in the generation of waste. These diagrams are used to decompose each waste type into specific waste streams. For example, a waste type may be sanitary waste, and waste streams within that type may be food waste, paper, and glass.

At level two, process diagrams are developed for each waste steam within a waste type. These diagrams depict the process flow at a greater level of detail. In addition to these waste stream process diagrams, new or modified procedures, processes, or technologies are identified which may reduce or eliminate the waste stream. The point in the process flow where the new technology can be deployed is identified, along with the likely impact of deployment.

For some high-priority waste streams, further detail is provided in a third level, including assessment of various options. The hierarchical structure described above is shown schematically in Figure 1.

As an example of roadmap structure and how the roadmap can be used, consider a path through Figure 1. The mission-level map element defines the N waste types. These waste types could be sanitary, hazardous, liquid effluent, or many others, depending on the nature of the operations at level zero.

At level one we define each waste stream within a waste type. In the example, we have associated five waste streams with the second waste type. The other waste types also have associated streams, not shown here for simplicity's sake.

At level two, a process map element is constructed to describe the processes that produce each waste stream. An adjunct to the process flow map element is the definition of procedure, process, or technology options for treating the object. The likely impact of each option is then described. Technical risk, schedule risk, cost, and health and safety impacts are assessed.

For high-priority or complex waste streams the options identified in the process flow map element are broken down in further detail, and a series of issues and attributes is developed to aid in comparing options. To clarify the construction process, we will show how each of the map elements at various levels is constructed.

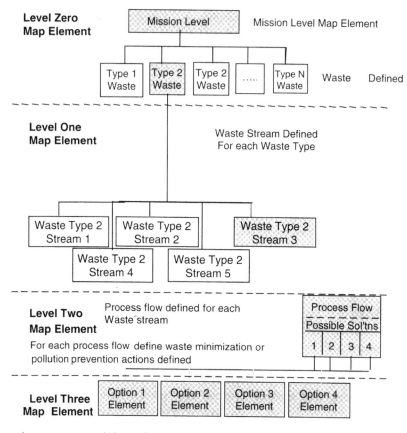

FIGURE 1   Roadmap hierarchy example.

3.1   Example: Los Alamos Environmental Stewardship Roadmap

To illustrate the techniques used in construction of the roadmap elements we will follow the specific path for sanitary waste through the Los Alamos Environmental Stewardship roadmap. The techniques can and should be generalized to other applications. We will start with a conceptual mission-level map for Los Alamos National Laboratory.

3.1.1   Level Zero or Mission-Level Map Element

Construct mission-level map element.
Define waste types.

This mission level map is constructed to represent the laboratory as a system with a series of material and energy flows, both into and out of the system. The first step in constructing the mission-level map is to decide the scope of the initial system. In this case, the system is the entire laboratory site. We can also choose to examine a smaller subset if we wish to focus on a particular area. Figure 2 shows the laboratory process map, which is a view of the laboratory from the local environmental perspective. The perspective can be important. If we had chosen a regional perspective, the resulting roadmap would have been quite different.

The map element is constructed by identifying inflows of materials and energy to the system, identifying the operations that use the materials and energy, and identifying system outflows, including all the products of the operations including wastes and pollutants. The wastes are accrued into a number of broad waste types. This is a critical step since it will form, in many cases, the foundation for all subsequent analysis. The waste types must be comprehensive and include all wastes generated from operations.

The laboratory performs work for government sponsors and private industry. In performing this work, the laboratory procures services, materials, equipment, new facilities, and commodities (electricity and natural gas). The laboratory also takes in water from the regional aquifer and air from the surrounding atmosphere. This series of inflows is shown at the left in Figure 2. Once in the laboratory, the inflows are used in the six different kinds of operations listed in Figure 2.

Most person-hours are spent conducting office operations. These involve office space, furniture, information processing equipment, paper, and office

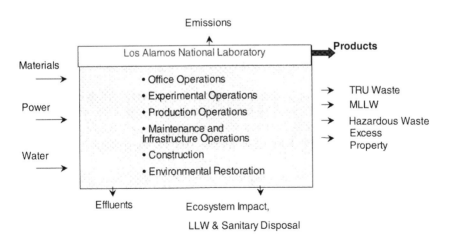

FIGURE 2 Laboratory process map.

supplies. Energy is expended to operate equipment and provide climate control. Water is used in evaporative cooling to transfer waste heat to the atmosphere.

Experimental operation includes bench-scale and large-scale research. Energy is expended to operate the equipment and provide climate control. Water is used in evaporative cooling to remove waste heat. Experimental operations typically procure large amounts of equipment but small amounts of chemicals and other materials.

Production operations include all the site production operations. Production operations consume material, water, and energy, but in this particular case energy and water usage is modest.

Maintenance and infrastructure operations include all maintenance activities, facility management activities, and site-wide infrastructure systems, such as the sanitary wastewater plant, on-site power plant, water influent system, and highway system. These operations consume large quantities of chemicals and produce most of the site's hazardous waste. They also consume significant amounts of energy and water.

Construction includes both smaller construction projects and major construction projects. Construction operations are important not only as a source of immediate environmental impact during construction activities; design decisions made during the construction process can lock in environmental impact for the lifetime of the facility.

Environmental remediation includes all remediation activities on the site. For purposes of this roadmap, only newly generated wastes and pollutants were considered, but that need not have been the case.

Because the products of the laboratory are mostly information, most material inflows become by-product or waste outflows. Identified outflows of waste and pollutants are divided into the eight categories shown in Figure 2. These include transuranic waste (TRU), mixed low-level waste (MLLW), low-level waste (LLW), hazardous waste, solid sanitary waste, excess property entering the salvage system for reuse or recycle, gaseous emissions, and liquid outfalls.

Another result of operation also occurs. The presence of laboratory facilities, infrastructure, operations, and land management affects local ecosystems. Much of this is unavoidable, and much of it is not necessarily harmful to the local environment. This local ecosystem impact can be minimized through wise operational choices.

Once the operations and outflows have been identified, a fundamental choice must be made. The subsequent lower-level maps can be organized and broken down according to either operation or waste type, depending on the specific goals of the mapping activity. Roadmaps based on operations are particularly good if one wishes to focus on organizational structure and its impact on pollution and waste generation and may include issues such as structure, funding, and customer base. Roadmaps based on waste type are generally more useful for devising pollution prevention and waste minimization strategies and choosing among technological alternatives. As an example, we will construct a map based on waste type and follow a specific waste type—sanitary waste—through the succeeding map elements. Janet Watson constructed the complete ESO sanitary roadmap, from which this was abstracted.

3.1.2 Level One or Waste Stream Definition Level

Construct waste type process flow diagram.
Define waste streams.
Define issues and constraints.
Prioritize waste streams.

Since we have chosen to follow the development of roadmap elements through the sanitary waste type, we first construct a sanitary waste process flow diagram. This diagram is constructed using the same principles that are used in all process flow diagrams: the inflows of materials and energy are identified, the process or operations are identified, and the outflows of waste material are identified. At this level, quantification becomes important, since it will be used to prioritize waste streams for waste reduction activities. The first-level map element that emerges from examination of the data is shown in Figure 3.

Nonhazardous, nonradioactive materials enter the laboratory as procured items, mail, food, and various other substances such as glass, brush, and construction materials. These items are used by the laboratory and are either recycled, reused, or salvaged, or are disposed in the county landfill. Materials disposed include such items as construction waste, food and food-contaminated wastes, paper products, glass, Styrofoam, and various other substances.

Material outflow pathways are shown at the right of the diagram. The composition of materials in those pathways is broken down in the pie charts at the bottom. It is important to carefully and completely identify the constituents of the waste and to quantify the volumes, since that information will form the foundation for the succeeding process flow diagrams.

In this case, we will examine the dumpster waste in greater detail. There are a number of waste streams to be considered. As a normal part of the roadmapping process, it is important to prioritize these waste streams for action based on some criteria. The basic question here is: Which waste streams should we attempt to minimize first? Other questions that need to be addressed are the cost of minimization and the return on investment for minimization activities.

Before choosing criteria, it is necessary to examine all issues and constraints associated with the waste streams. Issues might involve such considerations as lifetime of the landfill. If the landfill has a short lifetime and there are no easy alternatives to disposal at the current landfill, an overriding consideration

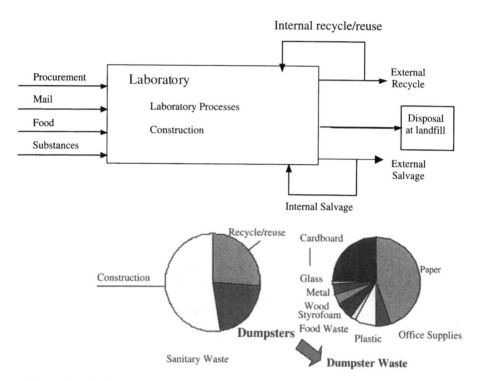

FIGURE 3 Sanitary waste streams.

might be to minimize the volume of material sent to the landfill, even if the cost savings for that action are small compared to other possible actions. Constraints might involve such things as regulatory requirements or operating policy. Actions designed to meet regulatory requirements will probably be placed high on the priority list even if the costs are high and the return on investment is low. In many pollution prevention and waste minimization programs, actions are prioritized based on Pareto analysis. The underlying assumption of Pareto analysis, generally surprisingly good, is that 80% of the waste comes from 20% of the operations and that 20% is where you should concentrate your efforts at prevention. If the operations are fully compliant with regulation and there are no overriding local issues, Pareto analysis is a very good way to prioritize activities. In a classic Pareto chart, the volumes of waste are plotted for each waste stream in a bar chart. The streams that contain 80% of the waste are then identified. If there are no overriding constraints, these are the streams that are selected for intervention, usually either in the order of total waste quantity or total waste cost. Robert Pojasek, in "Prioritizing P2 Alternatives" (*Pollution Prevention Review*, vol. 7, no. 1, pp. 105–112, 1997), discusses Pareto analysis and its major variants in detail.

At this point, it is usual to construct a table listing the waste streams, issues, constraints, the cost of treatment, handling, and disposal, the cost of regulatory support, and quantities for each stream. If the costs can be obtained on a unit quantity basis, calculating the cost associated with each stream is straight-forward. These factors all influence the final prioritization of the waste stream. As stated previously, issues and constraints may outweigh the more obvious cost arguments.

For this exercise we will not prioritize the sanitary waste streams but will choose to examine a subset of the paper waste stream—mail—in detail.

3.1.3 Level Two or Waste Stream Process Definition

Construct waste stream process flow diagram.
Identify intervention points.
Identify prevention options.

The process flow diagram for the "mail" waste stream is constructed like the previous process flow diagrams. However, at this level the operations are dia-gramed explicitly and in some detail rather than being simply listed. This is necessary because at this level we are trying to identify sources of waste and potential intervention points. Often, physical inspection of the operation is required to develop the necessary detail. For manufacturing or processing opera-tions, these diagrams can become very complex. In this diagram, we have diagramed all the major pathways for mail between receiving and final disposi-tion. Each box or node identifies a process or handling step.

Every year the laboratory receives and distributes 714 MT of mail. This mail includes junk mail, catalogs, telephone directories, and various documents, as well as business mail. The mail received by the laboratory includes a small amount of classified mail. The process flow diagram for the mail waste stream is shown in Figure 4.

Mail, including internally generated mail, is received by the laboratory and distributed. Any unwanted mail can be sent by the recipient to Mail Stop A1000 for sorting and recycle. Documents such as catalogs and directories that are glue-bound must first have the bindings sheared off before the paper is recycled. The bindings are sent to the landfill for disposal. Mail is also disposed by discarding in green desk-side containers or trash bins. The contents of the green containers are sent to recycle, while the contents of the trash bins are sorted for recyclable materials at the Material Recovery Facility (MRF). Classified material may not be disposed unless it has been security (crosscut) shredded. The strip-shredded material can be recycled, but crosscut shredded material currently goes to the landfill.

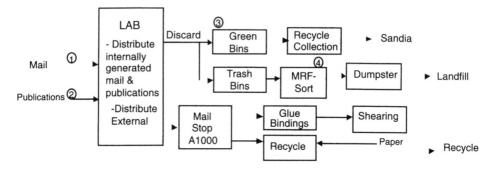

FIGURE 4 Mail and document distribution and disposal.

With the advent of MRF operations, the opportunity to recover nearly all the discarded recyclable mail is realized. The emphasis will then be on reducing the source of unwanted mail.

At this point, it is usual to identify possible intervention points. Examination of the process flow diagram shows limited opportunities to reduce the quantity of mail going to the landfill. The strategy has to be either to prevent mail from entering the laboratory or to increase the fraction of discarded mail that is recycled. Eliminating all incoming mail is, of course, not an option, but there may be ways to reduce particular incoming mail streams. Most notably, it is possible to reduce the volume of "junk mail" and certain print documents such as paper telephone books. This upstream approach has the advantage of preventing waste so that it never has to be handled. At the disposal end of the diagram, the only destinations for waste mail are recycle and the landfill, so if the quantity of discarded mail going to the landfill is to be reduced, the recycle fraction must be increased. Planning meetings with the personnel involved in recycle revealed two promising avenues to pursue.

With the intervention points identified, a set of initiatives are formulated, with the numbers keyed to process flow diagram.

> *Initiative 1: Reduction of "junk mail."* A substantial fraction of the mail consists of recurring, unwanted "junk" mail. A centralized stop-mail service for "junk mail" is currently in the pilot phase. Any laboratory employee who wishes to request removal of his or her name from a mailing list can use this service.
>
> *Initiative 2: Eliminate paper phonebooks.* Paper phonebooks are widely used and are difficult to recycle. US West directories, which are routinely distributed to all employees, will be eliminated as a source of waste by restricting delivery and asking employees to use the "on-line" directory

instead. Approximately 22 MT of waste per year can be avoided in this way.

Initiative 3: Additional items in paper recycle system. The current paper recycle program is limited to white and pastel paper; options for including other types of paper products (mail items) in this mix are being evaluated.

Initiative 4: Increase use of MS A1000. Although MS A1000 is widely used as a means of recycling various materials, many employees are still unaware of its existence. A publicity campaign will be developed to increase awareness; self-inking stamps (with the A1000 logo) will also be distributed to each mail stop within the laboratory to encourage use of this program.

These initiatives must be analyzed, of course, according to a number of relevant variables such as potential cost and effect on the waste stream. The variables that describe each of the initiatives are called factors and are sometimes tabulated in a spreadsheet. A very simple version of such a spreadsheet is shown in Table 1. In most cases, a cost is assigned to each of the initiatives. In this case, costs were not readily available or were being developed through pilot programs at the time the roadmap was constructed.

3.2   Differentiating Among Technology Solutions at Level Two

In the above example from the ESO roadmap, all the initiatives proposed were programs that did not require new technologies. Frequently, this is not the case and, in fact, some sets of initiatives feature competing technologies. Conceptually, the process flow diagram for such a project might look like Figure 5. This chart is representative of a many manufacturing processes, particularly those involving distillation or refining. Following an initial process step, the product stream is separated into two streams for further processing. Both of the streams have subsequent processing steps. Assume that at point 1 there is an opportunity to intervene and reduce the path B waste stream. Often in manufacturing processes, this intervention will involve deployment of a new technology, and there may be more than one candidate technology. In that case, the evaluation of the potential initiatives will be very different from the previous case involving mail. Now we are not dealing with administrative controls or modified processes alone, but must consider the relative merits of the competing technologies.

Some of the factors that must be considered might include the relative maturity of the technologies, which is usually understood to mean the scale of prior operation—i.e., bench, pilot, or full scale—schedule for deployment, cost to develop, cost to deploy, cost to operate, efficiency of the process or throughput, quantity of waste avoided, nature of any secondary waste streams from a

TABLE 1 Waste Minimization Initiatives

Initiative or project	Action/milestone	Status	Funding source	Waste avoided
Reduce "junk mail." Develop a centralized stop-mail service for "junk mail."	Evaluate pilot results; determine if results justify expense.	Ongoing Pilot test: completed	Base program	Source reduction, 4.4 MT/year
Eliminate paper phonebooks. Delivery of US West Telephone directories is restricted; employees are requested to use the "on-line" directory instead. Approximately 22 MT of waste per year can be avoided in this way.	Continue restricted delivery in future years.	Ongoing	Base program	Source reduction, 22 MT/year
Include additional items in paper recycle system. Include other paper products (mail items) in the program.	This option is being evaluated.	Not funded		Increased recycle
Increase use of MS A1000. Although MS A1000 is widely used as a means of recycling various materials, many employees are still unaware of its existence. This program within the laboratory will encourage use of this program.	A publicity campaign will be developed to increase awareness. Self-inking stamps (with the A1000 logo) will also be distributed to each mail stop.	Ongoing	Base program	Increased recycle

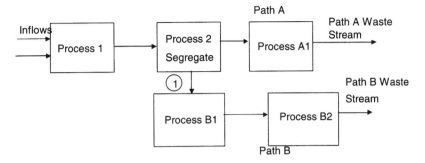

FIGURE 5 Conceptual technology process map.

particular process, and safety of the processes. A matrix similar to the one shown previously can be constructed and weights assigned to each of the factors. Issues frequently arise when trying to determine weights for the particular factors, and it is best to agree early about any constraints that must be applied. Typical constraints include the stipulation that the chosen process must be at least as safe and efficient as the process it will replace. Once weights are assigned to the various factors, the roadmapping team must meet with the technology advocates and the operations personnel to quantify the factors.

Since each technology is likely to have advocates and detractors, it is important to gather information on each technology from all concerned parties, including operators. Even then, it may be impossible to reach a consensus view with respect to all the relevant factors. For this reason, it is important to decide in advance how conflicts will be resolved. Normally, the roadmapping team resolves conflicts after gathering information from the technology advocates. After the factors have been quantified, one of several algorithms can be used to evaluate each of the competitive technologies. In this way technologies can be differentiated with regard to deployment in a particular process step and a basis for an action decision is established.

4 USING THE COMPLETED ROADMAP

To review briefly:

At level zero, the overall system operation was mapped and waste types were identified. Frequently this step is left out if waste types are well known or if only one waste type is of interest.

At level one, each of the waste types was broken down into waste streams. The size and nature of the waste streams was quantified and the waste streams were prioritized for minimization or prevention action.

At level two, detailed process maps for the waste streams were prepared, points of intervention were defined, and initiatives for minimization or prevention at these points were identified. Data were prepared for each of the initiatives to form a foundation for decision making.

At this point, a number of paths forward are possible. The zero level and level one maps are useful for many purposes, including education, training, and monitoring. The level two maps are normally used to enhance decision making and monitoring progress.

Part of the decision-making process involves developing an investment strategy. An investment strategy involves four items:

1. A decision about priorities and which waste streams should be addressed first with respect to minimization or prevention
2. A decision about which initiatives should be pursued first for the high-priority waste streams
3. An allocation of resources against the selected initiatives
4. Development of a fallback or contingency position for the initiatives, particularly those that require development and/or deployment of new technologies

Finally, a schedule for implementing the initiatives is developed and overlaid on the process map.

The schedule is normally prepared by redrawing the process map to represent the end state that will result from the implementation of selected initiatives. The redrawn map element includes an earliest start/latest finish date in the appropriate process nodes. A project control chart is frequently included as part of the revised process flow chart. The project control chart can include many or few schedule and control parameters such as start date, finish date, cost, and any other desired parameters. The redrawn process flow chart shown in Figure 5 would then look like Figure 6.

Clearly, if there are several initiatives in the same waste stream, the roadmap element can become complicated. In that case, it is usually easier to redraw a revised map element for each initiative so that the complete data on each initiative in a particular waste stream are located on its own map element. The redrawn map elements can be retained in one location for ease of review.

In addition, some roadmap developers include risk as part of the revised map element. The risk may be technical risk, programmatic risk, cost risk, or funding risk. The risk is usually specified as the risk of not being able to move successfully from one process node to the next. The risk is then associated with the link between nodes and aggregated along all pathways in the revised map element. In this way, risk to the project can be assessed, the sources of greatest risk can be identified, and contingency plans can be developed for those areas.

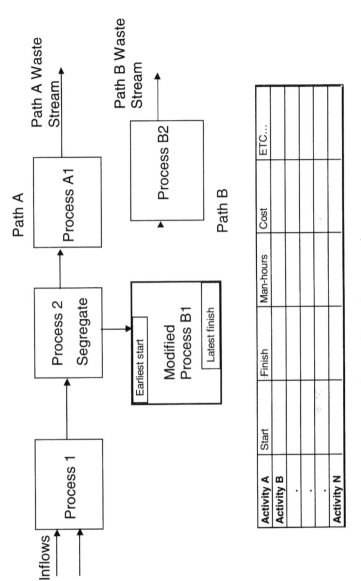

FIGURE 6 Redrawn process map element for project control.

Estimation of risk is necessarily subjective and cannot be taken too literally. The risk estimates serve simply as a guide to controlling risk.

Planning is a dynamic activity. Since pollution prevention operations change, hopefully in response to good planning, it is necessary to update the roadmaps periodically. The usual period for updates is yearly, but this can be adjusted to reflect the actual rate of changes in the system.

5 CONCLUSIONS

Roadmaps are useful tools for systematically evaluating the generation of waste and pollution in virtually any type of operation, large or small. For large systems like Los Alamos National Laboratory, the roadmap can be extensive. The Los Alamos ESO roadmap can be found online at http://emso.lanl.gov/publications.

Roadmaps provide a mechanism for evaluating the current state in detail, for deciding how to move toward a desired end state, for assessing the effectiveness of alternative options in moving toward the end state, for making investment decisions, and for controlling risk. More detailed information on the various aspects of roadmapping, as applied by a variety of institutions and industries, can be found in the bibliography that follows.

SELECTED BIBLIOGRAPHY

The following bibliography presents further information on roadmap construction and use and contains examples of different types of roadmaps. The Kostoff citation contains an exhaustive bibliography.

Aerospace Industries Association of America, Detailed Technology Roadmap for Superconductivity. Washington, DC: AIAA, Superconductivity Committee, 1992.

D. Barker, and D. Smith, Technology Foresight Using Roadmaps. *Long Range Planning*, vol. 28, no. 2, pp. 21–29, 1995.

Electronic Industry Environmental Roadmap, available from MCC Corporation, 3500 West Balcones Center Drive, Austin, TX 78759, 1998.

M. P. Espenschied, Graphical Status Monitoring System for Project Managers. Pretoria, South Africa: National Institute for Aeronautics and Systems Technology, Funder: National Aeronautics and Space Administration, Washington, DC, Report CSIRNIAST817, 1981.

J. H. Gurtcheff, US Strategic Nuclear Strategy and Forces: A Roadmap for the Year 2000. Study Project. Carlisle Barracks, PA: Army War College, 1991.

R. N. Kostoff, Science and Technology Roadmaps, http://www.dtic.mil/dtic/kostoff/Mapweb2I.html.

ORNL, *Oak Ridge National Laboratory Technology Logic Diagram. Volume 1, Technology Evaluation: Part A, Decontamination and Decommissioning.* Oak Ridge K-25 Site, TN, Report ORNLM2751V1PTA, 1993.

R. B. Pojasek, P2 Programs, Plans and Projects: Some Thoughts on Making Them Work. *Pollution Prevention Review*, vol. 9, no. 2, 1999.

U.S. Department of Energy, National TRU Waste Management Plan, DOE/NTP-96-1204, Revision 1, 1997.

REFERENCES

1. R. N. Kostoff, Science and Technology Roadmaps, http://www.dtic.mil/dtic/kostoff/Mapweb2I.html.
2. R. B. Pojasek, P2 programs, Plans and Projects: Some Thoughts on Making Them Work. *Pollution Prevention Review,* vol. 9, no. 2, 1999.

14

Pollution Prevention and DFE

Terrence J. McManus
Intel Corporation, Chandler, Arizona

1 BASIC PRINCIPLES OF POLLUTION PREVENTION AND WASTE MINIMIZATION

Beginning in the mid-1970s, environmental management of industrial air emissions and wastewater discharges focused on end-of-the-pipe or end-of-the-stack treatment technologies. Both the Clean Air Act of 1970 and the Federal Water Pollution Control Act of 1972 (now called the "Clean Water Act"), as well as the parallel regulatory structures set up at state and local levels, required new treatment technologies to be developed to manage air emissions and wastewater discharges. But none of these early statutes and regulations mandated that corporations minimize the amount of waste generated or prevent pollution during manufacturing.

With the passage of the Resource Conservation and Recovery Act (RCRA) in 1976, the government for the first time defined "hazardous waste" and began to focus on waste minimization, rather than just waste treatment. Large-quantity generators [producing more than 1000 kg (2200 lb) per month of hazardous waste] were required to ship waste to an approved treatment, storage, and disposal facility (TSDF), using a formal document known as a waste manifest. Because the new regulations were very strict, however, many off-site TSDFs had to close down, resulting in a sharp decrease in the supply of such facilities.

To reduce demand for the facilities, beyond the sharp rise in costs for TSDF services, Section 3000(b) of the RCRA requires that large-quantity generators who transport waste off-site must certify on the manifest that they have established a "program in place" to reduce the volume or quantity and toxicity of hazardous waste generated—to the extent economically practicable. For owners/operators who manage hazardous waste on-site in a permitted TSDF, Section 3005(h) similarly requires annual certification that a waste minimization program be in place and maintained in the facility's operating records. These two requirements put the burden of proof on generators or an owner/operators of TSDFs to show that they are implementing waste minimization strategies.

Small-quantity generators, who produce between 100 and 1000 kg per month of hazardous waste, are required to certify on their hazardous waste manifests that they have also "made a good faith effort to minimize" their waste generators (51 FR 35190; October 1, 1986).

Together, the large- and small-quantity generator requirements for waste minimization affect more than 95% of the hazardous waste generated in the United States. The primary mechanism for achieving such minimization is to identify the various hazardous waste streams and determine if it is possible to reduce the volume and/or toxicity of each (1).

The U.S. Environmental Protection Agency (EPA) also collects data, annually, on the emissions and disposal of a specific list of chemical compounds. Manufacturers who exceed certain thresholds have to inform the EPA as to whether the chemicals were released into the environment (air, water, or land) or transferred to another facility for management. The EPA, in turn, maintains a database known as the Toxics Release Inventory (TRI), which is one of the best data sources to review emissions performance on an industry sector basis. The first year for data reporting to EPA's TRI inventory was 1987. The database tends to be about two years behind in its reporting, however, as the reports are not due until July, and loading and analyzing the data takes about a year.

2   ROLE OF POLLUTION PREVENTION AND DESIGN FOR THE ENVIRONMENT

As methodologies for waste minimization improved in the 1980s, industries looked to more comprehensive approaches, such as pollution prevention (P2) and design for the environment (DFE). In 1990, the U.S. Congress passed the Pollution Prevention Act, which specifically required the evaluation of new opportunities and approaches to eliminate the generation of emissions and waste.

Under Section 6602(b) of the Pollution Prevention Act of 1990, Congress established a policy that:

Pollution should be prevented or reduced at the source wherever feasible.
Pollution that cannot be prevented should be recycled in an environmentally safe manner, wherever feasible.
Disposal and/or release into the environment should be employed only as a last resort and should be conducted in an environmentally safe manner.

The EPA established an operating definition for P2 as part of the agency's 1991 Pollution Prevention Strategy. That definition makes clear that prevention is the first priority within an environmental management hierarchy, which includes:

1. Prevention
2. Recycling
3. Treatment
4. Disposal or release

The EPA also recognized that any P2 strategy needs to be flexible. Any P2 option today, in fact, depends on three factors: legal requirements, levels of risk or toxicity reduction that can be achieved, and cost.

As with waste minimization, P2 typically focuses on existing manufacturing processes, by applying the prevention hierarchy to the various waste streams. When new manufacturing processes are developed, some corporations apply new environmental management techniques to reduce/eliminate waste generation as part of their manufacturing process design. This approach has been commonly called design for the environment (DFE).

Different people have defined DFE in different ways. For instance, the EPA defines a DFE program as "a voluntary partnership-based program that works directly with companies to integrate health and environmental consideration in business decisions (2). Intel Corporation has defined it as "a methodology to develop environmentally compatible products and processes, while maintaining desirable product price/performance and quality characteristics."

3 ENVIRONMENTAL FRAMEWORK

How do all these environmental components or programs work together to form a unified environmental management system (EMS)? Figure 1 presents a conceptual model of the environmental framework. This framework also demonstrates the evolution of environmental management over time, with waste treatment beginning at the center, as the earliest management technique, and current and future management approaches extending from there.

Indeed, waste treatment is the fundamental environmental management technology applied over many decades. The progression to each succeeding

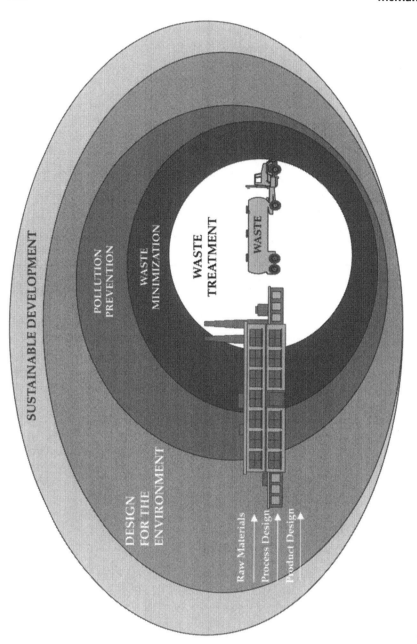

FIGURE 1 Environmental management evolution.

environmental management approach highlights the fact that both the number of choices and their scope and complexity are continually increasing.

4 SUSTAINABLE DEVELOPMENT

At the current outer edge of the framework is sustainable development. One way to achieve sustainable development is for many companies and their local communities to each adopt DFE strategies. In other words, sustainable development is the integration of many DFE programs, from many different entities, over a large geographic area or region. Such a unified approach is necessary because a single entity cannot provide all the components necessary to prevent pollution and recycle all recyclable materials. For example, a semiconductor manufacturing facility can establish programs to collect and recycle aluminum cans, paper materials, and chemicals. However, the company must rely on the aluminum industry, the paper recycle industry, and the chemical producers also to implement recycling in order to use the necessary recycle technology and/or effective and efficient economies of scale. When a region can develop such an integrated approach to the environment, the details can be presented in a "green plan."

In June 1993, President Clinton formed the President's Council on Sustainable Development (PCSD) to develop and recommend a national strategy for implementing sustainable development. This council consisted of leaders from industry, government, nonprofit organizations, and Native American groups. In 1996, the PCSD published the report, *Sustainable America*, which contained the following vision statement:

> Our vision is of a life sustaining earth. We are committed to the achievement of a dignified, peaceful and equitable existence. A sustainable United States will have a growing economy that provides equitable opportunities for satisfying livelihoods and a safe, healthy, high quality of life for current and future generations. Our nation will protect its environment, its natural resource base and the functions and viability of natural systems upon which all life depends (3, p. IV).

In support of this vision, the Council also recorded 16 beliefs that set the basis for implementing the strategy. The following four beliefs (3, pp. v–vi) refer specifically to industrial development:

> To achieve our vision of sustainable development, some things must grow—jobs, productivity, wages, capital and savings, profits, information, knowledge, and education—and others—pollution, waste, and poverty must not.
> The United States made great progress in protecting the environment in the last 25 years and must continue to make progress in the next 25 years.

We can achieve that goal because market incentives and the power of the consumers can lead to significant improvements in environmental performance at less cost.

Environmental progress will depend on individual, institutional, and corporate responsibility, commitment and stewardship.

Steady advances in science and technology are essential to help improve economic efficiency, protect and restore natural systems, and modify consumption patterns.

In May 1999, the PCSD published a second progress report, entitled *Towards a Sustainable America*. As part of this effort, the council focused specifically on the following issues (4, p. 3).

1. Policies to reduce greenhouse gas emissions
2. The next steps in building the new environmental management system for the twenty-first century
3. Policies and approaches to build partnerships to strengthen communities
4. Policies to foster U.S. leadership in international sustainable development policy, particularly in international capital flow

In the area of environmental management, it was very clear to the PCSD that most recent environmental reforms have not really focused on the objective of promoting sustainable development. This is partly because current definitions of environmental pollution, management, and protection are too narrowly scoped, with significant emphasis on point source emissions. Therefore, the solutions tend to be focused on single pollutants within a single media from a single source. Very little effort has been focused on aggregating and understanding environmental risks and impacts across a broader ecosystem basis. A similar conclusion was reached by the EPA Science Advisory Board's (SAB) Integrated Risk Project. In essence, the SAB believes that environmental management efforts to date have typically worked on targeted pollutants from single sources, and resulted in improvements in environmental performance over very localized areas. Specifically, SAB stated that the effort must be more holistic:

Concern for the environment has become an important part of the American value system. We care about the environment as it relates to human health, the viability of ecosystems, and our children's future. We care about the quality of life, today and in the future, and in the interconnected environmental conditions that play such an important role in determining life's quality (5).

Ultimately, the vision and environmental goals must be to protect the overall health of all people and the long-term viability of whole ecosystems. That means that risk reduction must be designed to control more than one pollutant at a time, to protect more than one human or ecological receptor, and thus to realize broader benefits of environmental improvement, at a lower cost.

The model for such integrated environmental decision making is presented in Figure 2. The three major components for the decision-making framework are

Problem formulation
Analysis and decisionmaking
Implementation and performance evaluation

Each of these steps requires constant feedback of information to both improve and optimize the specific measures implemented, yet some of the advantages of using an integrated decision-making model include:

An increase in the probability of focusing on the highest risks/most important issues
A methodology that includes both human health and ecological risks

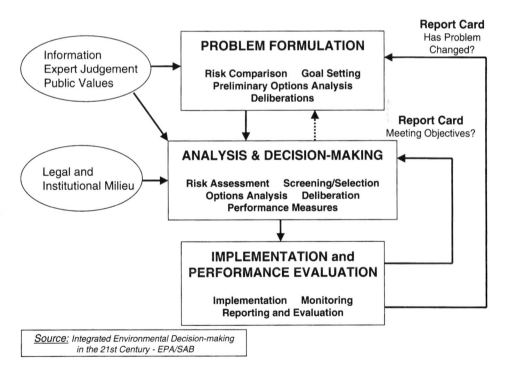

FIGURE 2   Integrated environmental decision-making framework.

Improvements in public accountability (because the framework is applied typically over a larger geographic area)

The inclusion of specific performance measurements

One of the biggest stumbling blocks to integrated environmental decision making is the fact that this approach is a marked departure from the current methodology for environmental management. That means it takes more time, in the beginning, to communicate with and educate the public so that they are willing to participate.

5 COMMON FOCUS AREAS AND DRIVERS FOR P2 AND DFE

Most industrial operations are committed to improvements in environmental performance and natural resource protection. The broad categories that industries focus on and try to manage are

Solid and hazardous waste generation
Chemical use
Air emissions
Water use
Wastewater discharge quantity and quality
Electrical use

Each category includes specific manufacturing operations that use natural resources or produce emissions and that therefore need the corporation's attention. Some of the reasons that corporations manage these processes in environmentally efficient ways are that they:

Reduce operating costs
Ensure compliance with environmental permits
Satisfy certification requirements of RCRA on waste manifests
Achieve a specific threshold, which eliminates reporting requirements or allows the corporation to achieve minor source status under an air emissions permit
Satisfy corporate commitments established in an environmental policy or a specific commitment to a local community
Ensure that emissions reporting under TRI demonstrates reduction from previous year's emission reporting data
Help the corporation develop a reputation as a "good corporate citizen"

The following example for waste solvent generation illustrates this approach for an individual manufacturing facility. The basic methodology for implementing either pollution prevention or design for the environment includes:

1. Identify which areas/manufacturing operations generate emissions and/ or consume natural resources (waste solvent generation is selected for this example).
2. Within the specific operation under evaluation, identify the various sources of waste solvent that are generated and identify the quantities per unit of time, quantities per unit of production, and characteristics of the waste solvent source(s).
3. Develop a list of options to recycle, reduce the volume, eliminate, or reduce the toxicity of the waste solvent that is generated.
4. Evaluate each potential option, using the following criteria as a minimum: Capital cost to implement
    Operating costs per year
    Quantities of waste solvent reduced or eliminated
    Reduction in toxicity
    Environmental impacts of the remaining materials/waste
    Health and safety advantages and disadvantages for implementing the proposed option
5. Select the best option for each solvent waste stream and implement.
6. Track performance of the implemented option and compare the data with original projections.
7. Reassess the option implemented and determine whether further improvements can be implemented, using the same basic methodology again.

6 INDUSTRY SECTOR APPROACH TO P2 AND DFE

Inevitably, each industrial manufacturing operation must determine how it will specifically implement its environmental management system and determine the proper leverage point for applying pollution prevention or design for the environment strategies. Within a specific industrial manufacturing sector, however, similarities in the manufacturing processes and chemistries frequently result in similarities in types of waste or emissions generated. Evaluations of P2 and/or DFE opportunities could be realized more efficiently by establishing specific design criteria or environmental performance goals at a sector level, rather than on a company or individual manufacturing facility basis. Both the semiconductor and metal finishing industries have embarked upon sector-wide approaches.

6.1 Semiconductor Manufacturing

More than three decades ago, Gordon Moore of Intel Corporation predicted that the number of transistors in a defined area of silicon would double every 18 months. This prediction, known as Moore's law, is presented in Figure 3. To keep

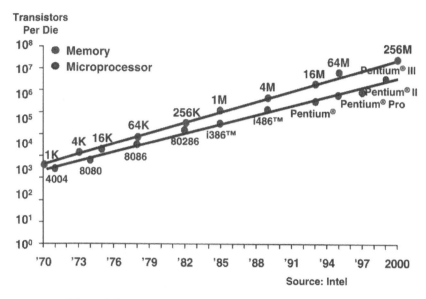

FIGURE 3 Moore's law.

pace with the increasing demand for higher performance and more electronic features, the semiconductor manufacturing process must continue to shrink the feature size on a chip. As the chip shrinks, the distance between transistors decreases, which in turn increases the speed of the device. Figure 4 presents a graphical picture of the rate of change of the semiconductor manufacturing process. Essentially, a new manufacturing process comes on line every two years. Currently, most of the manufacturing technology in the semiconductor industry is operating at a feature size of 0.25 µm and moving to 0.18 µm and on to 0.13 µm in 2002.

The rapid rate of change in manufacturing provides many opportunities for applying DFE strategies. Each new manufacturing tool set that is developed can be reviewed, and specific sources of air/water emissions or waste generation can be targeted for continuous improvement in the next generation.

To assist in DFE approaches, semiconductor companies leverage the trade association known as the Semiconductor Equipment and Materials International (SEMI) to establish environmental, health, and safety (EHS) design guidelines for new manufacturing equipment. The two primary guidelines are SEMI S2, "Safety Guidelines for Semiconductor Manufacturing Equipment," and SEMI S8, "Safety Guidelines for Ergonomic Engineering of Semiconductor Manufacturing Equipment." These two documents provide the basis for establishing design for EHS. Table 1 summarizes the environmental requirements contained in the SEMI S2 document.

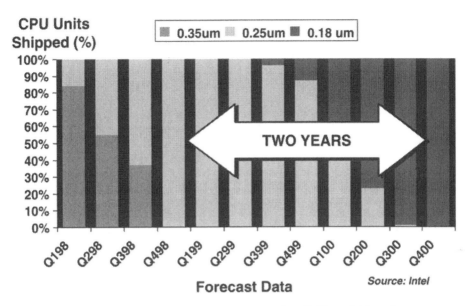

FIGURE 4 Semiconductor technology conversion (0.35 to 0.18 μm).

Since these are industry-wide EHS guidelines for developing new manufacturing tools, individual semiconductor manufacturers do not have to develop their own unique set of requirements. Nor do they suffer from the imposition of environmental design guidelines. It should be noted that this approach does not prevent individual semiconductor manufacturers from imposing additional standards that are stricter than those set forth in SEMI S2 or S8, on specific tools.

Implementation of the SEMI S2 guidelines by equipment manufacturers increased in acceptance globally in the early to mid-1990s. Through the implementation of the S2 guidelines and other approaches taken by the semiconductor industry, significant success in reducing emissions has been achieved. An example of this improvement is demonstrated in Table 2 and Figure 5, which presents the release of hazardous air pollutants (HAPs) by the U.S. semiconductor industry.

Using the TRI database, HAPs emissions data were generated for the semiconductor industry. For the period 1987–1990 (four years), the average annual HAPs emissions by the U.S. semiconductor industry was 4.09 million pounds. For the period 1994–1997, average annual HAPs emissions were 0.83 million pounds, which represents a fivefold decrease in HAPs emissions.

Figure 5 presents the HAPs emission data in pounds per million square inches of silicon wafers produced by U.S. semiconductor manufacturers. Table 2

TABLE 1 Summary of Semiconductor Manufacturing Equipment
International (SEMI) Environmental Design Considerations

The Environmental Design Guidelines apply across the full life of the equip-
ment, including decommissioning and disposal. The equipment manufac-
turer should consider resource conservation, including:
Water reuse/recycling
Reduce consumption of chemicals, energy, and water
Reduce resource requirements for equipment maintenance and reduc-
    tion in packaging requirements
The use of chemicals for processing, maintenance, and utilities must con-
sider chemical use effectiveness, environmental impacts, toxicity, waste
generation, and decommissioning.
Where practicable, the following chemicals should not be part of use or
operation:
Ozone-depleting substances as specified by the Montreal Protocol
Perfluorocarbons, including CF_4, C_2F_6, NF_3, C_3F_6, SF_6, and CHF_3
The equipment design must also reduce potential for unintended releases,
and include such features as overfill detectors and alarms, secondary
containment, chemical compatibility, and automatic shutoff of chemical
feed systems.
Segregation of effluents, wastes, and emissions needs to be provided to
prevent issues with chemical incompatibility, facilitation of recycle and
reuse, and to facilitate effective treatment technologies.
Equipment must also be designed to facilitate decommissioning.

provides the absolute quantity of HAPs emissions; Figure 5 reflects the fact that
production quantities of semiconductor wafers have increased each year since
1987. Both the absolute quantity of HAPs emissions and the quantity per unit of
production have continued to decrease.

Table 3 presents the total quantity of TRI chemical releases or transfers for
both the U.S. semiconductor industry and all U.S. industry. The data demonstrate
that the semiconductor industry produces less than 1% of the total TRI and that,
based on 1997 data, that sector has reduced TRI releases and transfers by a factor
of 3.5, compared to the first reporting year of 1987. During that same period, U.S.
industry as a whole reduced TRI releases by a factor of 2.7.

In summary, with the rapid change of manufacturing technology that
occurs within semiconductor manufacturing, the opportunities for engaging in
design for the environment are substantial. Use of SEMI S2 environmental
guidelines provides clear design guidelines for equipment suppliers on the types
of environmental improvements desired by the semiconductor industry. The key
to implementation is early engagement in the manufacturing development pro-

TABLE 2 Hazardous Air Pollutants
(HAPs) Emissions by U.S. Semiconductor
Industry (1987–1997)

Year	HAPs emissions (million pounds)
1987	3.85
1988	4.76
1989	4.30
1990	3.43
1991	2.82
1992	2.21
1993	1.39
1994	1.12
1995	0.88
1996	0.77
1997	0.55

Source: EPA Toxins Release Inventory (TRI) Database (1987–1997).

cess. Figure 6 presents a model for early engagement to achieve DFE within semiconductor manufacturing.

6.2 Metal Finishing

The metal finishing industry is characterized typically as a small manufacturing operation with about 10,000 job-shop or captive facilities across the United States. Under the Clean Water Act, the EPA defines metal finishing as follows:

Plants which perform any of the following six metal finishing operations on any basis material: Electroplating, Electroless Plating, Anodizing, Coating (chromating, phosphating and coloring), Chemical Etching and Milling and Printed Circuit Board Manufacture. If any of those six operations are present, then this part applies to discharges from those operations and also to discharges from any of the following 40 process operations: Cleaning, Machining, Grinding, Polishing, Tumbling, Burnishing, Impact Deformation, Pressure Deformation, Shearing, Heat Treating, Thermal Cutting, Welding, Brazing, Soldering, lame Spraying, Sand Blasting, Other Abrasive Jet Machining, Electric Discharge Machining, Electrochemical Machining, Electron Beam Machining, Laser Beam Machining, Plasma Arc Machining, Ultrasonic Machining, Sintering, Laminating, Hot Dip Coating, Sputtering, Vapor Plating, Thermal

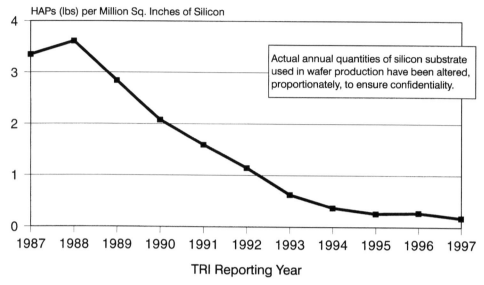

HAPs (lbs) per Million Sq. Inches of Silicon

Actual annual quantities of silicon substrate used in wafer production have been altered, proportionately, to ensure confidentiality.

TRI Reporting Year

SOURCE: EPA TRI Database (1987-97); Dataquest (June 1999)

FIGURE 5   HAPs release by U.S. semiconductor manufactures as a function of silicon substrate production (1987–1997).

Infusion, Salt Bath Descaling, Solvent Degreasing, Paint Stripping, Painting, Electrostatic Painting, Electropainting, Vacuum Metalizing, Assembly, Calibration, Testing and Mechanical Plating (6).

The primary pollutants generated by metal finishing are wastewaters containing heavy metals, waste solvent, and heavy metal-containing debris. The manufacturing technology changes less rapidly than that in the semiconductor industry, but is, in fact, more representative of the majority of U.S. manufacturing sectors. Typically, changes in metal finishing manufacturing are upgrades of specific tools within an existing line. One of the biggest improvements, which began about two decades ago, was the use of countercurrent rinsing techniques to reduce the volumes and heavy metal concentrations of wastewater generated during metal plating.

One of the concerns for this industry is that the heavy metals contained in the wastewater frequently discharge to a publicly owned treatment works (POTW) and accumulate in the municipal treatment plant's sludge or pass through the facility and discharge with the treated effluent. Therefore, significant effort has been undertaken under the Clean Water Act to reduce the emissions of metals from this industry sector.

TABLE 3 Total Quantity of TRI Chemical Releases or Transfers for U.S. Semiconductor Manufacturing and All U.S. Industry (1987–1997)

Year	U.S. semiconductor (million pounds)	All U.S. industry (million pounds)
1987	70.4	7,011.2
1988	38.7	6,489.0
1989	32.5	5,705.7
1990	24.5	4,999.2
1991	28.7	4,411.4
1992	26.0	4,299.2
1993	20.9	3,754.6
1994	15.3	3,134.9
1995	22.3	2,637.6
1996	18.1	2,545.3
1997	18.7	2,586.9

Source: EPA Toxics Release Inventory (TRI) Database, 1987–1997.

In 1993, EPA Administrator Carole Browner embarked on a new program, known as the Common Sense Initiative (CSI), to determine if a more effective approach to environmental regulations and management could be achieved through an industry sector-by-sector approach. The metal finishing industry was one of six selected for the CSI, with the goal to promote "cleaner, cheaper, smarter" environmental performance. One major outcome was a commitment by the metal finishing industry to reduce its environmental footprint. Table 4 presents the environmental commitments of the metal finishing industry developed under CSI.

The environmental goals set forth on both a facility and sector-wide basis in Table 4 provide the necessary direction for using quantitative performance goals to achieve design for the environment within the metal finishing industry. If successful, this industry will significantly reduce its emissions of metals into the environment and achieve significant reductions in consumption of natural resources, with specific emphasis on reducing water and energy use. Through the reductions in water and energy use, the industry should also reduce its operating costs and make it more competitive. Since this was a cooperative effort between the industry and the EPA, the agency has also agreed to reduce by 50% the administrative burden. This, in turn, reduces the operating costs of these types of metal finishing facilities.

In summary, P2 and DFE represent opportunities for industrial facilities to improve their environmental performance and, in many instances, reduce operat-

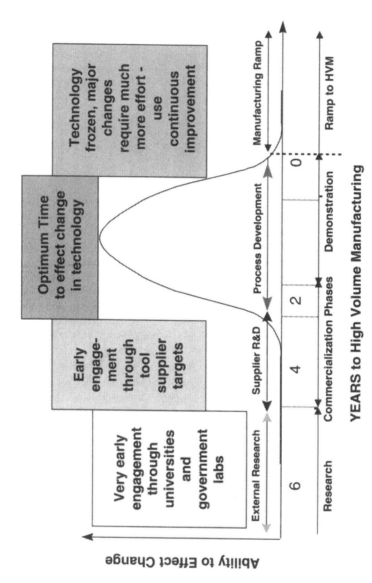

FIGURE 6   EHS technology engagement model for semiconductor manufacturing.

TABLE 4 Metal Finishing National Performance Goals: A Vision for a Cleaner, Cheaper, Smarter Future

Facility-based performance goals (by 2002)	Sector-wide performance goals (by 2002)

(1) Improved Resource Utilization ("Smarter")
  (a) 98% of metals ultimately utilized on product.
  (b) 50% reduction in water purchased/used (*from 1992 levels*).
  (c) 25% reduction in facility-wide energy use (*from 1992 levels*).
(2) Reduction in Hazardous Emissions and Exposures ("Cleaner")
  (a) 90% reduction in organic TRI emissions and 50% reduction in metals emissions to air and water (*from 1992 levels*).
  (b) 50% reduction in land disposal of hazardous sludges and a reduction in sludge generation (*from 1992 levels*).
  (c) Reduction in human exposure to toxic materials in the facility and the surrounding community, clearly demonstrated by actions selected and taken by the facility. Such actions may include, for example, pollution prevention, use of state-of-the-art emission controls and protective equipment, use of best recognized industrial hygiene practices, worker training in environmental hazards, and participation in a Local Emergency Planning Committtee.
(3) Increased Economic Payback and Decreased Costs ("Cheaper")
  (a) Long term economic benefit to facilities achieving Goals 1 and 2.
  (b) 50% reduction in costs of unnecessary permitting, reporting, monitoring and related activities (*from 1992 levels*), to be implemented through burden reduction programs to the extent that such efforts do not adversely impact environmental outcomes.
	(4) Industry-Wide Achievement of Facility Goals.
	  (a) 80% of facilities nationwide achieve Goals 1–3.
	(5) Industry-Wide Compliance with Environmental Performance Requirements.
	  (a) All operating facilities achieve compliance with Federal, State and Local environmental performance requirements.
	  (b) All metal finishers wishing to cease operations have access to a government sponsored "exit strategy" for environmentally responsible site transition.
	  (c) All enforcement activities involving metal finishing facilities are conducted in a consistent manner to achieve a level playing field, with a primary focus on those facilities that knowingly disregard environmental requirements.
	Note: At facilities where outstanding performance levels were reached prior to 1992, the percentage-reduction targets for Goals 1(b) and (c) and 2(a) and (b) may not be fully achievable, or the effort to achieve them may not be the best use of available resources. In these instances, a target should be adjusted as necessary to make it both meaningful and achievable.

Source: Ref. 8.

ing costs. Several industry sectors, such as metal finishing and semiconductor manufacturing, have moved to an industry sector-wide approach to help facilitate the implementation of P2 and DFE. It is anticipated that sector-wide approaches will provide additional encouragement for individual facilities to move forward in implementing new environmental management techniques.

REFERENCES

1. "Waste Minimization Requirements," http://yosemite.epa.gov/OSW/rcra.nsf/Documents/ F68B964A59054380852565DA006F08E1.
2. "Design for the Environment," http://www.epa.gov.
3. President's Council on Sustainable Development, *Sustainable America: A New Consensus for Prosperity, Opportunity and a Healthy Environment for the Future.* Washington, DC: U.S. Government Printing Office, 1996.
4. President's Council on Sustainable Development, *Towards a Sustainable America— Advancing Prosperity, Opportunity and Health Environment for the 21st Century.* Washington, DC: U.S. Government Printing Office, May 1999.
5. U.S. Environmental Protection Agency, Science Advisory Board, Integrated Environmental Decision Making in the 21st Century. Draft Document, May 3, 1999, p. 1.
6. *Code of Federal Regulations*, 1998 Version, 40 CFR Section 433.10.
7. *Safety Guidelines for Semiconductor Manufacturing Equipment*, Semiconductor Equipment and Materials International (SEMI) S2, 2000 Version, pp. 23–24.
8. EPA Common Sense Initiative, Metal Finishing Sector, "Strategic Goals Program— National Performance Goals and Action Plan," www.strategicgoals.org/plan/part1.htm.

15

Pollution Prevention and Life Cycle Assessment

Mary Ann Curran
U.S. Environmental Protection Agency, Cincinnati, Ohio

Rita C. Schenck
Institute for Environmental Research and Education,
Vashon, Washington

1 INTRODUCTION

Over the past 20 years, environmental management strategies in the United States, as well as in many other countries, have evolved through the development of laws and regulations that limit pollutant releases to the environment. For example, since its inception in 1970, the U.S. Environmental Protection Agency (EPA) has made important progress toward improving the environment in every major category of environmental impact caused by pollutant releases. Levels of emissions across the nation have stayed constant or declined, hundreds of primary and secondary wastewater treatment facilities have been built; land disposal of untreated hazardous waste has largely stopped, hundreds of hazardous waste sites have been identified and targeted for cleanup, and the use of many toxic substances has been banned. Together, these actions have had a positive effect on the nation's environmental quality and have set an example for nations everywhere. However, despite the combined achievements of the federal government, states, and industry in controlling waste emissions which have resulted in a healthier environment, the further improvement of the environment has slowed. This led to the realization that a new paradigm was needed for environmental protection.

Starting in the mid-1980s, pollution prevention was seen by visionaries as the way to go beyond such command and control approaches. Pollution prevention has received widespread emphasis internationally within multinational organizations and within the governments of both developing as well as developed nations. The European Union has designed some of its rules and programs based on the concepts behind pollution prevention. The United Nations Environmental Programme (UNEP) has a clean technologies program and has supported many workshops and meetings on various topics on pollution prevention as well as sustainable development.

Worldwide, the advancement of environmental protection strategies moving from end-of-pipe to pollution prevention and beyond has been steady. This evolution can be summarized by the chronology shown in Table 1. This chapter provides an overview of industrial pollution prevention, beginning with a brief definition of a pollution prevention opportunity assessment, followed by a discussion of life cycle assessment (LCA).

2 UNDERSTANDING OF INDUSTRIAL POLLUTION PREVENTION

Briefly stated, industrial pollution prevention is a term that is used to describe technologies and strategies that result in eliminating or reducing waste streams from industrial operations. The EPA defines pollution prevention as "the use of materials, processes, or practices that reduce or eliminate the creation of pollutants or wastes at the source." It includes practices that reduce the use of hazardous materials, energy, water, or other resources and practices that protect natural resources through conservation or more efficient use. The basic idea of

TABLE 1 Evolution of Environmental Protection

Chronology	Strategy
1970s to 1980s	End-of-pipe treatment
mid-1980s	Waste minimization
early 1990s	Pollution prevention
mid-1990s	ISO Certification/life cycle assessment
2000 and beyond	Agenda 21 for Sustainable Development

pollution prevention follows the axiom, "an ounce of prevention is worth a pound of cure." The U.S. Pollution Prevention Act of 1990 and pollution prevention experts conclude that it makes far more sense for a waste generator not to produce waste in the first place, rather than developing extensive, never-ending treatment schemes (1).

For industrial pollution prevention, two general approaches are used to characterize processes and waste generation. The first approach involves gathering information on releases to all media (air, water, and land) by looking at the output end of each process, then backtracking the material flows to determine the various waste sources. The other approach tracks materials from the point where they enter a facility, or plant, until they exit as wastes or products. Both approaches provide a baseline for understanding where and why wastes are generated, as well as a basis for measuring waste reduction progress. The steps involved in these characterizations are similar and include gathering background information, defining a production unit, general process characterization, understanding unit processes, and completing a material balance.

These steps, when performed systematically, provide the basis for a *pollution prevention opportunity assessment*. It begins with a complete understanding of the various unit processes and points in these processes where waste is being generated and ends with the implementation of the most economically and technically viable options. It may be necessary to gather information to demonstrate that pollution prevention opportunities exist and should be explored. Often, an assessment team is established to perform the steps along the way (2).

A preliminary assessment of a facility is conducted before beginning a more detailed assessment. The preliminary assessment consists of a review of data that are already available in order to establish priorities and procedures. The goal of this exercise is to target the more important waste problems, moving on to lower-priority problems as resources permit. The preliminary assessment phase provides information that is needed to accomplish this prioritization and to assemble the appropriate assessment team (3).

A subsequent detailed assessment focuses on the specific areas targeted by the preliminary assessment. Analyzing process information involves preparing a material and energy balance as a means of analyzing pollution sources and opportunities for eliminating them. Such a balance is an organized system of accounting for flow, generation, consumption, and accumulation of mass and energy in a process. In its simplest form, a material balance is drawn up according to the mass conservation principle:

Mass in = mass out − (generation + consumption + accumulation)

If no chemical or nuclear reactions occur and the process progresses in a steady state, the material balance for any specific compound or constituent is as follows:

Mass out = mass in

A process flow diagram may be helpful by providing a visual means of organizing data on the material and energy flows and on the composition of streams entering and leaving the system (see Figure 1). Such a diagram shows the system boundaries, all stream flows, and points where wastes are generated.

Boundaries should be selected according to the factors that are important for measuring the type and quantity of pollution prevented, the quality of the product, and the economics of the process. The amount of material input should equal the amount exiting, corrected for accumulation and creation or destruction.

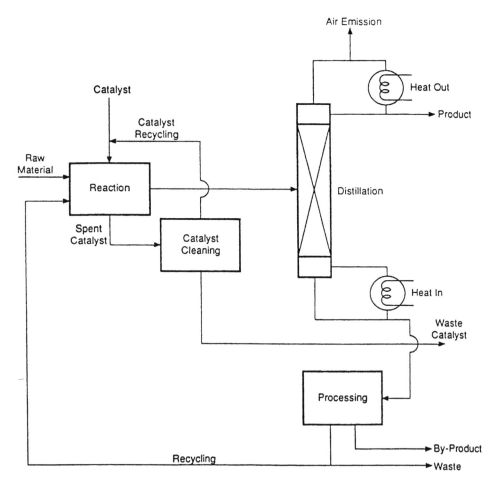

FIGURE 1 Example flow diagram (3).

A material balance should be calculated for each component entering and leaving the process, or other system being studied. A suggested approach for making these calculations is offered in Section 3.

Once the sources and nature of wastes generated have been described, the assessment team enters the creative phase. Pollution prevention options are proposed and then screened for feasibility. In this environmental evaluation step, pollution prevention options are assessed for their advantages and disadvantages with regard to the environment. Often the environmental advantage is obvious— the toxicity of a waste stream will be reduced without generating a new waste stream. Most housekeeping and direct efficiency improvements have this advantage. With such options, the environmental situation in the company improves without new environmental problems arising (3).

Along with assessing the technical and environmental effectiveness in preventing pollution, options are evaluated for the estimated cost of purchasing, installing, and operating the system. Pollution prevention can save a company money, often substantial amounts, through more efficient use of valuable resource materials and reduced waste treatment and disposal costs.

Estimating the costs and benefits of some options is straightforward, while for others it is more complex. If a project has no significant capital costs, the decision is relatively simple. Its profitability can be judged by determining whether it reduces operating costs or prevents pollution. Installation of flow controls and improvement of operating practices will not require extensive analysis before implementation. However, projects with significant capital costs require detailed analysis. Several techniques are available, such as payback period, net present value, or return on investment. These approaches are also described in Section 3.

At times, the environmental evaluation of pollution prevention options is not always straightforward. Some options require a thorough environmental evaluation, especially if they involve product or process changes or the substitution of raw materials. For example, the engine rebuilding industry is no longer using chlorinated solvents and alkaline cleaners to remove grease and dirt from engines before disassembly. Instead, high-temperature baking followed by shot blasting is being used. This shift eliminates waste cleaner but requires additional energy use for the shot blasting. It also presents a risk of atmospheric release because small quantities of components from the grease can vaporize. (3)

Others are moving toward the use of aqueous cleaners as substitutes for solvents in an attempt to avoid using toxic materials. However, while the less toxic aqueous cleaners offer a suitable substitution for chlorinated cleaning solvents from a performance standpoint, their use may be resulting in increased environmental impacts in other areas. Most obvious is the increased energy use that occurs from needing to heat the parts to be cleaned in order to get a satisfactory level of cleaning performance.

To make a sound evaluation, the team should gather information on all the environmental aspects of the product or process being assessed. This information would consider the environmental effects not only of the production phase but of the acquisition of raw materials, transportation, product use, and final disposal as well. This type of holistic evaluation is called a *life cycle assessment* (LCA). The stages that are included within the boundary of an LCA are shown in Figure 2. LCA's origins in mass and energy balance sheets have led to several important accounting conventions, including the following.

A *system-wide perspective* embodied in the term "cradle-to-grave" that implies efforts to assess the multiple operations and activities involved in providing a product or service. This includes, for example, resource extraction, manufacturing and assembly, energy supplies and transportation for all operations, use, and disposal.

A *multimedia perspective* that suggests that the account balance include resource inputs as well as wastes and emissions to most common environmental media, e.g., air, water, and land.

A *functional unit accounting* normalizes energy, materials, emissions, and wastes across the system and media to the service or product provided. Notably, this calculation allows the analysis of different ways to provide a function or service, for example, one can compare sending a letter via e-mail or via regular mail. Additionally, this approach entails *allocation* procedures so that only those portions or percentages of an operation specifically used to produce a particular product are included in the final balance sheet (4).

The functional unit approach of LCA takes the assessment beyond looking at the environmental impacts associated with a specific location or operation. The value of LCA lies in its broad, relative approach to analyzing a system and factoring in global as well as regional and local environmental impacts. This general, macro approach makes it theoretically feasible to frame numerous potential issues and environmental considerations, identify possible trade-offs between different parts of the life cycle, and make these possible issues and trade-offs apparent to decision makers. These attributes enable the user to understand complex and previously hidden relationships among the many system operations in the life cycle and the potential repercussions of changes in an operation on distant operations and other media. This is particularly true where unanticipated or unrecognized issues on the life cycle of a product or service are revealed to decision makers. This leads to a more complete and thorough evaluation for making decisions, including applications in strategic planning,

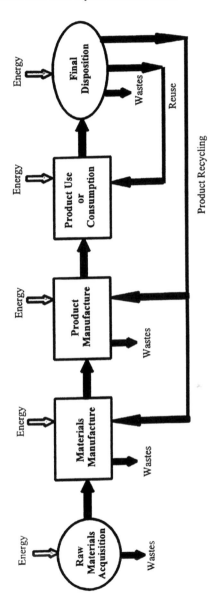

FIGURE 2 Input/output flows in a product life cycle.

environmental management, product development and R&D, and liability assessment, as well as pollution prevention.

3 CALCULATION

3.1 Measurement of Pollution Prevention

Accurate and meaningful measurement systems are vital to ensure long-term successful implementation of pollution prevention (5). To implement pollution prevention, industrial facilities must first measure the environmental impacts of their facilities, beginning with the accounting of the inputs and outputs across the facility's boundaries. This process is captured in a material and energy balance.

3.1.1 Material and Energy Balance

Analyzing process information involves preparing material and energy balances as a means of analyzing pollution sources and opportunities for eliminating them. Such a balance is an organized system of accounting for the flow, generation, consumption, and accumulation of mass and energy in a process. In its simplest form, a material balance is drawn up according to the mass conservation principle:

Mass in = mass out − (generation + consumption + accumulation)

The first step in preparing a balance is to draw a process diagram, which is a visual means of organizing data on the energy and material flows and on the composition of the streams entering and leaving the system. A flow diagram, such as Figure 1, shows the system boundaries, all streams entering and leaving the process, and points at which wastes are generated. The goal is to account for all streams so that the the mass equation balances.

The boundaries around the flow diagram should be based on what is important for measuring the type and quality of pollution prevented, the quality of the product, and the economics of the process. Again, the amount of material input should equal the amount exiting, corrected for accumulation and creation or destruction.

In addition to an overall balance, a material balance should be calculated for each individual component entering and leaving the process. When chemical reactions take place in a system, there is an advantage to performing the material balance on the elements involved.

Material and energy balances do have limitations. They are useful for organizing and extending pollution prevention data and should be used whenever possible. However, the user should recognize that most balance diagrams will be incomplete, approximate, or both (6).

Most processes have numerous process streams, many of which affect various environmental media.

The exact composition of many streams is unknown and cannot be easily analyzed.

Phase changes occur within the process, requiring multimedia analysis and correlation.

Plant operations or product mix change frequently, so the material and energy flows cannot be accurately characterized by a single balance diagram.

Many sites lack sufficient historical data to characterize all streams.

These are examples of the complexities that will recur in the analysis of real-world processes.

Despite the limitations, material balances are essential for organizing data and identifying data gaps and other missing information. They can help calculate concentrations of waste constituents where quantitative composition data are limited. They are particularly useful if there are points in the production process where it is difficult or uneconomical to collect or analyze samples. Data gaps, such as an unmeasured release, can also indicate that fugitive emissions are occurring. For example, solvent evaporation from a parts cleaning tank can be estimated as the difference between solvent added to the tank and solvent that is removed by disposal, recycling, or dragout (6). It is an essential characteristic of a mass balance that unmeasured flows are used to balance the equation.

3.1.2 Industrial Production and Waste Generation Tracking System

The Industrial Production and Waste Generation Tracking System shown in Figure 3 (7) establishes a framework for the determination of the main parameters for industrial production and waste generation. It is based on the following main production process variables:

1. Raw materials (rm)
2. Other materials entering production process (v)
3. Produced products (P)
4. Generated waste (y)

The generated waste may be:

1. Managed (g) by applying waste management
2. Released (z) into the environment, causing environmental pollution

Managed waste (g) may be further processed to be

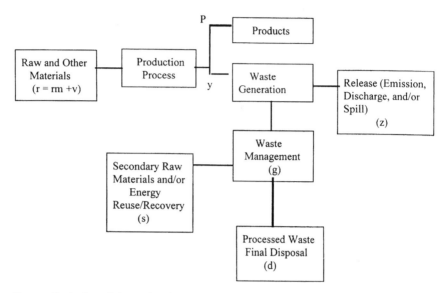

FIGURE 3 Industrial production and waste generation tracking system (7).

3. Used as secondary raw material and/or energy (*s*)
4. Finally disposed of as processed waste (residues) in a special (or secure) landfill (*d*).

The development of the Industrial Production and Waste Generation Tracking System model was based on the work of Baetz et al. (8). A model enabling calculation of quantities of waste generated in an industrial production was developed and defined as shown in Table 2.

During an industrial process at time *t*, a production factor *U*, correlating quantity of raw and other materials *r* and products *P*, has a value of $0 \leq U \leq 1$ and is defined as

$$U = \frac{P}{r}$$

Note that raw materials *r* includes "other materials" not typically defined as raw materials entering a production process. For example, paints and lacquers in "white goods" and furniture manufacture are usually not defined as raw materials but are still input materials.

Converting the last expression, the quantity of products may then be expressed as

$$P = Ur$$

TABLE 2 Definition of IPWGTS Model Parameters (7)

Parameter	Definition
t	Industrial production time in which system was observed
rm	Quantity of raw material entering industrial production in time t
v	Quantity of other materials (not defined as raw materials) entering industrial production in time t
$r = rm + v$	Quantity of raw and other materials entering industrial production in time t
$U = P/r$	Production factor after time t; $U = 1$ represents total production, while $U = 0$ represents zero level of production and therefore total waste generation
$P = Ur$	Quantity of products produced in time t
$y = (1 - U)r$	Quantity of solid, liquid, and/or gaseous waste generated in time t
$M = g/y = g/(1 - U)r$	Waste management factor after time t; $M = 1$ represents total waste management, while $M = 0$ represents zero level of waste management and therefore total release (emission, spill, and/or discharge) into the environment
$g = M(1 - U)r$	Quantity of solid, liquid, and/or gaseous waste managed by waste management (temporary storage, collection, transportation, processing, and final disposal)
$z = (1 - M)(1 - U)r$	Quantity of solid, liquid, and/or gaseous waste released to air, water, and/or soil/land, causing environmental pollution
$R = s/g = s/M(1 - U)r$	Waste recycling factor after time t; $R = 1$ represents total waste recycling by physical, chemical, thermal, and/or biological process as to recover secondary materials and/or energy, while $R = 0$ represents total waste processing by physical, chemical, thermal, and/or biological process for final disposal in the environment
$s = MR(1 - U)r$	Quantity of secondary raw materials and/or energy recovered from solid, liquid, and/or gaseous waste by waste recycling
$d = M(1 - R)(1 - U)r$	Quantity of processed waste for final disposal

while quantity of waste generated is

$$y = (1 - U)r$$

Managed waste is further quantified by a waste management factor M, which is defined as the ratio between waste managed and waste generated:

$$M = \frac{g}{y}$$

and can have a value $0 \leq M \leq 1$. The quantity of waste managed by storage, collection, transportation, processing, and final disposition is then

$$g = M(1 - U)r$$

while the quantity of waste released (emitted, discharged, and/or spilled) into the environment is

$$z = (1 - M)(1 - U)r$$

If waste is further processed by physical, chemical, thermal, and/or biological processing to recover secondary raw materials and/or energy, then processed waste is determined by the waste recycling factor R,

$$R = \frac{s}{g}$$

having the value $0 \leq R \leq 1$. The quantity of waste recycled into secondary raw materials and/or energy by waste processing is

$$S = MR(1 - U)r$$

while the quantity of waste to be finally disposed is

$$d = M(1 - R)(1 - U)r$$

Knowing quantities of raw and other materials ($rm + v$) entering the observed system and quantities of products (P) produced, quantities of waste generated (y) can be calculated. If the quantity of waste managed (g) by the waste generator is known, it is possible to predict quantities of material lost (z) through release (emission, discharge, and/or spill). Finally, if the waste generator recycles managed waste into secondary raw materials and/or energy (s), then the quantity of waste to be disposed (d) can be determined.

3.1.3 Production-Adjusted Pollution Prevention

After a pollution prevention activity has been implemented, adjusted figures from the process flow diagram should show a decrease in waste generation. This decrease is often approached in one of two ways (5). The first way is to look at the change in the quantity of chemicals or raw materials that are purchased or used for a process. This approach is based on the idea that materials that do not enter a production process in the first place cannot leave the process in the form of wastes or emissions, given that production is at a steady rate. The second, more common pollution prevention approach looks at how much the quantity of waste streams or emissions has been reduced over a given period of time. This results in a statement such as "Pollution prevention was achieved in a 20% reduction in the amount of spent chromium discharged compared to last year."

However, in both approaches, accounting for the varying levels of production is a key issue in more accurately capturing pollution prevention progress. If production drops off, a decrease in waste streams or emissions may be attributable to the decreased activity instead of to any particular pollution prevention effort. Production-adjusted measures of pollution prevention account for changes resulting from pollution prevention efforts. For production-adjusted pollution prevention measures, a unit of product is the factor used to adjust gross quantities of waste or chemical use to infer the amount of pollution prevention progress (see Table 3).

Using units of product to calculate pollution prevention improvements can filter out effects of change in production activity. For example, a firm generated 22,000 lb (1000 kg) of trichlorethylene (TCE) waste from a vapor degreasing operation used to remove oil for the 16,000 metal circuit boxes it manufactured. In 1994, the company implemented several pollution prevention activities which resulted in the generation of 15,000 lb (6800 kg) of TCE to manufacture 20,000 circuit boxes. At first, it would seem logical to express the achieved pollution prevention as the difference in the amount of TCE waste from 1993 to 1994, or 7000 lb (i.e., 20,000 lb − 15,000 lb). However, this way of measurement does not filter out the effect of increased production. Factoring in increased output, the pollution prevention progress can be calculated as follows:

TABLE 3   Typical Ways to Measure P2

Not production-adjusted
Change in quantity of emissions: "Reduced discharge of chromium by 20% last year."
Change in quantity of chemical or raw materials used: "Reduced plating solution purchases by 10% last year."
Production-adjusted
Change in quantity of chemical used per unit produced: "10% reduction in quantity of plating solution used per part shipped last year."
Change in quantity of chemical used per unit activity: "Reduced solvent use by 15% for every hour the degreaser ran last year."

Unit–of–product ratio: $\dfrac{\text{boxes in 1994}}{\text{boxes in 1993}} = \dfrac{20,000}{16,000} = 1.25$

The unit-of-product ratio is used to calculate the expected waste generation, given this year's level of production, if no pollution prevention changes had been made during the past year.

Expected waste generation in 1994 is calculated as follows:

(Production ratio) · (1993 waste generation) = (1.25) · (22,000 lb) = 27,500 lb

or = (1.25) · (10,000 kg) = 12,500 kg

Therefore,

(1994 adjusted waste generation) − (1994 actual waste generation)

= 27,500 lb (12,500 kg) − 15,000 lb (6,804 kg)

= 12,500 lb (5,680 kg) waste reduction

Another way to examine the effects of pollution prevention is to assess whether the amount of waste per "widget" produced has changed. Using the data from the preceding example, the calculations are as follows:

$\dfrac{\text{TCE waste generated in 1993}}{\text{number of widgets produced in 1993}} = \dfrac{22,000}{16,000}$

= 1.38 lb (0.626 kg) TCE per circuit box

$\dfrac{\text{TCE waste generated in 1994}}{\text{number of widgets produced in 1994}} = \dfrac{15,000}{20,000}$

= 0.75 lb (0.34 kg) TCE per circuit box

The two waste efficiencies can then be compared to conclude that the company made substantial waste reductions of:

1.38 lb − 0.75 lb = 0.63 lb of TCE per product

(0.626 kg − 0.34 kg = 0.29 kg of TCE per product)

These examples show the importance of finding and using a unit of product that is closely related to the waste or chemical usage being targeted. Suppose, however, that this facility has modified its degreasing operations and reduced solvent loss, but the loss of solvent is more related to the number of hours the degreaser was running than to the number of parts that were cleaned. In that case, calculating "solvent savings per part cleaned" has less meaning for indicating pollution prevention progress. Solvent saved per hour of degreasing operation, however, would provide a better picture of actual savings resulting from the change.

3.1.4 Calculating the Cost of Pollution Prevention

The "usual costs" of any process are commonly considered to be the costs associated directly with the polluting practice or proposed alternatives. Costs may then be categorized as either capital expenses that must be depreciated for tax purposes or other expenses that can be deducted from taxes in a single year. Other expenses are commonly calculated as "capital" costs because they are one-time costs that are needed before the process can be used. To collect the cost needed for determining the economic feasibility, the following cost items are needed.

Estimate Capital Cost Items

EQUIPMENT. This cost item represents the investment in new equipment needed to implement the pollution prevention option. The cost element should include the price (f.o.b. factory), taxes, freight, and insurance on delivery, and the cost for the initial spare parts inventory. Any additional equipment needed to support the pollution prevention alternative should be included, such as additional laboratory equipment.

MATERIALS. Materials costs include piping, electrical equipment, new instrumentation, and changes in the structure. These costs are those incurred in purchasing the materials needed to connect the new process equipment (or revise the use of existing equipment).

UTILITY CONNECTIONS. This item includes the costs for connecting new equipment (or for making new connections to existing equipment) as part of implementing the pollution prevention option. Typical utilities include electricity, steam, cooling water, process water, refrigeration, fuel (gas or oil), plant air (e.g., for process control), and inert gas.

SITE PREPARATIONS. This item includes the costs for any necessary site preparation, such as demolition, site clearing, paving, etc.

INSTALLATION. This item includes the costs incurred during the installation of the process equipment or process change, as well as charges by the vendor as well as by in-house staff.

ENGINEERING AND PROCUREMENT. This item includes the costs incurred to design the process equipment or process change and to purchase any new equipment. Charges for consultants used in designing and procuring equipment are also included.

Estimate Expenses. The costs in this category include both one-time costs and ongoing costs that are deductible for income tax purposes.

START-UP COSTS. Start-up costs include labor and material costs incurred during the start-up phase.

PERMITTING COSTS. These costs include both fees and the costs incurred by in-house staff in documenting the process change to meet permit requirements.

SALVAGE VALUE. Estimate the net amount (in today's dollars) that used equipment will be worth at the end of it useful lifetime. Include the value of working capital and catalysts and chemicals that will remain at the completion of the equipment's life.

TRAINING COSTS. Training costs include the costs for on-site and off-site training related to the use of the new equipment or for making sure the process change achieves its goal.

INITIAL CHEMICALS. The initial charges for chemicals and catalysts can be considered a capital item.

WORKING CAPITAL. This category includes all elements of working capital (required inventories of raw materials, in-process inventories, materials and supplies) not already included as charges for chemicals and catalysts for spare parts. Working capital may also include personnel costs for operations start-up.

DISPOSAL COSTS. The disposal cost includes all the direct costs associated with waste disposal, including solid waste disposal, hazardous waste disposal, and off-site recycling.

RAW MATERIAL COSTS. This category includes both the raw materials directly affected (e.g., chemicals for which more effective or less toxic substitutes are being found) and other raw materials affected by the change in the process (e.g., a change in a cleaning agent reduces the rejection rate of metal parts, thereby reducing total material costs).

UTILITIES COSTS. Utilities include electricity, process steam, water, compressed air, and heating oil or natural gas. It is important to consider whether a change causes downstream effects as well, e.g., recycling an aqueous waste stream may require energy to adjust the temperature of the stream to meet process requirements.

OPERATING COSTS. This cost element includes the labor needed to run the process or alternative.

Operation and maintenance (O&M) costs. This cost element includes supplies needed on a regular basis, such as glassware, buckets, cleaners, filters, protective equipment, etc.

Insurance and liability costs. In some cases, insurance rates may be adjusted accordingly, e.g., switching to a process that is know to be safer may lower insurance rates.

Other operating costs. This cost element includes other operating costs that have not been specifically mentioned above.

Estimate Operating Revenues. In some cases, implementing a pollution prevention option may lead to a change in the revenue from operations. The two main categories are revenues from products and revenues from marketable by-products.

PRIMARY PRODUCTS. If the process or procedural change changes the production rate of the process, then the revenues before and after the change should be included.

MARKETABLE BY-PRODUCTS AND RECOVERED MATERIAL. An increase in the amount of marketable by-products and materials that are recovered and reused should be included.

According to Humphreys and Wellman (9), the most widely used methods for calculating potential savings are

Payback (or payout) time
Payback time with interest
Return on investment
Return on average investment
Discounted cash flow (also known as interest rate of return)
Venture worth analysis (also known as incremental present worth)

Each of these techniques offers both advantages and disadvantages, possibly resulting in different order of profitability. All these methods are used to compare alternatives. In that sense they give several alternatives with relative position of preference to each other. For this reason, when pollution prevention alternatives are being evaluated, more than one method of calculation should be used.

The following section evaluates an example investment of $1 million using two methods: payback time and rate-of-return using discounted cash flow.

Payback time is the time required for all cash flows to equal the original investment. In other words, it is the time it takes to recover the original investment. The estimated cash flow and savings for a proposed $1 million project with a projected life of 5 years is shown in Table 4. Through interpolation between the accumulated savings, we see that the savings will reach $1 million between the second and third year:

$$\frac{1,000,000 - 825,000}{1,105,000 - 825,000} + 2 = 2.6$$

TABLE 4 Estimated Cash Flow for a $1 Million Project with a Projected Life of 5 Years

Year	Savings per year	Cumulative savings
1	$525,000	$ 525,000
2	300,000	825,000
3	280,000	1,105,000
4	200,000	1,305,000
5	125,000	1,430,000

Therefore the payback period is 2.6 years.

While it is very simple to calculate, the shortcoming of this method is that it does not consider the value of money, i.e., interest. The *discounted cash flow* (DCF), or interest-rate-of-return, method is the most widely used of all these types of calculations. It is also probably the most valid technique, since it considers all cash flows in and out of a project as well as the time value of money. DCF is based on the realization that a dollar in the future is not worth as much as today's dollar, which is available for reinvestment. In effect, DCF applies the expected rate of return on an investment made today over the life of the investment.

The equation to calculate DCF is

$$0 = -I + \sum_{1}^{n} \frac{CF_n}{(1 + i)^n}$$

where

CF_n = cash flow in year n

i = interest rate of return

I = initial capital investment

Again using the previous example, DCF is calculated as follows:

$$0 = -1,000,000 + \frac{525,000}{(1 + i)^1} + \frac{300,000}{(1 + i)^2} + \frac{280,000}{(1 + i)^3} + \frac{200,000}{(1 + i)^4} + \frac{125,000}{(1 + i)^5}$$

At $i = 20\%$, the equation yields $-45,449$; and at $i = 15\%$, the equation yields $+43,980$. Interpolating between the two values and setting the equation equal to zero, $i = 17.5\%$. Therefore the DCF rate of return is 17.5% per year.

3.2 Life Cycle Assessment

Every day, both individual consumers and industry make choices that affect the environment. Manufacturers choose from among different materials, suppliers, or production methods. Consumers decide on the need for a product and make purchasing choices. Environmentally responsible choices need reliable information based on the life cycle characteristics of the alternative products or processes being considered.

LCA considers the environmental aspects and the potential impacts of a product or service system throughout its life—from raw material acquisition through production, use, and disposal. This information has many potential uses: it can help identify ways to improve environmental aspects of a product at various stages in its life cycle; it can support decision making in industry, governmental, or nongovernmental organizations; it can aid in the selection of relevant indicators of environmental performance; and (with proper precautions) it can support the marketing of products or services.

The environment is complex, with many interrelationships, and a major challenge in any LCA study is to isolate the impacts of a single product or service system. Comparability between LCA studies is also an issue, as products that perform the same function may be made of widely varying materials. There is a clear need for neutral, scientifically oriented, consensus-based guidance on the conduct of LCA. Toward this end, various research organizations and practitioners are working to develop LCA methodology and tools. Especially, the International Standards Organization (ISO) has developed a series of standards and technical reports that cover the various stages involved in LCA, from scoping through impact assessment and interpretation (10–13). Using agreed-upon principles, an LCA study can be done responsibly, transparently, and consistently.

Before the ISO efforts in LCA, early research conducted by the U.S. EPA in LCA methodology along with efforts by the Society of Toxicology and Chemistry (SETAC) led to the four-part approach to LCA that is widely accepted today (14):

Goal and scope definition: identifying the purpose for conducting the LCA, the boundaries of the study, assumptions, and expected output

Life cycle inventory: quantifying the energy use and raw material inputs and environmental releases associated with each stage of the life cycle

Life cycle impact assessment: assessing the impacts on human health and the environment associated with the life cycle inventory results

Improvement analysis/interpretation: evaluating opportunities to reduce energy, material inputs, or environmental impacts along the life cycle.

LCA is not strictly a technical process. Various simplifying assumptions and value-based judgments must be used throughout the process. The key is to keep these to a minimum and be explicit in the reporting phase about what assumptions and values were used. Readers of the study can then recognize the judgments and decide to accept, qualify, or reject them and the study as a whole.

Finally, it should be recognized that the results of an LCA can provide much of the information needed to make a decision. In most, if not all, cases LCA should be integrated with other assessment tools and techniques, such as the financial tools described above, to make sound decisions.

In addition to comparing the environmental soundness of products, LCA is also being used to assess applications within industrial processes, such as pollution prevention activities. The following two examples demonstrate how LCA has been used to evaluate options for material substitution and raw material sourcing.

Aqueous cleaners. While aqueous cleaners offer a suitable substitution for chlorinated cleaning solvents, they generally require pretreatment prior to discharge to a POTW to adjust the pH, remove oil, grease, and solids, and to precipitate phosphates and inactive chelating agents. Another impact is that energy use is often higher than that required for chlorinated solvents (15).

Butanediol (BDO). An alternative to natural gas-derived feedstock to produce 1,4-butanediol (BDO) is a feedstock process that is based on the fermentation of corn-derived glucose to succinic acid, followed by catalytic reduction to BDO. The higher energy use of the alternative process indicates that the overall environmental consequences would be greater than the conventional process. Because electricity generation is inefficient, and energy production in the United States is mostly coal-based, the alternative process was projected to have a greater potential for impact in multiple impact categories, including global warming, acid rain, smog, water use, particulates, and solid waste (coal ash) disposal (16).

LCA, while comprehensive in theory, encounters practical limitations and barriers that are slowing its widespread adoption. A major barrier is the lack of knowledge of the life cycle concept. Producers need to be made aware of the life cycle impacts that their activities carry and the importance of going beyond meeting compliance. More important, government offices that issue media-based or industry-focused regulations and policies need to begin using life cycle thinking. There are numerous instances where life cycle thinking is potentially beneficial in making public policy. Introducing LCA concepts into the rule-making process extends the regulatory analysis upstream and downstream and across all media to account for the effects of the proposed standard, which may otherwise escape a traditional regulatory impact analysis.

Another key barrier is the lack of reliable data. Lack of data has hindered, perhaps prevented, many applications. Several efforts are underway in North America and Europe to make data more easily accessible. Another fundamental barrier to performing LCAs at this time is the lack of a generally agreed-on impact assessment method. This seems to be more of a barrier in the United States than in Europe, where several attempts at life cycle impact assessment have been published. Figure 4 depicts the historical development of LCA practice.

3.2.1 Goal and Scope Definition

The goal and scope of an LCA study should be clearly defined and consistent with the intended application of the results. The goal should be stated unambiguously, together with the reasons for carrying out the study. In defining the scope of the study, the following items should be considered and clearly described: the system(s) being studied, the methodology and interpretation approach to be used,

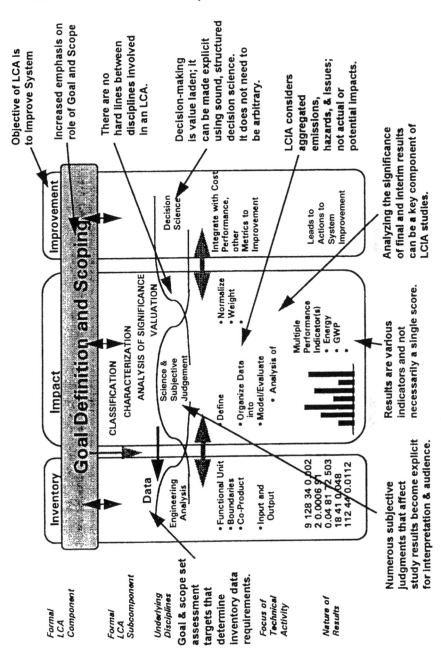

FIGURE 4 Emerging LCA practice (4).

and data needs. Specifically, the decisions made during the planning phase include the following:

The purpose of the study
The audience, and whether publication will occur
The system function and functional unit
The study boundaries
What types of study to perform
Allocation methods
Environmental impact categories, and model and impact indicators
Data quality requirements

Scoping decisions make for the success or failure of the study. For example, drawing the systems boundaries tends to be a balance between time and money resources and practical limitations. Sometimes, in their attempt to save costs, practitioners have drawn the boundaries of the study too narrowly, thus yielding a study that leaves important questions unanswered. On the other hand, LCA boundary conditions sometimes expand to the point that a full data set is never accomplished (17).

Scoping an LCA is the most critical part of the LCA Study. Studies that are not appropriately scoped are rarely successful.

3.2.2 Inventory—Calculating Energy and Material Inputs and Environmental Releases

The second activity of a life cycle assessment is the identification and quantification of energy and resource use and environmental releases to air, water, and land (18). This inventory component is a technical, data-based process with a goal of achieving a mass and energy balance for the life cycle system being studied. In the broadest sense, inventory begins with raw material extraction and continues through final product consumption and disposal. The boundaries for the system being studied are determined in the goal definition and scoping phase and should be broad enough to allow the study to quantify resource use and environmental releases.

The quality of a life cycle inventory depends on an accurate description of the system to be analyzed. The necessary data collection and interpretation is contingent upon proper understanding of where each stage life cycle begins and ends. The general scope of each stage can be described as follows.

Raw materials acquisition. This stage of the life cycle of a product includes the removal of raw materials and energy sources from the earth, such as the harvesting of trees or the extraction of crude oil. Land disturbance as well as transport of the raw materials from the point of acquisition to the point of raw materials processing are considered part of this stage.

Manufacturing. The manufacturing stage produces the product from the raw materials and delivers it to consumers. Three substages or steps are involved in this transformation: materials manufacture, product fabrication, and filling/packaging/distribution.

Materials manufacture. This step involves converting raw material into a form that can be used to fabricate a finished product. For example, several manufacturing activities are required to produce a polyethylene resin from crude oil: the crude oil must be refined; ethylene must be produced in an olefins plant and then polymerized to produce polyethylene. Transportation between manufacturing activities and to the point of product fabrication should also be accounted for in the inventory, either as part of materials manufacture or separately.

Product fabrication. This step involves processing the manufactured material to create a product ready to be filled, or packaged—for example, blow molding a bottle, forming an aluminum can, or producing a cloth diaper.

Filling/packaging/distribution. This step includes all manufacturing processes and transportation required to fill, package, and distribute a finished product. Energy and environmental wastes caused by transporting the product to retail outlets or to the consumer are accounted for in this step of a product's life cycle.

Use/reuse/maintenance. This is the stage consumers are most familiar with, the actual use, reuse, and maintenance of the product. Energy requirements and environmental wastes associated with product storage and consumption are included in this stage.

Recycle/waste management. Energy requirements and environmental wastes associated with product disposition are included in this stage, as well as postconsumer waste management options such as recycling, composting, and incineration.

The following general issues apply across all life cycle stages.

Energy and transportation. Process and transportation energy requirements are determined for each stage of a product's life cycle. Some products are made from raw materials, such as crude oil, which are also used as sources for fuel. Use of these raw materials as inputs to products represents a decision to forego their fuel value. The energy value of such raw materials that are incorporated into products typically is included as part of the energy requirements in an inventory. Energy required to acquire and process the fuels burned for process and transportation use is also included.

Environmental waste aspects. Three categories of environmental wastes are generated from each stage of a product's life cycle: atmospheric emissions, waterborne wastes, and solid wastes. These environmental wastes are generated by both the actual manufacturing processes and the use of fuels in transport vehicles or process operations.

Waste management practices. Depending on the nature of the product, a variety of waste management alternatives may be considered: landfilling, incineration, recycling, and composting.

Allocation of waste or energy among primary and co-products. Some processes in a product's life cycle may produce more than one product. In this event, energy and resources entering a particular process and all wastes resulting from it are allocated among the product and co-products. Allocation is described below in more detail.

3.2.3   Function and Functional Unit Determination

Since any system is a collection of processes connected by flows of intermediate products, the system is defined by the function that has been selected for study. This in turn determines the scope of the study, which sets the boundaries and determines which unit processes will be included. And, because the system is a physical system, it obeys the law of conservation of mass and energy. Mass and energy balances are the goal of a life cycle inventory and perform a useful check on the completeness and validity of the data.

The selection of inputs and outputs to model the system are based on the functional unit. Comparisons between systems are made on the basis of the same function, quantified by the same functional unit. For example, one might compare sending a letter via e-mail and via regular mail. The system function is the transfer of the document. The system function is measured in functional units, such as thousands of pages transferred and the distance of their transfer. *Functional units* permit direct comparisons between disparate systems and processes.

As another example, the *function* of drying hands in a public facility can be evaluated comparing a paper towel system to an air-dryer system. The *functional unit* may be expressed in terms of the identical number of pairs of hands dried for both systems. For each system, it is necessary to determine the mass and energy flows through the product system based on the functional unit, e.g., the average mass of paper used or the average volume of hot air required to dry hands each time. In cases like this, where the use is small, it is common to multiply the product use in order to get to numbers that are easier to work with (for example, 1000 times to dry hands).

3.2.4   Co-Product Allocation Procedures

Allocation procedures are needed when dealing with systems involving multiple products (i.e., co-products and by-products). The materials and energy flows as well as associated environmental releases and resource uses must be allocated to the different products in the inventory process. A basis for co-product allocation needs to be selected, with careful attention paid to the specific items calculated. Allocation can be made based on mass, or energy content, or economic considerations (the market value of the different products). Allocation on a mass basis is most often used and is appropriate in many circumstances.

A mass-based allocation calculation begins with the identification of the functional unit that is being used to assess the product system. As an example, 1000 units (e.g., pounds or kilograms) of a final product are being produced in a process that generates 30 units of waste as depicted in Figure 5.

The final product is produced in a two-step process. In the first step, 600 units of raw material A and 420 units of raw material B are reacted to form the intermediate product. The process results in the production of a co-product along with 20 units of waste. In the second step, the intermediate product is mixed with 150 units of raw material C and 60 units of additive to form the final product. This step results in the formation of 10 units of waste. The question then is how much of the 20 units of waste from the first step should be allocated to the release inventory for the production of the final product? First completing the mass balance, we see that 800 units of the intermediate product is being produced along with 200 units of the co-product. Using these output numbers to calculate the mass ratio, 80% of the 20 units of waste (or 16 units) in the first step should be allocated to the final product. This gives a total amount of waste of 26 units (16 + 10 units) for this two-step process based on the functional unit of 1000 units of final product.

Amount allocated to:	Final product	Intermediate product
Raw material A	480	120
Raw material B	336	84
Process waste 1	16	4
Raw material C	150	
Additive	60	
Process waste 2	10	

There are exceptions to a mass-based calculations which may be appropriate at times, such as the use of the market value of the products and co-products. For example, in the process of mining a high-value ore, e.g., gold, a large quantity of mined waste material is produced. Often this waste can be sold as a by-product to be used as roadbed material. Using a mass-based allocation rule would result in the majority of the environmental releases and resource use being allocated to

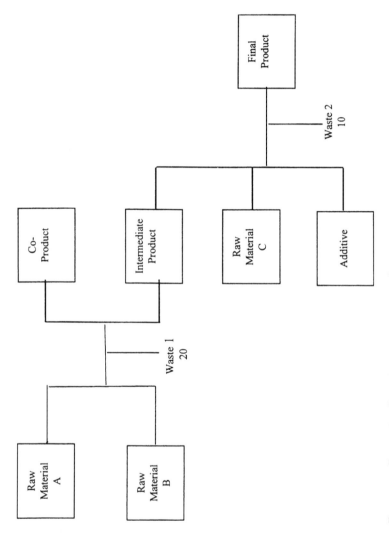

Figure 5 Example allocation among mass flows.

the roadbed material. However, this allocation rule is inappropriate, because the intent of mining is to produce gold, not roadbed material. In this case, since gold is sold at a much higher cost per weight than roadbed material, the market value of the co-products would result in a better allocation of environmental releases and resource use.

For policy analysis, the functional unit should be set at a level that is politically significant, e.g., the number of products produced within a state or within an airshed. A "higher"-level functional unit such as these allows the analyst to include geographic and time considerations which may be significant to the interpretation of the results. However the functional unit is set, it must be consistent with the goal that was defined for the study.

3.2.5 Life Cycle Impact Assessment

The impact assessment component of LCA is a technical, quantitative and/or semiqualitative process which characterizes and classifies the effects of the resource requirements and environmental loadings that are identified in the inventory phase (19). LCIA has three steps:

1. Identification of the impact categories, category indicators, and the models pertaining to them
2. Classification of inventory data into impact categories
3. Characterization of the inventory to yield numerical values of indicators for each impact category.

Impact categories represent the issues of concern for a particular product system, and indicators are readily calculated entities that are believed to be well correlated to the environmental endpoints. Thus, for example, global climate change might be an impact category, and one classifies all greenhouse gas emissions into this category. Then, typically one calculates the category indicator called radiative forcing. The radiative forcing indicator is a mathematical combination of the atmospheric behavior and optical properties of the separate greenhouse gases. Radiative forcing is believed to be well correlated to the endpoints of global climate change, e.g., droughts and floods, or changes in growing season or increased storm activity. However, the links between radiative forcing and these endpoints are not well understood, while the link between the emissions of greenhouse gases and radiative forcing is very well understood and thoroughly modeled by the Intergovernmental Panel on Climate Change.

A uniform method for impact assessment has not yet been agreed to, but the analysis should address all relevant environmental impacts: ecological and human health impacts as well as resource use. Some effects, such as low-level radiation or heat and noise pollution that are not easily modeled, because the level of understanding of the environmental mechanism is low, or because there is poor

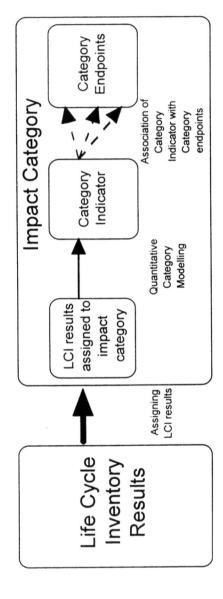

FIGURE 6 Impact categories and indicators.

linkage between the inventory results and effects in the environment. Others (e.g., habitat modification) are known to play a critical role in environmental impacts of products (e.g., agricultural products), but are difficult to model quantitatively. Life cycle impact assessment practice is moving more and more toward using sophisticated fate and transport models to evaluate indicators of environmental impacts.

The choice of impact categories and category indicators and models can drive the collection of inventory data. For example, one might choose to evaluate only minerals whose reserves are predicted to be depleted within 100 years, or some other reasonable time frame. This would eliminate the need to gather data on such materials as bauxite, clay, or iron ore, and would decrease the cost of inventory collection and management.

To date, no "standardized" listing of impact categories to be used in LCA has been established, but several categories are employed in common practice, as shown in Table 5.

The Classification Step. Inventory data need to be classified into the relevant impact categories for modeling. Some emissions have influence on more than one environmental mechanism and must be classified into more than one category. The classic example oif this is oxides of nitrogen, or NO_x, which acts as catalyst in the formation of ground-level ozone (smog), but also is a source of acid precipitation. These substances must be characterized into both categories. One form of NO_x (nitrous oxide, N_2O) is also active as a greenhouse gas. The classification rules for any LCIA must be clearly reported, so that readers of a study understand what exactly was done to the inventory data.

The Characterization Step The goal of life cycle impact assessment is to convert collected inventory inputs and outputs into indicators for each category (aggregates can be system-wide, by life cycle stage, or by unit operation).

TABLE 5 Typical Impact Categories

1.	Stratospheric ozone depletion
2.	Global warming
3.	Human health
4.	Ecological health
5.	Smog formation
6.	Nonrenewable resource depletion
7.	Land use/habitat alteration
8.	Acidification
9.	Eutrophication
10.	Energy: processing/transportation

These indicators do not represent actual impacts, because the indicator does not measure actual damage, such as loss of biodiversity. However, together, they do constitute an ecoprofile for a product or service.

While there is no universally accepted "right" list of impact categories or indicators, basic objectives have been set by the Society of Toxicology and Chemistry (SETAC) that help define categories:

1. Category definition begins with a specific relevant endpoint. Ideally, the endpoint can actually be observed or measured in the natural environment.
2. Inventory data are correctly identified for collection. In principle, those inventory inputs and outputs which relate to the particular impact are identified.
3. An indicator describes the aggregated loading or resource use for each individual category. The indicator is then a representation of the aggregation of the inventory data.

Figure 7 compares the real-world causes and effects (the environmental mechanism) with the modeled world of LCIA. There are many differences between the two. In an LCI, for example, the inventory information is typically modeled as a constant and continuous flow, while in the real world, emissions typically occur in a discontinuous fashion, varying from minute to minute.

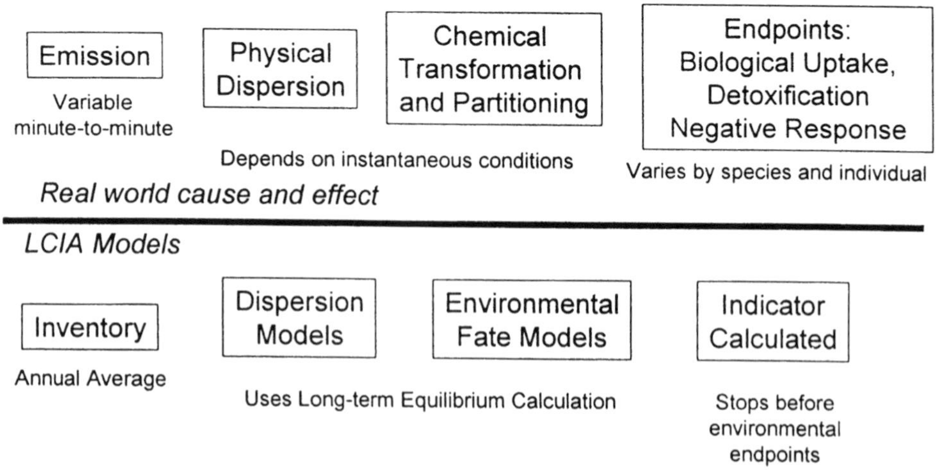

FIGURE 7 Comparison of "real-world" endpoints to LCIA indicators.

Both natural and anthropogenic flows act physically, chemically, and biologically to produce real impacts on the biota (see Figure 8). This series of events is called the environmental mechanism.

In the virtual reality of the environmental model, many assumptions and simplifications are made to yield indicators. Even the best current air dispersion models are accurate only within a factor of two to three, but the level of accuracy is getting better all the time. The principle methodological issue in life cycle impact assessment is the modeling management of often very complex, extended environmental mechanisms. A listing of all possible endpoint impacts is quite long and can look like the following suggested list.

I. Toxicity issues
    A. Human health considerations
        1. Acute human occupational
        2. Chronic human by consumer
        3. Chronic human by local population
        4. Chronic human by occupational
        5. Human health
        6. Human toxicity by ingestion
        7. Human toxicity by inhalation/dermal exposure
        8. Inhalation toxicity
    B. Ecological considerations
        1. Aquatic toxicity
        2. Biodiversity decrease
        3. Endangered species extinction
        4. Environmental toxicity
        5. Landfill leachate (aquatic) toxicity
        6. Species change
        7. Terrestrial toxicity
        8. Eutrophication (aquatic and terrestrial)
II. Global issues
    A. Atmospheric considerations
        1. Acid deposition
        2. Acidification potential
        3. Global warming potential
        4. Stratospheric ozone depletion potential
        5. Photochemical oxidation potential
        6. Tropospheric ozone
    B. Resource considerations
        1. Energy use
        2. Net water consumption
        3. Nonrenewable resource depletion

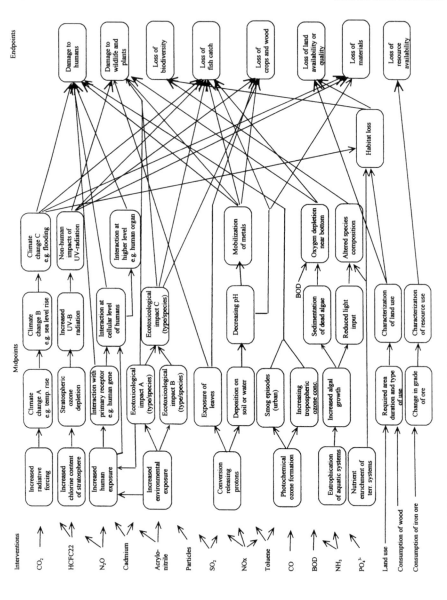

FIGURE 8 Midpoints versus endpoints (20).

    4. Preconsumer waste recycle percent
    5. Product disassembly potential
    6. Product reuse
    7. Recycle content
    8. Recycle potential for postconsumer
    9. Renewable resource depletion
   10. Resource depletion
   11. Resource renewability
   12. Source reduction potential
   13. Surrogate for energy/emissions to transport materials to recycler
   14. Waste-to-energy value

III. Local issues

  A. Waste considerations
    1. Airborne emissions
    2. Hazardous waste
    3. Incineration ash residue
    4. Material persistence
    5. Particulates
    6. Toxic content
    7. Toxic material mobility after disposal
    8. Solid waste generation rate
    9. Solid waste landfill space
   10. Waterborne effluents

  B. Public relation considerations
    1. Esthetic (e.g., odor)
    2. Habitat alteration
    3. Heat
    4. Industrial accidents
    5. Noise
    6. Radiation

  C. Environment considerations
    1. Local land
    2. Local water quality
    3. Physical change to soil
    4. Physical change to water
    5. Regional climate change
    6. Regional land
    7. Regional water quality

Clarifying the environmental mechanism can help determine when impacts may be additive or when they are independent and non-additive. Two illustrative examples are global climate change and stratospheric ozone depletion.

Example 1: Global climate change. The conversion of various greenhouse gases into radiative equivalents is universally applicable based on a scientifically supported mechanism (once a judgment has been made to select a time frame for analysis.)

Example 2: Stratospheric ozone depletion. Stratospheric ozone depletion is caused by the interaction of halogenated free radicals in the upper atmosphere directly reducing concentrations of ozone. However, many ozone-depleting agents are effective greenhouse gases as well. In addition, recent research indicates that greenhouse effects on the lower atmosphere have led to trapping of energy near the earth, and consequent cooling of the upper atmosphere. The stratospheric cooling tends to exacerbate the effects of ozone depleters.

Nevertheless, for the purposes of LCIA models, these two mechanisms are treated separately. This simplification helps develop an overall view of the environmental impacts of industrial systems at a first-order level. In fact, although LCIA modeling tends to be technically complex, one can view LCIAs as extended back-of-the-envelope calculations of realistic worst-case potential impacts.

The goal in assigning LCI results to the impact indicator categories is to highlight environmental issues associated with each. Assignment of LCI results should:

First assign results which are exclusive to an impact category and

Then identify LCI results that relate to more than one impact category, including

Distinguishing between parallel mechanisms (where a given molecule is "used up" in its actions), and serial mechanisms, where a molecule can act in one mechanism, and then in a second mechanism without losing its potency. SO_x acts in parallel mechanisms of allocated between human health and acidification, while NO_x acts in a serial mechanism as a catalyst in photochemical smog formation and then in acidification.

Typically, in impact assessment a "nonthreshold" assumption is used. That is, inventory releases are modeled for their potential impact regardless of the total load to the receiving environment from all sources or consideration of the assimilation capacity of the environment. However, there is a trend, particularly in Europe, to consider thresholds in evaluating indicators. For example, ground-level ozone formation is often calculated as an indicator for photochemical smog. Background levels of ozone are about 20 ppb, while some vegetative damage has been observed at 40 ppb, and human health effects at 80 ppb. All these levels, as well as intermediate levels, have been used in determining indicators for photo-chemical smog.

If LCI results are unavailable or of insufficient quality to achieve the goal of the study, then either iterative data collection or adjustment of the goal is required.

The following sections offer descriptions of current approaches that are being applied to model some of the impact category indicators listed in Table 4. The most simplistic models are described in order to offer insight into the types of approaches that are being considered useful from both a practical aspect as well as least cost.

Stratospheric Ozone Depletion. Ozone depletion is suspected to be the result of the release of man-made halocarbons, e.g., chlorofluorocarbons, that migrate to the stratosphere. For a substance to be considered as contributing to ozone depletion, it must (a) be a gas at normal atmospheric temperatures, (b) contain chlorine or bromine, and (c) be stable within the atmosphere for several years (21).

The most important groups of ozone-depleting compounds (ODCs) are the CFCs (chlorofluorocarbons), HCFCs (hydrochlorofluorocarbons), halons, and methyl bromide. HFCs (hydroflourocarbons) are also halocarbons but contain fluorine instead of chlorine or bromine, and are therefore not regarded as contributors to ozone depletion.

The ozone depletion potential (ODP) is calculated by multiplying the amount of the emission (Q) by the equivalency factor (EF)

$$ODP = Q \cdot EF$$

Current status on reporting equivalency factors uses CFC11 as the reference substance. The equivalency factor is defined as

$$EF_{ODP} = \frac{\text{contribution to stratospheric ozone depletion from } n \text{ over \# years}}{\text{contribution to stratospheric ozone depletion from CFC11-\# years}}$$

General LCA practice uses values that represent ODC's full contribution, but Table 6 also shows factors for 5, 20, and 100 years for some gases. The ozone depletion potential (ODP) is calculated by multiplying a substance's mass emission (Q) by its equivalency factor. These individual potentials can then be summed to give an indication of projected total ODP for substances 1 through n in the life cycle inventory that contribute to ozone depletion:

$$ODP = \sum_{n}^{1} (Q \cdot EF_{ODP})$$

Global Warming. The most significant impact on global warming has been attributed to the burning of fossil fuels, such as coal, oil, and natural gas.

TABLE 6   Equivalency Factors for Ozone Depletion (21)

Substance	Formula	ODP g CFC11/g substance			
		5 years	20 years	100 years	∞
CFC11	$CFCl_3$	1	1	1	1
CFC12	CF_2Cl				0.82
CFC113	$CF_2ClCFCl_2$	0.55	0.59	0.78	0.90
CFC114	CF_2ClCF_2Cl				0.85
CFC115	CF_2ClCF_3				0.40
Tetrachloromethane	CCl_4	1.26	1.23	1.14	1.20
HCFC22	CHF_2Cl	0.19	0.14	0.07	0.04
HCFC123	CF_3CHCl_2				0.014
HCFC124	CF_3CHFCl				0.03
HCFC141b	$CFCl_2CH_3$	0.54	0.33	0.13	0.10
HCFC142b	CF_2ClCH_3	0.17	0.14	0.08	0.05
HCFC225ca	$CF_3CF_2CHCl_2$				0.02
HCFC225cb	CF_2ClCF_2CHFCl				0.02
1,1,1,-Trichlorethane	CH_3CCl_3	1.03	0.45	0.15	0.12
Methyl chloride	CH_3Cl				0.02
Halon 1301	CF_3Br	10.3	10.5	11.5	12
Halon 1211	CF_2ClBr	11.3	9.0	4.9	5.1
Methyl bromide	CH_3Br	15.3	2.3	0.69	0.64

Several compounds, such as carbon dioxide (CO_2), nitrous oxide (N_2O), methane (CH_4), and halocarbons, have been identified as substances that accumulate in the atmosphere, leading to an increased global warming effect.

For a substance to be regarded as a global warmer, it must (a) be a gas at normal atmospheric temperatures, and (b) either be able to absorb infrared radiation and be stable in the atmosphere with a long residence time (in years) *or* be of fossil origin and converted to CO_2 in the atmosphere (21).

Table 7 is a list of substances that are considered to contribute to global warming. Equivalency factors, based on carbon dioxide as 1, are shown for each substance over 20-, 100-, and 500-year spans. The choice of time scale can have considerable effect on how global warming potential is calculated. The 100-year time frame is often selected, unless reasons exist that indicate otherwise.

$$EF_{GWP} = \frac{\text{contribution from } n \text{ to global warming over \# years}}{\text{contribution from } CO_2 \text{ to global warming over \# years}}$$

TABLE 7 Equivalency Factors for Global Warming (21)

Substance	Formula	GWP g CO_2/g substance		
		20 years	100 years	500 years
Carbon dioxide	CO_2	1	1	1
Methane	CH_4	62	25	8
Nitrous oxide	N_2O	290	320	180
CFC11	$CFCl_3$	5000	4000	1400
CFC12	CF_2Cl_2	7900	8500	4200
CFC113	$CF_2ClCFCl_2$	5000	5000	2300
CFC114	CF_2ClCF_2Cl	6900	9300	8300
CFC115	CF_2ClCF_3	6200	9300	13000
Tetrachloromethane	CCl_4	2000	1400	500
HCFC22	CHF_2Cl	4300	1700	520
HCFC123	CF_3CHCl_2	300	93	29
HCFC124	CF_3CHFCl	1500	480	150
HCFC141b	$CFCl_2CH_3$	1800	630	200
HCFC142b	CF_2ClCH_3	4200	2000	630
HCFC225ca	$CF_3CF_2CHCl_2$	550	170	52
HCFC225cb	CF_2ClCF_2CHFCl	1700	530	170
1,1,1-Trichloroethane	CH_3CCl_3	360	110	35
Chloroform	CH_3Cl	15	5	1
Methylene chloride	CH_2Cl_2	28	9	3
HFC 134a	CH_2FCF_3	3300	1300	420
HFC 152a	CHF_2CH_3	460	140	44
Halon 1301	CF_3Br	6200	5600	2200
Carbon monoxide[a]	CO	2	2	2
Hydrocarbons (NMHC)[a]	Various	3	3	3
Partly oxidized hydrocarbons[a]	Various	2	2	2
Partly halogenated hydrocarbons[a]	Various	1	1	1

[a]Contributes indirectly due to conversion into CO_2. Only compounds of petrochemical origin.

The global warming potential (GWP) is calculated by multiplying a substance's mass emission (Q) by its equivalency factor. These individual potentials can then be summed to give an indication of projected total GWP for substances 1 through n in the life cycle inventory that contribute to global warming:

$$GWP = \sum_{n}^{1} (Q \cdot EF_{GWP})$$

Nonrenewable Resource Depletion. This impact category models resources that are nonrenewable, or depletable. The subcategories include:

Fossil fuels
Net non-fuel oil and gas
Net mineral resources
Net metal resources

Some models also include the energy that is inherent in a product that is made from a petroleum feedstock in order to reflect the amount of stock that was diverted and is no longer available for use as an energy source.

This category can also reflect land use as a resource. Land that has been disturbed directly due to physical or mechanical disturbance can be accounted for as a resource that is no longer available either for human use or for ecological benefit (such as providing habitat for a certain species). Other subcategories under the resource category include:

Net marine resources depleted
Net land area
Net water resources
Net wood resources

Scientific Certification Systems (SCS) proposes the following approach in their Life-Cycle Stressor Effects Assessment (LCSEA) model for calculating net resource depletion (22). The LCSEA model is based on (a) the relative rates of depletion of the various resources and (b) the relative degree of sustainability of the resources.

The model considers the key factors that affect resource depletion and includes consideration of recycled material as supplementing raw material inputs. It also takes into account materials that are part of the standing reserve base, i.e. materials, such as steel in a bridge, that will become available as a recovered reserve at some future time. Recycling of metals has great significance for the depletion calculation (see Figure 9).

The elements to be considered in factoring resource depletion include:

Current world reserves
Raw material input (i.e., the amount used)
Amount recycled (both direct and standing stock)
Waste generation
Natural accretion

The reserve base-to-use ratio can be calculated as follows:

$$\frac{\text{Reserve base } (R)}{\text{Use } (U)} = \text{number of years of remaining use left (at current use rate)}$$

$$\frac{\text{Use } (U)}{\text{Reserve base } (R)} = \% \text{ of reserve base used}$$

The recycled resource is linked to the original virgin material use and corresponding reserve base. Emissions are not spatially or temporally lined to the original virgin unit operation. Accounting for all reserve bases:

$$\frac{\text{Waste } (\sum W)}{\text{Reserve base } (R) + \text{recyclable stock } (\sum S)}$$

The current assumption is that only one iteration of recycling and material integrity is sustained. If natural accretion is accounted for, the following formula results:

$$\frac{\text{waste } (\sum W) - \text{natural accretion } (N)}{\text{reserve base } (R) + \text{subsequent uses } (\sum S)}$$

Including the time period in the equation, we get:

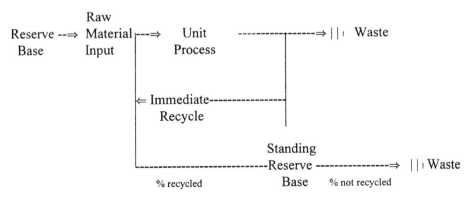

FIGURE 9 Flow of metals, including standing reserve.

$$\frac{\left(\sum W - N\right) \Delta T}{R + \left(\sum S\right) \Delta T} \quad \text{Current assumption: } \Delta T = 50 \text{ years}$$

And accounting for baseline reserve bases,

$$\frac{\left(\sum W - N\right) \Delta T + (R_b - R)}{R + \left(\sum S\right) \Delta T} \quad R_b = \text{a reserve base baseline}$$

Therefore,

$$\text{Resource depletion factor (RDF)} = \frac{\left(\sum W - N\right) \Delta T + (R_b - R)}{R + \left(\sum S\right) \Delta T}$$

Resource depletion of fossil fuels represents a simple application. Accretion is zero and recycling is nil. Thus, wasted resource equals resource used, or

$$\text{RDF} = \frac{(W) * T}{R}$$

The impact for the resource depletion category can then be calculated according to the formula:

Resource depletion indicator (RD) = resource use ×
resource depletion factor (RDF)

For net resources depleted (or accreted), the units of measure express the equivalent depletion (or accretion) of the identified resource. All of the net resource calculations are based on RDFs.

Indicator—net resource	Units of measure
Water	Equivalent cubic meters
Wood	Equivalent cubic meters
Fossil fuels	Tons of oil equivalents
Non-fuel oil and gas	Tons of oil equivalents
Metals	Tons of (metal) equivalents
Minerals	Tons of (mineral) equivalents
Land area	Equivalent hectares

Acidification. For acidification, an equivalency approach is typically applied and the stressor flows are converted into SO_2 or H^+ equivalents. For example, NO_2 is multiplied by $64/(2 * 46) = 0.70$, since this is the molar proton release potency of NO_2 compared to SO_2. Table 8 shows sample calculations using potency factors for an inventory with SO_2, NO_2, and HCl releases. The LCSEA approach takes the calculation one step further and includes an emission loading factor to reflect how much of the inventory release is expected to reach the receiving environment.

Eutrophication. Eutrophication occurs in aquatic systems when the limiting nutrient in the water is supplied, thus causing algal blooms. In fresh water, it is generally phosphate which is the limiting nutrient, while in salt waters it is generally nitrogen which is limiting. In general, addition of nitrogen alone to fresh waters will not cause algal growth, and addition of phosphate alone to salt waters will not cause significant effects. In brackish waters, either nutrient can cause algal growth, depending on the local conditions at the time of the emissions.

Eutrophication is generally measured using the concentration of chlorophyll-a in the water. Waters with less than 2 mg of chlorophyll-a per cubic meter (2 mg chla m^{-3}) are considered "oligotrophic," while those with 2–10 mg chla m^{-3} are considered "mesotrophic," and those with more than 10 mg chla m^{-3} are termed "eutrophic." Waters over 20 mg chla m^{-3} are considered "hypereutrophic."

As waters become mesotrophic, their species assemblages change, favoring species that grow rapidly in the presence of nutrients ("weed" species) over those which grow more slowly. There is some indication that eutrophication in salt waters is the source of the red tides that are a worldwide problem.

Under eutrophic conditions, the algae in the water significantly block light passage, while in hypereutrophic conditions the amount of biomass produced is so high that anoxic conditions occur, leading to fish kills. There are some indications that similar sorts of effects occur in terrestrial systems as well.

The ratio of carbon to nitrogen to phosphorus in aquatic biomass is 106:16:1 (23), on an atomic basis. This ratio is the basis of combining nitrogen and phosphorus in calculating the eutrophication potential of emissions.

(Molar quantity of nitrate + nitrite + ammonia) × Redfield ratio
+ molar quantity of phosphate × [endpoint characterization factor
(fresh, salt water)] = eutrophication indicator

Eutrophication is typically measured in PO_4 equivalents. The EPA has set a concentration of 25 μg PO_4 L^{-1} as the level needed to protect fresh-water aquatic ecosystems from eutrophication.

Energy. While inventory analyses involves the collection of data to quantify the relevant inputs and outputs of a product system, the accounting of electricity as a flow presents a unique challenge. The use of energy audits makes the idea of balancing energy flows around a process a familiar one. However, in LCA the reporting of energy flows is in itself insufficient to perform a subsequent

TABLE 8 Calculating Acidification "Emission Loading" (22)

Unit operation	Inventory emission	LCI result (ton/30a)	Potency factor	Molar equivalent (ton/30a)	Characterization factor	Emission loading (ton/30a)
Coal mining/transport	SO₂	31,620	1	31,620	0.5	15,810
	NO₂	9,660	0.7	6,762	0.3	2,029
	HCl	270	0.88	238	0.5	119
CaO product/transport	SO₂	240	1	240	0.15	36
	NO₂	1,260	0.7	882	0.075	66
Coal use	SO₂	50,190	1	50,190	0.15	7,529
	NO₂	36,480	0.7	25,536	0.075	1,915
	HCl	15,210	0.88	13,385	0.15	2,008
Total				128,853		29,512

TABLE 9 Equivalency Factors for Acidifiers (21)

Formula	Conversion	M_w g · mol	n	EF kg SO_2/ kg substance
SO_2	$SO_2 + H_2O \rightarrow H_2SO_3 \rightarrow 2H^+ + SO_3^{2-}$	64.06	2	1
SO_3	$SO_3 + H_2O \rightarrow H_2SO_4 \rightarrow 2H^+ + SO_4^{2-}$	80.06	2	0.8
NO_2	$NO_2 + \frac{1}{2}H_2O + \frac{1}{4}O_2 \rightarrow 2H^+ + SO_3^{2-}$	46.01	1	0.7
NO	$NO + O_3 + \frac{1}{2}H_2O \rightarrow H^+ + NO_3^- + \frac{3}{4}O_2$	30.01	1	1.07
HCl	$HCl \rightarrow H^+ + Cl^-$	36.46	1	0.88
HNO_3	$HNO_3 \rightarrow H^+ + NO_3^-$	63.01	1	0.51
H_2SO_4	$H_2SO_4 \rightarrow 2H + SO_4^{2-}$	98.07	2	0.65
H_3PO_4	$H_3PO_4 \rightarrow 3H^+ + PO_4^{3-}$	98	3	0.98
HF	$HF \rightarrow H^+ + F^-$	20.01	1	1.6
H_2S	$H_2S + \frac{3}{2}O_2 + H_2O \rightarrow 2H^+ + SO_3^{2-}$	34.03	2	1.88
NH_3	$NH_3 + 2O_2 \rightarrow H^+ + NO_3^- + H_2O$	17.03	1	1.88

impact assessment. Ideally, the environmental impacts associated with energy generation should be captured in the approach. That is, the generation of electricity from fossil fuels should also show the contribution to the emission of global warming gases, solid waste (especially coal ash), etc. This type of detail also allows for the consideration of the use of waste materials in energy recovery operations. Also, the calculation of energy flow should take into account the different fuels and electricity sources used, the efficiency of conversion and distribution of energy flows, as well as the inputs and outputs associated with generation and use of that energy flow. In addition, a more robust assessment may consider an evaluation of the specific sources of electrical power that are contributed to the national energy grid on a more regional approach. This type of consideration is important in determining local impacts. For example, electricity that is produced in Maine is not used in California. Therefore, the impacts of electricity generation based on a national average may not be appropriate.

In the absence of a readily available model that can convert energy-related inventory data into potential impacts based on the fuel source, a fallback position can be to look at the source of the total energy used and identify what percentage is obtained from the national energy grid (which is mainly fossil fuels) and what percentage comes from other sources, such as the burning of waste materials. At this high-level decision point, this information is appropriate and the approach fits the indicator-by-indicator comparison framework.

3.2.6 Weighting

Weighting, also called valuation, assigns relative weights to the different impact indicator categories based on their perceived importance. Since there are various ways in which different individuals consider things to be important, formal valuation methods should make this process explicit and be representative of the individual or group making the final decision.

ISO 14042 requires that weighting of individual categories only be done after fully disclosing unweighted indicators. When comparing two systems, the trade-offs between impacts often require a judgment call to be made in order to arrive at a decision.

Table 10 shows the partial results of an evaluation that was conducted at Fort Eustis, Virginia, as part of ongoing efforts to reduce waste generation from chemical agent-resistant coating (CARC) depainting/painting operations (24). This example focuses on a portion of the evaluation that compared the baseline CARC system with an alternative system using a different primer and thinner combination. The proposed switch to the alternative primer/thinner system was identified as a possible way to reduce the facility's air releases and potential contribution to global climate change.

TABLE 10 Environmental Impact Scores for Baseline and Alternative CARC Systems (24)

Spatial scale	Impact category[a]	Baseline	Alternative
Global	ODP	1.090	0.367
	GLBLWRM	1.013	0.984
	FSLFUELS	1.263	1.180
Regional	ACIDDEP	1.198	1.175
	SMOG	1.114	0.992
	WTRUSE	[b]	[b]
Local	Toxicity:		
	HUMAN	2.150	1.793
	ENVTERR	3.799	2.862
	ENVAQ	1.280	3.540
	LANDUSE	1.577	1.585

[a]ODP = ozone depletion potential; GLBLWRM = global warming potential; FSLFUELS = fossil fuel & mineral depletion potential; ACIDDEP = acid deposition potential; SMOG = smog creation potential; WTRUSE = water use; HUMAN = human health toxicity potential; ENVTERR = terrestrial wildlife toxicity potential; ENVAQ = aquatic biota potential; LANDUSE = land use for waste disposal.
[b]Water use was not reported as an impact because water availability is plentiful where CARC operations are located, and because water is typically treated and reused or released to the environment.

After life cycle inventory data for the raw materials, painting, and disposal of the baseline CARC and alternative system were collected, additional impact information was then included to complete the LCA. A valuation process was conducted on nine selected impact categories using the analytical hierarchy process (AHP) in order to assign weights to the categories. AHP is a recognized methodology for supporting decisions based on relative preferences of pertinent factors.

It should be recognized that valuation is inherently a subjective process. In the CARC study, the results of the valuation process indicated that *relative to this particular group*, the greatest potential environmental concern is ozone depletion (weight = .332). Water use was included in the valuation process, but it was not included in the impact assessment since water is plentiful near CARC operations, and because water is treated and reused or released to the environment. The weights of all the impact categories (in order of decreasing importance) were determined to be as follows:

Ozone depletion	.332
Acidification	.189
Global warming potential	.124
Human health	.099
Photochemical smog formation	.097
Land use	.058
Fossil fuel use	.037
Water use	.025
Terrestrial toxicity	.020
Aquatic toxicity	.020

These weights were multiplied by the normalized inventory data to arrive at the scores shown in Table 10. For most of the impact categories, the difference is not great enough to conclude that there is a preference between these systems. However, for ozone depletion (ODP) and aquatic toxicity (ENVAQ), some differences can be noted. While the ozone depletion score appears to decrease (1.090 to 0.367), showing potential improvement, the environmental aquatic toxicity score appears to increase (1.280 compared to 3.540). Looking back at the inventory data, it is noted that the increased aquatic toxicity is due to increased cadmium and chlorine releases to the wastewater associated with manufacturing the ingredients for the alternative primer.

If the decision is made in favor of selecting the alternative system because of its potentially lower impact on the ozone layer, it is now clear that this decision may result in an increased burden on the wastewater system. The benefit of using life cycle data to support the decision-making process is that the decision is being made in a broader context and with recognition of how the production of the alternative product can be factored in. If, on the other hand, concerns are more immediate and focused on the local aquatic environment, with a higher weight being assigned to aquatic toxicity, the final decision could go the other way, with a preference for the baseline system, depending on whether the inventory data are sufficient to influence the results in addition to an increased weight being placed on aquatic toxicity. In either case, such weighting schemes should be made very explicit in the final analysis.

3.2.7 Interpretation

In the interpretation step of LCA, the results of the inventory and impact modeling are analyzed, conclusions are reached, and findings are presented in a transparent manner. It is critical that the report that results from this activity is clear, complete, and consistent with the goal and scope of the study. ISO 14043 lists key features of life cycle interpretation as follows:

> The use of a systematic procedure to identify, qualify, check, evaluate, and present the conclusions based on the results of an LCA or life cycle interpretation (LCI), in order to meet the requirements of the application as described in the goal and scope of the study;
>
> The use of an iterative procedure both within the interpretative phase and with the other phases of an LCA or LCI
>
> The provision of links between LCA and other techniques for environmental management by emphasizing the strengths and limits of an LCA study in relation to its defined goal and scope

Transparency throughout the interpretation phase is essential. Whenever preferences, assumptions, or value choices are used in the assessment or in reporting, these need to be clearly stated in the final report. The goal of life cycle interpretation is to give credibility to the results of the LCA in a way that is useful to the decision maker.

3.3 Life Cycle Costing

Over 30 years ago, the U.S. Department of Defense recognized that operation and maintenance (O&M) costs were substantial components of the total costs of owning equipment and systems. In fact, ownership costs can far outweigh the costs of procurement. By considering the full costs over the life cycle of the system and the time value of money (e.g., discounting), better choices can be made.

The broader practice of environmental accounting now uses words such as *total cost analysis/assessment* and *life cycle costing* to emphasize that traditional approaches overlook important environmental costs (and potential cost savings and revenues). A firm's cost accounting system traditionally serves as a way to track and allocate costs to a product or process for operational budgeting, cost control, and pricing. In life cycle costing, accurate allocation serves to identify environmental impacts in order to achieve pollution prevention across the entire life cycle.

Life cycle costing has not yet achieved a single functional definition and has been used to mean different things. However, the concept behind it refers to the management application of environmental accounting (e.g., cost accounting, capital budgeting, process/product design) across the life span of a product or process. It is difficult to discern life cycle costing from total cost assessment (TCA), because TCA is sometimes used to refer to a specific application of environmental accounting, such as the life span of a technology or process. TCA is often used to refer to the act of adding environmental costs into capital budgeting, whereas life cycle costing is used more frequently when incorporating environmental accounting into the entire design of a process or product (25).

It is essential to determine the scope of environmental costs to be included in a life cycle costing evaluation, including not only a firm's private costs only (i.e., those that directly affect the firm's bottom a line), but also private and societal costs, some of which do not show up directly or even indirectly in the firm's bottom line. An expanded accounting approach is described in the EPA's *Pollution Prevention Benefits Manual* (26). The manual distinguishes among four levels of costs:

Usual costs (Tier 0): Equipment, materials, labor, etc.
Hidden costs (Tier 1): Monitoring, paperwork, permit requirements, etc.
Liability costs (Tier 2): Future liabilities, penalties, fines, etc.
Less tangible costs (Tier 3): Corporate image, community relations, consumer response, etc.

Further, there is an important distinction between costs for which a firm is accountable and costs resulting from a firm's activities that do not directly affect the firm's bottom line:

Private costs are the costs incurred by a business or costs for which a business can be held responsible. These are the costs that directly affect a firm's bottom line. Private costs are sometimes termed *internal* costs.
Societal costs are the costs of activities, anywhere within the life cycle, which impact on the environment and on society for which the product manufacturer is not directly held financially responsible. These costs do not directly affect the company's bottom line. Societal costs are also referred to as *external costs* or *externalities.* They may be expressed qualitatively, in physical terms (e.g., tons of releases, exposed receptors), or quantitatively, in dollars and cents. Societal costs can be divided as being either *environmental costs* or *social costs.*

Life cycle costing includes all internal plus external costs incurred throughout the life cycle of a product or process. External costs are not borne directly by the company (or the ultimate consumer of the company's goods or services) and do not typically enter the company's decision-making process. The use of electricity can be used to demonstrate the difference between internal and external costs. The generation of electrical power imposes various environmental impacts and costs. Facility construction, operation, and maintenance are costs that are incurred by the electrical generators, who recover the costs through the prices they set to sell their electricity. Other impacts are not borne by the generator and are not reflected in the price. For example, fossil fuel plants emit sulfur dioxide and nitrogen oxides, precursors to acid rain. Life cycle costing would attempt to describe qualitatively or place a dollar value on those impacts to reflect the overall cost to society and the environment, such as human health risk, damage to buildings and other structures (e.g., statues), damage and loss of trees and other plant life, alteration of habitat and resulting animal species loss, etc.

Uncovering and recognizing environmental costs associated with a product, process, system, or facility is an important goal for making good management decisions. Attaining such goals as reducing environmental expenses, increasing revenues, and improving future environmental performance requires paying attention to current and potential future environmental costs. Whether or not a cost is "environmental" is not critical; the goal is to ensure that relevant costs receive appropriate attention.

Inherent in life cycle costing are the same considerations that were discussed in conducting a life cycle inventory: costs that are omitted may skew the results. Also, life cycle costing cannot be used to compare disparate products, but it is a tool for assessing comparable products or processes. Further, the function of the products being compared should be equivalent.

4 CONCLUSIONS

Pollution prevention is a valuable concept for facility managers tasked with environmental protection. It is a method that allows them to think about their operations and identify opportunities to improve their operations. The main goal of pollution prevention is to reduce or eliminate the creation of pollutants and wastes at the source in order to reduce costs and to meet or exceed federal and state regulations on environmental discharges and emissions. Over the years, significant work has been done by various government offices, universities, and industry to demonstrate pollution prevention techniques and effectively transfer this information to wider audiences for implementation. A wealth of material on case studies for many different industrial sectors can be found in the open literature on this subject.

A life cycle perspective in combination with pollution prevention elevates the concept by looking beyond a single process or facility to encompass the environmental aspects that may be affected somewhere else within the entire system. This type of holistic approach to identifying secondary consequences leads the thought process toward sustainability rather than simple environmental protection. It is LCA's key message and the reason why LCA is becoming widely accepted as the basis for approaches to environmental management. The systematic application of life cycle thinking in all aspects of decision making, including process improvement, product selection, and end-of-life management, provides a stronger model for environmental management than does simple pollution prevention. The information that an LCA provides allows for better-informed decision making to occur. As a result, LCA is an environmental management tool and model that is quickly being adopted at the international level. The LCA provides information that is useful not only to the individual facility or corporation but to environmental policy makers in governments.

LCA is a relatively recent technique in environmental protection and sustainability, but much has been learned in the relatively short period of time that has been dedicated to this subject. It is increasingly obvious that life cycle-based approaches are needed to fully evaluate environmental impacts in all our decisions and choices. The wide spectrum of activities involved in a product or process requires that practitioners and method developers test new ways to model LCA and exchange information among themselves and share it with potential users (i.e., environmental decision makers) in order to advance the understanding and application of LCA. LCA is an evolving tool that continues to improve as better site-specific models become coupled with simpler ways to communicate results to the users of life cycle data.

REFERENCES

1. Harry Freeman, Pollution Prevention. In Harry M. Freeman (ed.), *Industrial Pollution Prevention Handbook*. New York: McGraw-Hill, 1995.

2. L. Case, L. Mendicino, and D. Thomas, Developing and Maintaining a Pollution Prevention Program. In Harry M. Freeman (ed.), *Industrial Pollution Prevention Handbook*. New York: McGraw-Hill, 1995.

3. U.S. Environmental Protection Agency, *Facility Pollution Prevention Guide,* EPA/600/R-92/088. Cincinnati, OH: Risk Reduction Engineering Laboratory, May 1992.

4. The Society of Environmental Toxicology and Chemistry, *Life-Cycle Impact Assessment: The State-of-the-Art,* Larry Barnthouse, Jim Fava, Ken Humphreys, Robert Hunt, Larry Laibson, Scott Noesen, James Owens, Joel Todd, Bruce Vigon, Keith Weitz, John Young (eds.), Pensacola, FL: SETAC Foundation, 1997.

5. U.S. Environmental Protection Agency, *Developing and Using Production-Adjusted*

Measurements of Pollution Prevention, EPA/600/R-97/048. Cincinnati, OH: National Risk Management Research Laboratory, September 1997.

6. David P. Evers, Facility Pollution Prevention Planning. In Harry M. Freeman (ed.), *Industrial Pollution Prevention Handbook.* New York: McGraw-Hill, 1995.

7. U.S. Environmental Protection Agency, *Development of Computer Supported Information System Shell for Measuring Pollution Prevention Progress,* EPA/600/R-95/130, NRMRL, Cincinnati, OH: National Risk Management Research Laboratory, August 1995.

8. B. W. Baetz, E. I. Pas, and P. A. Vesiland, Planning Hazardous Waste Reduction and Treatment Strategies: An Optimization Approach. Waste Manage. Res., vol.7, no. 2, pp. 153–163, 1989.

9. Kenneth Humphreys and Paul Wellman, *Basic Cost Engineering.* New York: Marcel Dekker, 1996.

10. International Standards Organization, Environmental Management—Life Cycle Assessment—Principles and Framework, ISO 14040, 1997.

11. International Standards Organization, Environmental Management—Life Cycle Assessment—Goal and Scope Definition and Inventory Analysis, ISO 14041, 1998.

12. International Standards Organization, Environmental Management—Life Cycle Assessment—Life Cycle Impact Assessment, ISO 14042, 2000.

13. International Standards Organization, Environmental Management—Life Cycle Assessment—Life Cycle Interpretation, ISO 14043, 2000.

14. J.A. Fava, R. Denison, B. Jones, M. A. Curran, B. W. Vigon, S. Selke, and J. Barnum (eds.), *A Technical Framework for Life Cycle Assessments.* Pensacola, FL: The Society of Environmental Toxicology and Chemistry, 1991.

15. K. Stone and J. Springer, Review of Solvent Cleaning in Aerospace Operations and Pollution Prevention Alternatives, Environ. Prog., vol. 14, no. 4, pp. 261–272, 1995.

16. U.S. Environmental Protection Agency, *Streamlined Life-Cycle Assessment of 1,4-Butanediol Produced from Petroleum Feedstocks versus Bio-Derived Feedstocks,* in Cincinnati, OH: National Risk Management Research Laboratory, September 1997.

17. The Society of Environmental Toxicology and Chemistry, *Streamlined Life Cycle Assessment,* Joel Ann Todd and Mary Ann Curran (eds.), Pensacola, FL, June 1999.

18. U.S. Environmental Protection Agency, *Life Cycle Assessment: Inventory Guidelines and Principles,* EPA/600/R-92/245. Cincinnati, OH: Risk Reduction Engineering Laboratory, February 1993.

19. U.S. Environmental Protection Agency, *Life-Cycle Impact Assessment: A Conceptual Framework, Key Issues, and Summary of Existing Methods.* prepared by the Research Triangle Institute (RTI), July 1995.

20. U. de Haes, O. Jolliet, G. Finnveden, M. Hauschild, W. Krewitt, and R. Mueller-Wenk, Best Available Practice Regarding Impact Categories and Category Indicators in Life Cycle Impact Assessment, SETAC Life Cycle Impact Assessment Workgroup discussion paper, February 1999.

21. Henrik Wenzel, Michael Hauschild, and Leo Alting, *Environmental Assessment of Products,* London, UK: Chapman & Hall, 1997.

22. S. Rhodes, F. Kommonen, and R. Schenck, Evolution of Life-Cycle Assessment as an Environmental Decision-Making Tool: ISO 14042 and Life Cycle Stressor Efforts Assessment (LCSEA), Workshop booklet, February 1998.

23. A. C. Redfield, The Process of Determining the Concentration of Oxygen, Phosphate, and Other Organic Derivatives within the Depths of the Atlantic Ocean. *Pap. Phys. Ocean. Meteor. 9,* 1942.
24. U.S. Environmental Protection Agency, Life Cycle Assessment for Chemical Agent Resistant Coating, EPA/600/R-96/104, prepared by Battelle and Lockheed-Martin for the National Risk Management Research Laboratory, Cincinnati, OH, 1996.
25. Allen White, D. Savage, and K. Shapiro, Life Cycle Costing: Concepts and Applications. In M. A. Curran (ed.), *Environmental Life Cycle Assessment.* New York: McGraw-Hill, 1996.
26. U.S. Environmental Protection Agency, *Pollution Prevention Benefits Manual,* EPA230/R-98/100, October 1989.
27. U.S. Environmental Protection Agency, *Pathway to Product Stewardship: Life-Cycle Design as a Business Decision-Support Tool,* EPA/742/R-97/008. Office of Pollution Prevention and Toxics, December 1997.

GLOSSARY

Functional unit   The measure of a life cycle system used to base reference flows in order to calculate inputs and outputs of the system

Inventory   See *Life cycle inventory.*

ISO   International Standards Organization (or International Organization of Standardization).

Life   (1) Economic: that period of time after which a product, machine, or facility should be discarded because of its excessive costs or reduced profitability. (2) Physical: that period of time after which a product, machine, or facility can no longer be repaired in order to perform its designed function properly.

Life cycle assessment   Evaluation of the environmental effects associated with any given activity from the initial gathering of raw materials from the earth to the point at which all materials are returned to the earth; this evaluation includes all releases to the air, water, and soil.

Life cycle cost   The sum of all discounted costs of acquiring, owning, operating, and maintaining a project over the study period (i.e., the life of the product or process). Comparing life cycle costs among mutually exclusive projects of equal performance has been used as a way to determine relative costs.

Life cycle impact assessment   A scientifically based process or model which characterizes projected environmental and human health impacts based on the results of the life cycle inventory.

Life cycle inventory   An objective, data-based process of quantifying energy and raw material requirements, air emissions, waterborne effluents, solid waste, and other environmental releases throughout the life cycle of a product, process, or activity.

Pollution prevention   The use of materials, processes, or practices that reduce or eliminate the creation of pollutants or wastes at the source.

Pollution prevention opportunity assessment   The systematic process of identifying areas, processes, and activities which generate excessive waste streams or waste by-products for the purpose of substitution, alteration, or elimination of the waste.

POTW (Publicy Owned Treatment Works)   Any device or system used to treat (including recycling and reclamation) municipal sewage or industrial wastes of a liquid nature that is owned by a state, municipality, intermunicipality, or interstate agency [defined by Section 502(4) of the Clean Water Act].

RCRA (Resource Conservation and Recovery Act of 1976)   Amending the Solid Waste Disposal Act (SWDA), the RCRA established a regulatory system to track the generation of hazardous substances from the time of generation to disposal. The U.S. Congress declares it to be the national policy of the country that, whenever feasible, the generation of hazardous waste is to be reduced or eliminated as expeditiously as possible. Waste that is nevertheless generated should be treated, stored, or disposed of so as to minimize the present and future threat to human health and the environment (40 USC 6902).

Waste minimization   Approaches or techniques that reduce the amount of RCRA-regulated wastes generated during industrial production processes; the term applies to recycling and other efforts to reduce waste volume.

16

Application of Life Cycle Assessment

W. David Constant
Louisiana State University and A&M College, Baton Rouge, Louisiana

1 INTRODUCTION

Application of life cycle assessment (LCA) ensures that environmental impact is explicitly included in the design process, yielding the "best" alternative. The "best" choice can be difficult in the final analysis to assess, as many factors may cause an objective LCA to fall into a gray or subjective area. Graedel (1) presents an excellent methodology for streamlined LCA with a matrix approach, yielding objective results, and explores the approaches used by major manufacturers in obtaining data for the matrices. Others (2,3) present basic studies and applications, and there are many articles and texts available between basics and detailed methods. The objective of this chapter is to explore the application of LCA for waste site remediation, as an example of the extension of LCA to areas beyond manufacturing goods and consumer products.

2 BASICS FROM CHEMICAL ENGINEERING

Application of LCA to assess a waste site remedy is making use of the basics of chemical engineering and related fields, the material and energy balances, with a few other topics included, such as economics, eco- and/or health risk assessment, and regulatory constraints. Key components for assessment in the material and energy balance approach are (a) the problem basis, (b) boundaries, (c) generation and accumulation, and (d) time. These terms and their components must be consistent in analysis of, for example, technology comparisons in remediation or waste management methods for treatment of a process stream or contaminant-media matrix. In terms of a general equation, we may write for any LCA assessment:

$$\text{Input} = \text{output} + \text{generation} + \text{accumulation} \tag{1}$$

It should be noted that the generation term may be positive or negative (if material is consumed). It is critical that the terms are handled in a consistent fashion for use in LCA. The basis of "measurement" is key—should we look at, as with paper cups versus ceramic cups, the number of cups or the number of cups per use? The decision depends on what we are investigating in the LCA. Also, in extension of LCA to site remediation, the value of the source term, for example, the mass of waste within the source of a groundwater contaminant plume, may be very difficult to obtain without significant uncertainty associated, making the LCA an estimate with significant error. The accumulation term is critical for understanding the burden with time for unsteady-state processes. Since time is included in the accumulation term, the length of the LCA, or time line, must be selected initially to satisfy stakeholders in the application (regulators, landowners, industry, public, etc.). One should determine what interval for study is appropriate, as regulations use years, decades, and in some cases centuries as windows into model predictions of the fate and transport. Generation components will include transformation, reaction, and other terms that change the physical or chemical properties of the material of interest within the boundaries being explored. Boundaries must be carefully defined in order to ease the solution of the mathematical portion of the problem, be representative of the system under study, and also to reflect the focus of the LCA being performed. Equations may be written from this basic form for any process wherein boundaries can be determined, in either differential form or integral manner. Further, those without accumulation terms are considered steady state, and of course, many processes have numerous streams to consider. The matrix approach is typically utilized in such process balances, due to the complexity found when processes are linked for a final solution, to ease algebra. Herein we obtain the overall balances and individual component balances for overall matrix solutions. Energy balances are handled in a similar fashion, incorporating thermodynamic properties and allowing for both open and closed systems, determined by the transfer of mass across a boundary. A batch process would be considered closed by definition and continuous systems are defined as open for energy balance purposes. Basics of material and energy balances many be found in a number of chemical engineering texts, such as Felder and Rousseau (4) and others.

3 ADDITION OF CHEMODYNAMICS

In order to take LCA beyond its usual scope to assess a waste site remedy, we need not only to include the mass and energy balances for environmental compartments, but also the models and mechanisms of inter- and intramedia transport. The concepts and models of chemodynamics are provided in significant detail by Thibodeaux (5), and by Reible and Choi (6). For the purposes of this discussion, it is assumed that appropriate models of the fate and transport of contaminants for a case are known, estimated, or available in some fashion. Since the objective of the LCA in remediation is to assess the burden being placed on the environment by various remedies, the application of realistic and appropriate models is essential and found in references above (5,6) for certain cases.

4 APPLICATION

With the objective being assessment of the environmental (eco- and human health) burden from various remedies in a waste site problem, we may employ, as with LCA, risk assessment in the reverse, following the methodology of Hwang (7) as extended by Constant et al. (8). The exposure is summed from different routes to get total exposure, E_t, which, to have an acceptable risk, should not exceed the reference dose level (RL). Exposures are calculated as in any other assessment, based on concentration, time, body weight, etc. Key here is the concentration at the exposure point (receptor), which is a function of the concentration in the soil or other medium. This function is the subject of much modeling work as described above, as it is essential to have accurate representations of the fate and transport of the contaminants in and across media under investigation. Thus, one may apply the above exposure method for an acceptable risk, combined with LCA, as follows (8). First, land use or other resource needs are established, followed by assuming a chemical management or remediation scheme or treatment train concept. Then, incorporating regulations, laws, and liabilities, chemicals are traced from "cradle to grave" throughout the process as in a typical LCA. Next, risk is assessed for the scheme (7) with all material and energy balances incorporated, along with eco-risk as appropriate. With these balances completed including appropriate fate and transport models, economics of the process(es) being considered are established. If the economics are acceptable, then the technology or waste processing is appropriate for the present level of available technology considered (or assumed) in the first step. If the economics are unfavorable, another treatment, manufacturing, or remediation scheme should be applied to this integrated management methodology for comparison to the first attempt. Thus, methods can be evaluated for risk-based remediation or manufacturing as in LCA for the most economical approach. It should be noted here that technology is on a moving time line, and what is successful and economical today will need to be reassessed several times in the future as chemical fate and transport modeling, our understanding of the environment, technologies, and regulations all change. In this manner, "new" technologies such as monitored natural attenuation (MNA) can be assessed alongside and with active remedies as described in the example below.

5 EXAMPLE OF ASSESSMENT

The Petro Processors, Inc. (PPI), site is one of the most significant Superfund sites in the United States. While it is not directly under Superfund, due to agreement among the parties and existing consent decree in the U.S. District Court, Middle District of LA, it provides an excellent example, via hindsight, of the utility of LCA for risk-based remediation methodology as described by Constant et al. (8). Details of the site(s), being named Brooklawn and Scenic, for the nearby roadways, may be found at EPA's Region VI Web site (9), and are not described here. The site(s) contain approximately 400,000 tons of chlorinated hydrocarbon wastes, including hexachlorobenzene, hexachlorobutadiene, TCE, DCE, lower chlorinated products, and numerous other petrochemical wastes of similar nature. The mixture forms a dense, nonaqueous, viscous phase, which is currently covered and maintained by hydraulic containment of the groundwater and recovery of the organic phase (both of which are treated on-site) under strict regulatory requirements. Louisiana State University (LSU) has served as the Court's Expert for the last 10 years regarding cleanup of this site, with the role being research into advanced technologies and assisting the court in monitoring and assessing the remediation.

Just prior to LSU's involvement, the remedy was to remove the waste, stabilize the material, and place it in a lined 1×10^6 yd^3 vault, with some hydraulic containment and recovery of site areas found difficult to remove due to the high water table. However, due to volatile emissions exceeding fence-line limits, the removal method was abandoned. The next remedy to be put in place was active hydraulic containment (expansion of the previous pumping layout), with several hundred wells to remove organics and contaminated water over the long term in order to protect the underlying aquifer and prevent migration. To date, only about 1% of the free-phase material has been removed, but containment appears successful, at the expense of treating millions of gallons of water containing only trace contamination levels. As this pumping method was being installed at Brooklawn, via ongoing research by LSU and site personnel for augmentation of the pumping operation by both passive and active technologies, intrinsic biodegradation and hence natural attenuation was studied and then incorporated into the remedy. The addition of biodegradation, sorption, dispersion, etc. as required within the lines of evidence of MNA as presented by EPA (10), has had a significant impact on the remedy of the PPI site.

It appears at this time that active pumping of significant water volumes may be greatly reduced, while still focusing on source removal (NAPL), with the understanding that significant residuals remain bound in these systems. The MNA component is proposed to contain and reduce the size of contaminated groundwater over time, with acceptable risk. Thus, using the same receptors and risk issues, incorporation of MNA into the PPI remedy promises to reduce significantly the cost of the remedy, without increasing eco- or human health risk. In a recent comparison, starting with the same source and endpoints, in the same time frame, MNA with limited hydraulic containment (a few source removal wells) was evaluated against active hydraulic containment and recovery (many pumping wells across the site) for the Scenic site of PPI. This assessment included monitoring and other costs associated with these technologies, and a 70% cost savings was found by incorporation of MNA into the remedy. This was due mainly to the reduced number of wells to be installed and the significant reduction in treatment of produced water. While this example case of remedy comparison has occurred during the remediation of PPI, it clearly shows that we have the tools of LCA, risk-based remediation, and MNA at our disposal. When properly integrated, one can provide acceptable waste management schemes prior to initiation of a manufacturing project or site remediation. However, as technology changes, as stated earlier, this assessment must be ongoing.

REFERENCES

1. T. E. Graedel, *Streamlined Life-Cycle Assessment*, pp. 97–98. Englewood Cliffs, NJ: Prentice Hall, 1998.
2. P. C. Schulze (ed.), *Measures of Environmental Performance and Ecosystem Condition*. Washington, DC: National Academy Press, 1999.
3. R. J. Walter, *Practical Compliance with the EPA Risk Management Program*. New York: Center for Chemical Process Safety of the AIChE, 1999.
4. R. M. Felder and R. W. Rousseau, *Elementary Principles of Chemical Processes*, pp. 81–86. New York: Wiley, 1978.
5. L. J. Thibodeaux, *Chemodynamics: Environmental Movement of Chemicals in Air, Water and Soil*. New York: Wiley, 1979.
6. B. Choy and D. D. Reible, *Diffusion Models of Environmental Transport*. Boca Raton, FL: Lewis Pub., CRC Press, 2000.
7. S. T. Hwang. *J. Environ. Sci. Health, A,* vol. A27, no. 3, pp. 843–861, 1992.
8. W. D. Constant, L. J. Thibodeaux, and A. R. Machen, Environmental Chemical

Engineering: Part I—Fluxion; Part II—Pathways. *Trends Chem. Eng.*, vol. 2, pp. 525–542, 1994.

9. U.S. Environmental Protection Agency (EPA) Region VI Website, Information on Superfund sites, Louisiana, Petro Processors, Inc., http://www.epa.gov/earth1r6/6sf/6sf-la.htm.

10. Robert S. Kerr Environmental Research Center (U.S. EPA), Natural attenuation short course materials, Ada, OK, December 2–4, 1997.

17

Risk-Based Pollution Control and Waste Minimization Concepts

Gilbert J. Gonzales

Los Alamos National Laboratory, Los Alamos, New Mexico and
New Mexico State University, Las Cruces, New Mexico

1 INTRODUCTION

Ecological risk assessment is defined as "the qualitative or quantitative appraisal of impact, potential or real, of one or more stressors (such as pollution) on flora, fauna, or the encompassing ecosystem." The underlying principles behind risk reduction and integrated decision making that are detailed in U.S. Environmental Protection Agency (EPA) strategic initiatives and guiding principles include pollution prevention (1). Pollution control (PC) and waste minimization (WM) are probably the most effective means of reducing risk to humans and the environment from hazardous and radioactive waste. Pollution control can be defined as any activity that reduces the release to the environment of substances that can cause adverse effects to humans or other biological organisms. This includes pollution prevention and waste minimization. Waste minimization is defined as pollution prevention measures that reduce Resource Conservation and Recovery Act (RCRA) hazardous waste (2). Reduced risk is one benefit of these practices, and it results most directly from lower concentrations of contaminants entering the environment from both planned and accidental releases.

Although it is difficult to quantify the reductions in the release of contaminants to the environment that have resulted from reductions in waste, the reductions have most assuredly reduced risk to humans and the environment posed by toxicants. Using examples from Los Alamos National Laboratory (LANL), we will discuss the interrelatedness of pollution prevention/waste minimization with risk reduction and how risk assessment can be generally applied to the field of pollution prevention and waste minimization. While emphasis in this chapter is on "ecological risk assessment," the concepts and principles can also be applied to human health risk assessment. For purposes of this chapter, distinction is not made between pollution prevention, waste minimization, recycling or other waste management techniques, and other related terms. Rather, emphasis is on source reduction, which includes any practice that reduces the amount of contaminant entering a waste stream or the environment.

It is evident that the EPA's hierarchy of general pollution prevention and waste minimization methods is implicitly related to risk reduction (2). The preferred method, source reduction, results in the greatest reduction in human and ecological risk from contaminants. Source reduction is followed in effectiveness by recycling, treatment, and disposal. While it is possible, depending on the technology used, that recycling and treatment may increase risk to worker health because of increases in contact handling of waste, the prioritized list (source reduction → recycling → treatment → disposal) generally results in a decrease in risk to the public and the environment as one progresses from the least preferred to the most preferred method. The reason for this is simple. The preferred method results in little, if any, release of contaminants into the environment compared to the less preferred methods; with the less preferred methods, not only does the quantity of contaminants potentially released into the environment increase, the potential to release them increases. With this premise, it is then important to realize the magnitude of risk reduction achieved by employing pollution prevention and waste minimization at facilities such as LANL.

2 SITE DESCRIPTION AND CHARACTERIZATION

LANL is located in north-central New Mexico, approximately 60 miles northwest of Santa Fe (Figure 1). LANL is a U.S. Department of Energy-owned complex managed by the University of California that was founded in 1943 as part of the Manhattan Project to create the first nuclear weapon. Since then, LANL's mission to design, develop, and test nuclear weapons has expanded to other areas of nuclear science and energy research.

The Laboratory comprises dozens of individual technical areas located on 43 square miles of land area; about 1400 major buildings and other facilities are part of the Laboratory. The Laboratory is situated on the Pajarito Plateau, which consists of a series of fingerlike mesas separated by deep east-to-west–oriented

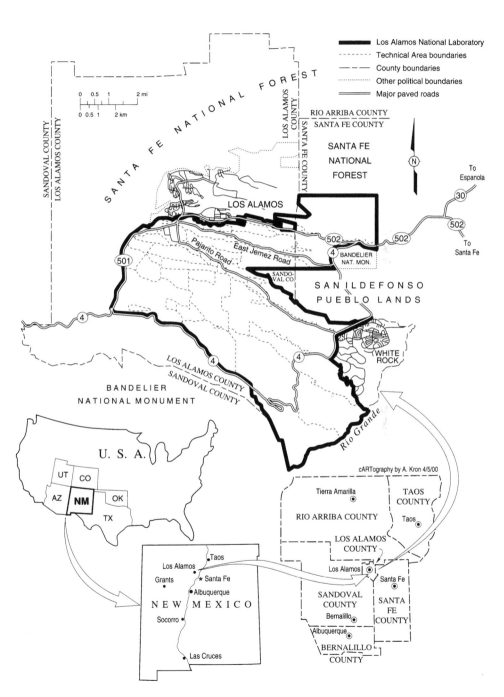

FIGURE 1 Location of the Los Alamos National Laboratory.

canyons cut by intermittent streams (Figure 2). Mesa tops range in elevation from approximately 7800 ft on the flanks of the Jemez Mountains to about 6200 ft at their eastern termination above the Rio Grande.

Researchers at Los Alamos work on initiatives related to the Laboratory's central mission of enhancing global security as well as on basic research in a variety of disciplines related to advanced and nuclear materials research, development, and applications; experimental science and engineering; and theory, modeling, analysis, and computation. As a fully functional institution, LANL also engages in a number of related activities including waste management; infrastructure and central services; facility maintenance and refurbishment; environmental,

FIGURE 2 Overhead view of the topography in and around the Los Alamos National Laboratory.

ecological, cultural, and natural resource management; and environmental restoration, including decontamination and decommissioning. As result of the scientific and technical work conducted at Los Alamos, the Laboratory generates, treats, and stores hazardous, mixed, and radioactive wastes.

About 2120 contaminant potential release sites (PRSs) have been identified at LANL. The LANL PRSs are diverse and include past material disposal areas (landfills), canyons, drain lines, firing sites, outfalls, and other random sites such as spill locations. Categorizing contaminants into three types—organics, metals, and radionuclides—Los Alamos has all three present. The contaminants include volatile and semivolatile organics, polychlorinated biphenyls (PCBs), asbestos, pesticides, herbicides, heavy metals, beryllium, radionuclides, petroleum products, and high explosives (3). The primary mechanisms for potential contaminant release from the site is surface-water runoff carrying potentially contaminated sediments and soil erosion exposing buried contaminants. The main pathways by which released contaminants can reach off-site residents are through infiltration into alluvial aquifers, airborne dispersion of particulate matter, and sediment migration from surface-water runoff. Like many other sites, the predominant pathway by which contaminants enter terrestrial biological systems is the ingestion of soil, intentional or not.

Diverse topography, ecology, and other factors make the consideration of issues related to contamination in the LANL environment complex. "Since 1990, LANL's environmental restoration project has conducted over 100 cleanups. The environmental restoration project has also decommissioned over 30 structures and conducted three RCRA closure actions during this period. Schedules have been published for the planned cleanup of approximately 700 to 750 additional sites. This schedule encompasses a period of about 10 years, beginning with fiscal year 1998. The number of cleanups per year varies from approximately 100 in fiscal year 2002 to 18 in fiscal year 2008. An important and integral part of this pollution prevention technology and of identifying interim protection measures is ecological risk assessment" (3).

3 POLLUTION PREVENTION AND WASTE MINIMIZATION AT LANL

The pollution prevention program at the Laboratory has been successful in reducing overall LANL wastes requiring disposal by 30% over the last 5 years. The program is site wide but has facility-specific components, especially for the larger generators of radioactive and hazardous chemical wastes. Past reductions indicate that waste generation in the future should be less than that projected. The Site Pollution Prevention Plan for Los Alamos National Laboratory (4) describes the LANL Pollution Prevention and Waste Minimization Programs, including a general program description, recently implemented actions, specific volume reductions resulting from recent actions, and current development/demonstration efforts that have not yet been implemented.

More specifically, LANL has achieved reductions in the generation of hazardous waste, low-level radioactive waste (LLW), and mixed LLW. These and two other waste types are defined as follows:

> Hazardous waste—Any substance containing waste that is regulated by RCRA, the Toxic Substances Control Act, and New Mexico as a Special Waste.
>
> Low-level radioactive waste (LLW)—Radionuclide-containing substances with a radionuclide activity (sometimes referred to as concentration) of less than 100 nCi/g.
>
> Mixed LLW—Substances containing both RCRA constituents and LLW.
>
> Transuranic radioactive (TRU) waste—Radionuclide-containing substances with a radionuclide activity (concentration) equal to or greater than 100 nCi/g.
>
> Mixed TRU waste—Substances containing both RCRA and TRU waste.

Estimated reduction rates of chronically generated waste, by waste type, over a 6-year period are (D. Wilburn, personal communication, 2000)

> Hazardous waste: 11%/yr (65% from 1993 to 1999)
> LLW: 11%/yr (67% from 1993 to 1999)
> Mixed LLW: 12%/yr (72% from 1993 to 1999)

Production of TRU and mixed TRU waste combined has increased by an average of 38%/yr (228% from 1993 to 1999).

The Laboratory has dozens of pollution prevention projects ongoing and planned. Some of the efforts are pollution prevention and waste minimization in the strictest sense and some (e.g., separation of waste types or satellite treatment followed by centralized treatment) are pollution control in the broadest sense, including efforts where cost savings is the primary goal and pollution control is a secondary benefit. Three examples of pollution control/waste minimization at LANL follow.

> *Generator Set-Aside Fee-Funded (GSAF) Plutonium Ingot Storage Cubicle Project.* "An aliquot casting and blending technique is under imple- mentation at LANLs Plutonium Facility. The aliquot process allows out-of-specification plutonium to be blended with other plutonium so that the final mixed batch meets specifications and is uniform. This avoids the cost and waste generation related to reprocessing out-of-specification plutonium ingots through the nitric acid line. In addition the more uniform product will reduce the reject rate and will avoid reprocessing and remanufacturing wastes. This project will fabricate a storage system

with 20 cubicles. It is expected that 12.5 m^3 of TRU waste can be avoided over the 25 year life of the cubicle system. This project has an estimated 100% return on investment when prior investments are included in the base project cost. This project is the final step necessary to achieve aliquot blending. The GSAF program is funding purchase of the storage cubicle and the operating group has programmatic support for installing and qualifying the cubicle."

GSAF Reduction of Acid Waste and Emissions. "The Laboratory's Analytical Chemistry Sciences Group's performance of analytical services at one of the Laboratory's technical areas requires the dissolution of up to 20,000 samples per year. The current process is hot plate digestion which requires large quantities of chemicals, mostly acids, and results in the volatilization and release to the atmosphere of 90% of those chemicals. The yearly consumption of HNO$_3$, HCl, HF, HClO$_4$ and NH$_4$OH is estimated to be 3395 kgs. About 3095 kgs of these chemicals are volatilized and become air emissions from the stacks; the balance is diluted and discharged to the Laboratory's Radioactive Liquid Waste Treatment Facility via the LLW acid line. The stack emissions constitute about 30% of the Laboratory's annual hazardous air pollutant discharges. By switching to a microwave and muffle furnace oven process, annual consumption of the chemicals listed above can be reduced to about 370 kgs, an 89% reduction. It is estimated that implementing the new process will reduce the Laboratory's hazardous air pollution discharges by three tons. The estimated return on investment for this project is 82%."

Waste Minimization and Microconcentric Nebulization for Inductively Coupled Plasma-Atomic Emission Spectroscopy. "One of the most popular techniques for multi-element analysis is inductively coupled plasma atomic emission spectroscopy (ICP-AES). The most common method of introducing samples into the ICP torch is pneumatic nebulization. The standard nebulizers, in conjunction with typical spray chambers, exhibit very low transport efficiency. Because of this inefficiency the ICP-AES generates almost a liter per day of rinse water that consists primarily of dilute nitric acid mixed with other contaminants. The other contaminants can be TRU waste, LLW, mixed LLW or hazardous waste depending on the samples analyzed. Under this project an existing microconcentric nebulizer will be deployed with the ICP-AES and optimized for analysis of trace elements in a variety of plutonium-containing matrices. It is expected that using an optimized microconcentric nebulizer will reduce the daily rinse water from the spray chamber drain to about 50 ml. This represents a 95% or 200 l/yr reduction in the volume of waste requiring disposal. A 50 liter per year reduction in LLW plastic sample containers is expected. Reducing the volume of samples in the glovebox will simplify the ICP-AES operation, reducing various risk parameters. The return on investment ranges from 10–50%."

4   ECOLOGICAL RISK ASSESSMENT

Ecological risk assessment is defined as "the qualitative or quantitative appraisal of impact, potential or real, of one or more stressors (such as pollution) on flora, fauna, or the encompassing ecosystem." The Laboratory is employing the EPA's iterative and tiered approach to ecological risk assessment whereby less complex assessments with greater conservatism and uncertainty are employed first, followed by the advancement of potential problem areas and contaminants to progressively more complex and realistic assessments with lower uncertainty. The purpose of this section is not to present ecological risk assessment methods, as there are many excellent sources on this subject (5–8). Rather, the purpose is to introduce concepts in ecological risk assessment that could be considered in designing pollution control and waste minimization programs.

4.1   History at the Los Alamos National Laboratory

Ecological risk assessment ("ecorisk") at the Laboratory is a work in progress. Methods for ecorisk screening and "tier 2" assessments have been in development and implementation since approximately 1993. Most screenings and assessments completed at the Laboratory thus far are based on the U.S. EPA hazard quotient method, whereby hazard quotient values are calculated for receptors for each contaminant by area and may be thought of as a ratio of a receptor's exposure at the site to a safe limit or benchmark:

$$HI_i = \sum_{j=1}^{n} HQ_{ij} \tag{1}$$

where

HI$_i$ is the hazard index for receptor I to n contaminants of potential concern (COPCs), and

$$HQ_{ij} = \frac{\text{exposure}_{ij}}{\text{safe limit}_{ij}} \tag{2}$$

where

HQ$_{ij}$ is the hazard quotient for receptor I to COPC j
exposure$_{ij}$ is the dose to receptor i for COPC j
safe limit is an effects or no effects level, such as a chronic no-observed-adverse-effects level (NOAEL)

4.1.1 Screening

In 1995 a very conservative method for screening contaminants and contaminated areas was developed (9). The method involved the selection of ecological endpoints that focused on animal feeding guilds, a listing of candidate contaminants, exposure/dose–response estimation, estimation of food and soil ingestion, and risk characterization. A comparison value or safe limit was based on foraging mode, behaviors, types of food consumed, the amount consumed, and NOAELs or radiation dose limits when available. The safe limits for radionuclides were largely laboratory-derived rat-based values used in human risk assessments. As such, this and the use of other conservative factors or assumptions resulted in a method that likely overestimated potential risk by orders of magnitude, especially for radionuclides. At least one application of the method is known to have occurred (10).

Currently at LANL, ecorisk screening encompasses qualitative "scoping evaluation" as the basis of problem formulation and "screening evaluation" to identify contaminants of potential concern (COPCs) by exposure media. The screening evaluation focuses on identifying sites that require further investigation and risk characterization (11). A key component to the screening evaluation, and one of interest to PC/WM, is the ecological screening level (ESL) concept. An ESL is basically a contaminant safe limit, or acceptable effects level, below which measurable effects are not expected. ESLs are most useful in units of contaminant amount per quantity of soil, so that measurements of contaminant levels in soil can be quickly and directly compared. The ESLs are developed for each ecological receptor of interest, each chemical, and are media specific. They are determined so that if an area has levels of a chemical above the ESL in any medium, then the area is deemed to pose a potentially unacceptable risk to ecological receptors. Calculations of ESLs require toxicity information, including toxicity reference values (TRVs), preferably chronic NOAELs, and knowledge of transfer coefficients including bioconcentration and bioaccumulation factors. Details on this information and on the process for calculating and selecting ESLs are documented in a LANL report (11); however a summary is provided here.

Nonradionuclides. "Although soil ESLs are based on exposure of terrestrial receptors—plants, invertebrates (earthworms), and wildlife—they are determined differently for each receptor. The different approaches are required because of the different ways that toxicological experiments are performed for these organisms. For plants, earthworms and other soil-dwelling invertebrates, effects are based on the concentration of a COPC in soil. Therefore, ESL values are directly based on effects concentrations and modeling is not required. For plants and invertebrates the soil ESL was 'back-calculated' from No-Observed-Effect-Concentrations. Exposure to wildlife, however, is dependent on exposure of the organism to a chemical constituent from a given medium (such as soil or foodstuff) through direct and indirect means (i.e., ingestion, inhalation, and dermal). Wildlife ESL values were based on the dietary regimen of the receptor, including consumption of plants, invertebrates, and vertebrate flesh, with some incidental soil ingestion (11)." The soil ESL was "back-calculated" as the soil concentration of a COPC resulting from exposure to the NOAEL. Starting with the EPA's general terrestrial wildlife exposure model (12), inverting to convert soil concentration to dose, relating food intake to soil intake, and solving for COPC- and receptor-specific ESLs, models such as Eq. (3) for omnivores were used to compute ESLs for nonradionuclides:

$$ESL_{ij} = \frac{NOAEL_{ij}}{I_i \cdot (fs_i + fp_i \cdot TF_{plant,j} + fi_i \cdot TF_{invert,j}}\tag{3}$$

where

ESL$_{ij}$ is the soil ESL for omnivore i and COPC j (mg/kg)
NOAEL$_{ij}$ is the NOAEL for omnivore i and COPC j (mg/kg/day)
fs$_i$ is the fraction of soil ingested by omnivore i, expressed as a fraction of the dietary intake
I_i is the normalized daily dietary ingestion rate for omnivore i (kg/kg/day)
fp$_i$ is the fraction of plants in diet for omnivore i
TF$_{plant,j}$ is a unitless transfer factor from soil to plants for COPC j
fi$_i$ is the fraction of invertebrates in diet for omnivore i
TF$_{insect,j}$ is a unitless transfer factor from soil to insects for COPC j

For any given COPC, the soil ESL used for initial screening was often the lowest receptor-specific soil ESL value among plants, invertebrates, robin, kestrel, shrew, mouse, cottontail, and fox. Models were developed for sediment and water. More detail can be found in Ref. 11.

Radionuclides. Radionuclide dose limits are the equivalent of the NOAELs used to develop nonradionuclide ESLs (11). For screening at LANL, radionuclide ESLs were based on the International Atomic Energy Agency (IAEA)-recommended dose limit of 0.1 rad per day (13). At LANL the radionuclide dose to terrestrial biota was taken as the sum of the dose from internally deposited radionuclides and external dose from gamma emitting radionuclides in soil [Eq. (4)].

Total acceptable dose $= 0.1$ (rad/day) = internal dose
+ external dose (4)

Detailed formulas can be found in Ref. 11. Conservative assumptions about the size of the organism, its diet, the geometry of the contaminated source, and the location of the receptor relative to the contaminated source are used by LANL for estimating internal and external doses for screening purposes. The internal dose to wildlife is calculated by multiplying the effective energy of a radionuclide by the body burden of that radionuclide in an organism. The external dose is the dose coefficient for skin exposed to contaminated water or soil. Calculations for estimating internal and external doses from radionuclides in soil can be found in literature by Higley and Kuperman (14). At LANL, ESLs are developed separately for plants and invertebrates versus wildlife, and decay and metabolic elimination rates are incorporated (11).

4.1.2 Integration with Safety Analyses

In 1994 the Laboratory began a pilot project to develop an ecological risk assessment method that could be used to augment hazards and safety analyses (15). Specifically, the Laboratory explored the possiblity of using an ecological risk assessment approach to explicitly incorporate environmental consequence analysis into hazards analyses and probabilistic risk assessment processes. Investigators used a two-phase approach: (a) Using five COPCs, conservative ecological screening was interacted with worst-case accident analysis results to determine whether ecological impacts could be excluded from further investigation or would require more detailed assessment; (b) using a dynamic ecological transport model (BIOTRAN), a more detailed and realistic assessment was conducted of the land area that the screening process identified as most susceptible to potential adverse impacts from the five contaminants. BIOTRAN (now BIOTRAN.2) is an ecological transport model used for predicting the flow of organic and inorganic contaminants, including radionuclides, through simulated ecosystems—which can include plant, animal, and human populations—using biomass vectors (16). Twenty-two subroutines include variables of meteorology, soil physics, hydrology, geology, plant and animal physiology, limnology, radiation biology, ecology, epidemiology, risk analysis, and health physics. "The results of the detailed assessment were consistent with the findings of the screening assessment, thus validating the screening assessment approach in the context of hazards analysis."

4.1.3 Tier 2 Assessments

From 1997 to 1999, Gonzales et al. completed tier 2 preliminary ecological risk assessments of federally protected threatened and endangered species (17–19). Risk assessments of the Mexican spotted owl (*Strix occidentalis lucida*), the American peregrine falcon (*Falco peregrinus*), the bald eagle *(Haliaeetus leucocephalus)*, and the Southwestern willow flycatcher (*Empidonax traillii extimus*) were performed using a custom FORTRAN model, ECORSK.5, and the geographic information system (GIS) (17–19). Using Eq. (5), estimated doses resulting from soil ingestion and food consumption were compared against TRVs for nonradionuclides and radiation dose limits for radionuclides. This generated hazard indices (HIs) that included a measure of additive effects from multiple contaminants (radionuclides, metals, and organic chemicals) and included bioaccumulation and biomagnification factors.

$$
\mathrm{HI} = \frac{I_f/\mathrm{BW} \sum\limits_{i=1}^{\mathrm{ncs}} O_i\, \mathrm{ENH}_i \sum\limits_{j=1}^{\mathrm{ncoc}} [F_s + (1 - F_s)\, \mathrm{BMF}_j]\, \mathrm{Cs}_{i,j}}{\mathrm{TRV}} \tag{5}
$$

where

HI = hazard index, cumulative for all summed COPC hazard quotients
I_f = food consumption, kgdwt/day
BW = body weight of organism, kgfwt
ncs = total number of nest sites
O_i = occupancy factor for the ith feeding site
ENH_i = enhancement factor for the ith feeding site
ncoc = total number of different COPCs found over all nesting sites
F_s = % soil in diet
BMF_j = biomagnification factor via food chain transport for the jth contaminant
$\mathrm{Cs}_{i,j}$ = COPC concentration of soil/sediment, mg/kgdwt, for the ith feeding site and the jth COPC
TRV_j = toxicity reference value ("safe limit"), mg/kg-body weight/day, for the jth COPC

Other terms were sometimes included, such as calculating sediment intake from prey data and inferring BMFs through differential equations.

ECORSK.5 was used to integrate environmental contaminant data, species-specific biological, ecological, and toxicological information, and GIS spatial information as depicted in Figure 3. Using ECORSK.5, the risk assessor is able to scale home ranges (HRs) so that a rectangular HR simulates foraging patterns based on natural features such as topography or prey distribution. ECORSK.5 is also coded to allow the option of weighting the simulated foraging process on an exponential basis as based on the distance from randomly selected nest sites such that a species forages more in the vicinity of its nest (or any other focal point) than it does at the extremity of its HR. A second manner in which the simulated foraging process is weighted in the tier 2 process is based on animal distribution and other population data. For example, a GIS elk habitat use model based on positions fed from radio collars via a Global Positioning System was used to create an occupancy scheme whereby the amount of time spent foraging on any one area is specified. Coding schemes identify home ranges, habitat core areas, species, and occupancy weights.

ECORSK.5 generates information on risk by specific geographic location for use in management of contaminated areas, species habitat, facility siting,

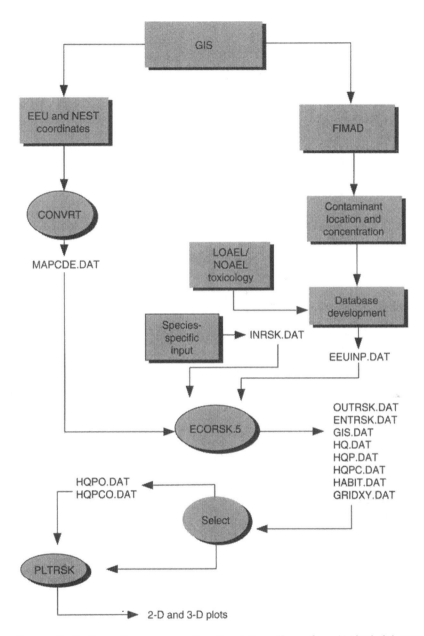

FIGURE 3 Schematic respresenting the integration of ecological risk assessment parameters by ECORSK.5.

and/or facility operations in order to maintain risk from contaminants at acceptably low levels. Several different levels of risk breakdown, from very summarized values to very detailed values, are generated by ECORSK.5. Figure 4 shows one level of breakout, where the risk contribution by Aroclor-1254 from different discrete locations in a HR is presented graphically. Thus, the strength of this custom FORTRAN model is for animals with large individual home ranges or populations with large distributions. ECORSK.5 is documented in an internal report (20).

4.2 Ecorisk Assessments Applied

Results of the tier 2 assessments described above indicated a small potential for impact to the peregrine falcon, but the studies concluded no appreciable potential impact was expected to the Mexican spotted owl, the bald eagle, or the Southwestern willow flycatcher. With all the hazardous substances used at the Laboratory over the years, including radionuclides, an application of a pesticide unrelated to Laboratory operations 37 years ago resulted in the contaminant that dominated the risk contribution for the peregrine falcon. In 1963, dichlorodiphenyltrichloroethane (DDT) was applied to one-half million acres of forest west of the Rio Grande at an application rate of about 140,000 ppm (21). DDT and dichlorodiphenylethelyne (DDE), a DDT breakdown product, were two of the three top contaminants in several modeling scenarios for the peregrine falcon. Since this is currently a widely diffuse source that largely originated in 1963, PC/WM cannot be applied to this chemical, raising the point that not all risk issues can be addressed in the design of a PC/WM program. Nevertheless, the risk assessments summarized above also provide an example of how risk information can be used to augment PC/WM activities as follows. The top three contaminants, DDT, DDE, and a PCB mixture—Aroclor-1254—contributed a total of 81% of the risk under conditions of one of several modeled foraging scenarios. In part this resulted in a charter at the Laboratory to give added emphasis to PCB pollution control (A. Jackson, personal communication, 2000). The increased effort aims to

> Adopt a PCB-free program for authorized sources as an institutional sponsored program (consistent with Laboratory policy of zero environmental incident)
> Incorporate into the yearly budgetary exercise a schedule to replace PCB items
> Place greater division-level organizational emphasis on PCB-free facilities
> Gain senior management endorsement of an approach to address PCB vulnerabilities based on an ecological-risk concept as a coordinated institutional program, i.e., as a Laboratory taking a leadership and proactive role in working with regulators and the Department of Energy

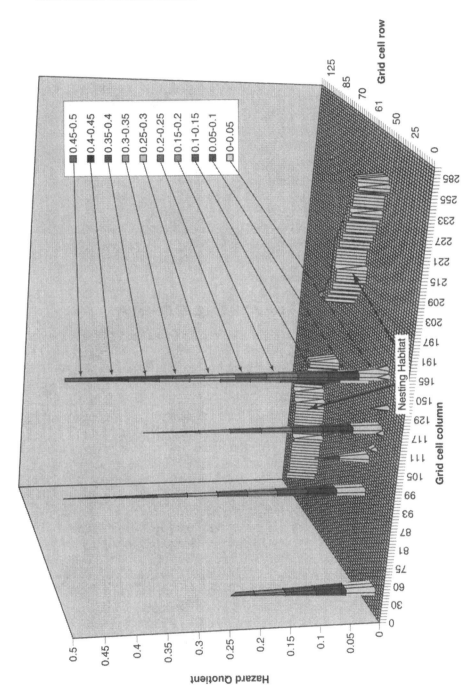

FIGURE 4   Three-dimensional plot of ecological risk hazard quotient contributions for Aroclor-1254 by discrete location.

in arriving at a cost-effective and scientifically sound risk management framework for decision making that meets performance goals and is protective of the environment).

Implementing pollution controls such as those listed above is no small task at a sprawling 43-mi^2 industrial site, and costs can be substantial. In the example above, risk assessments helped focus field studies such that in the first year following the completion of the assessments approximately $122,000 was saved by targeting specific contaminants and omitting the analysis of those deemed inconsequential from a risk perspective. Thus, the consideration of risk issues in the design of PC/WM programs can improve the effectiveness and efficiency of the programs.

As mentioned earlier, the Laboratory is employing the iterative and tiered approach to ecological risk assessment whereby less complex assessments with greater conservatism and uncertainty are employed first, followed by the advancement of potential problem areas and COPCs to progressively more complex and realistic assessments with lower uncertainty.

4.3 Risk-Based Prioritization of PC/WM COPC Targets

As discussed in the previous section, development of the ecorisk screening method at LANL included the calculation of screening values termed ESLs. Table 1 is a list of soil ESLs by contaminant in decreasing order (22). For each contaminant ESLs were developed for eight terrestrial species. The values in Table 1 are for the species that usually had the most stringent (lowest) ESL. ESL values are still undergoing revision at LANL, but the relative rank is not expected to change much.

Ecological screening levels are related directly to toxicity and exposure, therefore ranking contaminants on the basis of ESL value gives some indication of the contaminants, which if eliminated or reduced by PC/WM techniques, would most effectively reduce risk to flora and fauna. However, before prioritizing contaminants from a PC/WM perspective, one must also consider other factors, some practical, such as the ubiquity of contaminants in the environment because if a contaminant were quite toxic but was absent or present at very low levels in the environment relative to its ESL, elimination or reduction of the contaminant would gain little in terms of risk reduction. For example, since 2,3,7,8-tetrachlorodibenzodioxin is ranked first in Table 1, thus can be considered relatively very toxic, a large degree of risk reduction could be achieved by PC/WM efforts that minimize emissions or releases of it into the environment. However, its presence in the environmental contaminant database used in the threatened and endangered species assessments discussed above was sparse, therefore it is not likely a good candidate for PC/WM consideration.

TABLE 1 Examples of Ecological Screening Levels (ESLs) in Soil Used for Screening-Level Ecological Risk Assessments at the Los Alamos National Laboratory

Type[a]	Name	CAS no.	ESL[b]	Rank
dioxin/furan	Tetrachlorodibenzodioxin[2,3,7,8-]	1746-01-6	1.80E-06	1
pest	DDE[4,4'-]	72-55-9	1.80E-03	2
pest	DDT[4,4'-]	50-29-3	2.80E-03	3
pcbs	Aroclor-1248	12672-29-6	3.00E-03	4
pest	Endrin	72-20-8	3.40E-03	5
inorg	Mercury (methyl)	22967-92-6	1.60E-02	6
inorg	Thallium	7440-28-0	3.30E-02	7
pest	BHC[gamma-]	58-89-9	3.40E-02	8
pest	Dieldrin	60-57-1	4.10E-02	9
inorg	Mercury (inorganic)	7439-97-6	5.00E-02	10
pcbs	Aroclor-1260	11096-82-5	5.00E-02	11
pcbs	Aroclor-1254	11097-69-1	1.20E-01	12
svocs	Di-n-butylphthalate	84-74-2	1.30E-01	13
he	m-Dinitrobenzene	99-65-0	1.60E-01	14
pcbs	Aroclor-1242	53469-21-9	1.60E-01	15
pest	Kepone	143-50-0	1.60E-01	16
pest	Heptachlor	76-44-8	1.80E-01	17
Inorg	Silver	7440-22-4	2.00E-01	18
pahs	Naphthalene	91-20-3	2.00E-01	19
svocs	Bis(2-ethylhexyl)phthalate	117-81-7	2.40E-01	20
inorg	Vanadium	7440-62-2	2.50E-01	21
he	Trinitrotoluene[2,4,6-]	118-96-7	3.00E-01	22
svocs	Pentachlorophenol	87-86-5	3.20E-01	23
inorg	Chromium VI	18540-29-9	3.50E-01	24
pest	Endosulfan	115-29-7	3.50E-01	25
inorg	Antimony	7440-36-0	5.00E-01	26
inorg	Selenium	7782-49-2	5.00E-01	27
inorg	Cobalt	7440-48-4	5.10E-01	28
inorg	Arsenic	7440-38-2	5.70E-01	29
he	dinitrotoluene[2,6-]	606-20-2	7.40E-01	30
pest	BHC[beta-]	319-85-7	9.70E-01	31
inorg	Beryllium	7440-41-7	1.00E+00	32
inorg	Cadmium	7440-43-9	1.00E+00	33
inorg	Cyanide	57-12-5	1.00E+00	34
he	HMX	2691-41-0	1.00E+00	35
vocs	Xylenes	1330-20-7	1.10E+00	36
he	dinitrotoluene[2,4-]	121-14-2	1.20E+00	37
svocs	Chlorophenol[2-]	95-57-8	1.20E+00	38
pahs	Benzo(a)pyrene	50-32-8	1.80E+00	39

TABLE 1 Continued

Type[a]	Name	CAS no.	ESL[b]	Rank
vocs	Acetone	67-64-1	1.80E+00	40
vocs	Trichloroethene	79-01-6	1.90E+00	41
pest	Chlordane[alpha-]	5103-71-9	2.10E+00	42
pest	Chlordane[gamma-]	5103-74-2	2.10E+00	43
he	Tetryl	479-45-8	2.20E+00	44
pahs	Dibenz(a,h)anthracene	53-70-3	2.30E+00	45
pahs	Acenaphthene	83-32-9	2.50E+00	46
inorg	Chromium, total	7440-47-3	3.10E+00	47
pcbs	Aroclor-1016	12674-11-2	3.20E+00	48
pahs	Benzo(a)anthracene	56-55-3	3.30E+00	49
pahs	Chrysene	218-01-9	3.30E+00	50
vocs	Trichlorobenzene[1,2,4-]	120-82-1	3.50E+00	51
svocs	Chloro-3-methylphenol[4-]	59-50-7	3.60E+00	52
vocs	Tetrachloroethene	127-18-4	3.80E+00	53
he	4-amino-2,6-dinitrotoluene	19406-51-0	4.20E+00	54
inorg	Aluminum	7429-90-5	5.00E+00	55
inorg	Manganese	7439-96-5	5.00E+00	56
inorg	Uranium	7440-61-1	5.00E+00	57
he	o-Nitrotoluene	88-72-2	5.00E+00	58
he	RDX	121-82-4	5.10E+00	59
inorg	Boron	7440-42-8	5.70E+00	60
pahs	Methylnaphthalene[2-]	91-57-6	5.90E+00	61
he	2-amino-4,6-dinitrotoluene	35572-78-2	6.10E+00	62
he	m-Nitrotoluene	99-08-1	6.10E+00	63
vocs	Methylene chloride	75-09-2	6.60E+00	64
he	sym-Trinitrobenzene	99-35-4	7.00E+00	65
pahs	Benzo(b)fluoranthene	205-99-2	7.40E+00	66
svocs	Phenol	108-95-2	7.90E+00	67
svocs	Benzoic acid	65-85-0	8.20E+00	68
pest	Methoxychlor[4,4'-]	72-43-5	8.40E+00	69
svocs	Pentachloronitrobenzene	82-68-8	8.80E+00	70
inorg	Zinc	7440-66-6	1.00E+01	71
he	p-Nitrotoluene	99-99-0	1.10E+01	72
pest	Toxaphene	8001-35-2	1.10E+01	73
pahs	Benzo(g,h,i)perylene	191-24-2	1.20E+01	74
pahs	Indeno(1,2,3-cd)pyrene	193-39-5	1.20E+01	75
vocs	Dichlorobenzene[1,4-]	106-46-7	1.20E+01	76
inorg	Copper	7440-50-8	1.30E+01	77
pahs	Benzo(k)fluoranthene	207-08-9	1.30E+01	78
svocs	Nitrobenzene	98-95-3	1.40E+01	79
pahs	Pyrene	129-00-0	1.50E+01	80
rad	Americium-241	86954-36-1	1.70E+01	81

Type[a]	Name	CAS no.	ESL[b]	Rank
rad	Plutonium-238	13981-16-3	1.70E+01	82
rad	Thorium-228	14274-82-9	1.70E+01	83
rad	Plutonium-239/240	15117-48-3	1.80E+01	84
rad	Neptunium-237	13994-20-2	1.90E+01	85
rad	Thorium-229	15594-54-4	1.90E+01	86
inorg	Fluoride	16984-48-8	2.00E+01	87
inorg	Lead	7439-92-1	2.00E+01	88
inorg	Nickel	7440-02-0	2.00E+01	89
rad	Radium-226	13982-63-3	2.00E+01	90
rad	Thorium-230	14269-63-7	2.00E+01	91
rad	Uranium-233	13968-55-3	2.00E+01	92
rad	Uranium-234	13966-29-5	2.00E+01	93
rad	Uranium-235	15117-96-1	2.10E+01	94
rad	Uranium-236	13982-70-2	2.10E+01	95
rad	Uranium-238	7440-61-1	2.20E+01	96
inorg	Barium	7440-39-3	2.30E+01	97
rad	Thorium-232	7440-29-1	2.30E+01	98
svocs	Chlorobenzene	108-90-7	2.40E+01	99
pahs	Fluoranthene	206-44-0	2.60E+01	100
pahs	Fluorene	86-73-7	2.90E+01	101
vocs	Chloroform	67-66-3	3.20E+01	102
vocs	Tetrachloroethane[1,1,2,2-]	79-34-5	3.90E+01	103
svocs	Dibenzofuran	132-64-9	6.10E+01	104
vocs	Benzene	71-43-2	6.60E+01	105
vocs	Toluene	108-88-3	7.10E+01	106
inorg	Titanium	7440-32-6	7.20E+01	107
vocs	Dichloroethylene[1,1-]	75-35-4	7.60E+01	108
vocs	Dichloroethylene[1,2-]	540-59-0	9.20E+01	109
pahs	Phenanthrene	85-01-8	1.10E+02	110
svocs	Dimethylphthalate	131-11-3	1.30E+02	111
he	Nitrogylcerin	53-63-0	1.50E+02	112
pahs	Acenaphthylene	208-96-8	1.60E+02	113
svocs	Di-n-octylphthalate	117-84-0	1.60E+02	114
inorg	Strontium	7440-24-6	1.90E+02	115
pahs	Anthracene	120-12-7	2.20E+02	116
rad	Strontium-90 + Yttrium-90	10098-97-2	2.80E+02	117
svocs	Butyl benzyl phthalate	85-68-7	3.40E+02	118
rad	Plutonium-241	14119-32-5	3.70E+02	119
rad	Cobalt-60	10198-40-0	4.10E+02	120
rad	Europium-152	14683-23-9	4.20E+02	121
rad	Cesium-134	13967-70-9	4.90E+02	122
rad	Lead-210	14255-04-0	5.10E+02	123

TABLE 1 Continued

Type[a]	Name	CAS no.	ESL[b]	Rank
rad	Cesium-137 + Barium-137		5.50E+02	124
vocs	Dichloroethane[1,1-]	75-34-3	7.30E+02	125
vocs	Butanone[2-]	78-93-3	9.50E+02	126
rad	Sodium-22	13966-32-0	1.00E+03	127
vocs	Trichloroethane[1,1,1-]	71-55-6	2.80E+03	128
he	PETN	78-11-5	1.40E+04	129
rad	Tritium	10028-17-8	2.10E+04	130
rad	Radium-228	15262-20-1	3.50E+04	131
inorg	Calcium	7440-70-2	n/a	132
inorg	Chloride	16887-00-6	n/a	133
inorg	Iron	7439-89-6	n/a	134
inorg	Lithium	7439-93-2	n/a	135
inorg	Magnesium	7439-95-4	n/a	136
inorg	Molybdenum	7439-98-7	n/a	137
inorg	Potassium	7440-09-7	n/a	138
inorg	Sodium	7440-23-5	n/a	139
svocs	Carbazole	86-74-8	n/a	140
vocs	Dichloroethane[1,2-]	107-06-2	n/a	141

[a]he, high explosives; inorg, inorganic; pahs, polycyclic aromatic hydrocarbons; pcbs, polychlorobiphenyls; pest, pesticide; rad, radionuclide; vocs, volatile organic carbons; svocs, semivolatile organic carbons.

Another practical consideration is the source type of a PC/WM candidate constituent, i.e., a point source versus diffuse. Obviously, point-source contaminants are subject to PC/WM and diffuse sources typically are not. For example, DDT and DDE rank high in terms of toxicity as measured by its safe limit, or ESL (Table 1), and as measured in potential risk to threatened and endangered species at Los Alamos (17); however, the predominant source at the Laboratory was aerial spraying in the 1960s (20).

Seven of the 10 most stringent ESLs in Table 1 are for organic contaminants. This is consistent with the results of ecorisk screening and ecological risk assessments and related studies that have been completed to date at LANL. The assessments and studies have begun to demonstrate that organic contaminants dominate risk to nonhuman organisms, and radionuclides are of less concern (17–19, 23). Thus, for purposes of reducing ecological risk from contaminants, it may be that the largest gains in risk reduction (ignoring other factors) can be achieved by targeting a category or group of contaminants as a priority.

In summary, risk-related factors that should be considered in designing a PC/WM program include

Contaminant risk rank
Contaminant ubiquity
Contaminant source and date of origin

There are factors additional to the three above that can influence the effectiveness of prioritization of PC/WM efforts but that are inherent in the above three. For example, in addition to toxicity, contaminant risk ranks that are based on ESLs inherently consider the potential for a contaminant to bioaccumulate (soil-to-first-level transfer) or biomagnify (food-chain transfer). This is true because ESLs are derived from safe limits using transfer coefficients that include bioaccumulation and biomagnification factors for those contaminants that exhibit this tendency. Another example of a factor inherently included in ESL-based risk ranking is the amount of soil ingested by an organism. Soil ingestion is an important route of exposure to environmental contaminants for wildlife (24).

5 CONCLUSION

Pollution control and waste minimization can be used to effectively minimize or eliminate risk to humans and the environment. Using the Los Alamos National Laboratory as a case study, concepts have been introduced at the rudimentary level that may serve as "seeds" that enable practitioners to further increase the effectiveness of pollution control and waste minimization programs in minimizing or eliminating risk. Numerous factors, some practical and some technical, must be considered when attempting to design holistic pollution control and waste minimization programs. Indices of risk used to rank contaminants that are targeted for elimination or reduction by pollution control/waste minimization techniques can be used effectively to reduce risk to flora and fauna.

REFERENCES

1. CENR (Committee on Environment and Natural Resources of the National Science and Technology Council), *Ecological Risk Assessment in the Federal Government*, CENR/5-99/001, Washington, DC, 1999.
2. P. A. Reinhardt, K. L. Leonard, and P. C. Ashbrook, *Pollution Prevention and Waste Minimization in Laboratories*. Boca Raton, FL: CRC Press, Lewis Publishers, 1996.
3. DOE (U.S. Department of Energy), Site-Wide Environmental Impact Statement for Continued Operation of the Los Alamos National Laboratory, DOE/EIS-0238 Albuquerque, NM: Albuquerque Operations Office, 1999.
4. LANL (Los Alamos National Laboratory), Site Pollution Prevention Plan for Los Alamos National Laboratory, Los Alamos National Laboratory Report LA-UR-97-1726, Los Alamos, NM, 1997.

5. EPA (U.S. Environmental Protection Agency), Guidelines for Ecological Risk Assessment; Final, U.S. Environmental Protection Agency Report EPA/630/R-95/002F, Washington, DC, 1998.

6. E. J. Calabrese and L. A. Baldwin, *Performing Ecological Risk Assessments*. Chelsea, MI: Lewis Publishers, 1993.

7. S. M. Bartell, R. H. Gadner, and R. V. O'Neill, *Ecological Risk Estimation*. Chelsea, MI: Lewis Publishers, 1992.

8. G. W. Suter II, *Ecological Risk Assessment*. Chelsea, MI: Lewis Publishers, 1993.

9. M. H. Ebinger, R. W. Ferenbaugh, A. F. Gallegos, W. R. Hansen, O. B. Myers, and W. J. Wenzel, Screening Methodology for Ecotoxicological Risk Assessment, Los Alamos National Laboratory Report LA-UR-95-3272, Los Alamos, NM, 1995.

10. G. J. Gonzales and P. G. Newell, Ecotoxicological Screen of Potential Release Site 50-006(d) of Operable Unit 1147 of Mortandad Canyon and Relationship to the Radioactive Liquid Waste Treatment Facilities Project, Los Alamos National Laboratory Report LA-13148-MS, Los Alamos, NM, 1996.

11. LANL (Los Alamos National Laboratory), Screening Level Ecological Risk Assessment Methods for the Los National Laboratory's Environmental Restoration Program, R. Ryti, E. Kelly, M. Hooten, G. Gonzales, G. McDermott, L. Soholt, Los Alamos National Laboratory Report LA-UR-99-1405, Rev. 1, Los Alamos, NM, December, 1999.

12. EPA (U.S. Environmental Protection Agency). Wildlife Exposure Handbook; Final, U.S. Environmental Protection Agency Report EPA/630/R-95/002F, Washington, DC, 1998.

13. IAEA (International Atomic Energy Agency), Effects of Ionizing Radiation on Plants and Animals at Levels Implied by Current Radiation Standards, Technical Report Series No. 332, Vienna, Austria, 1992.

14. K. A. Higley and R. Kuperman, Ecotoxicological Benchmarks for Radionuclide Contaminants at Rocky Flats Environmental Technology Site. Appendix C, Argonne National Laboratory Report RF/ER-96-0039, Argonne, IL, 1996.

15. A. Gallegos, E. Kelly, M. Kramer, C. McDaniel, P. Newell, K. Sasser, and D. Stack, Integrating Ecological Risk Assessment and Hazards Analysis, A Pilot for the Proposed Los Alamos Waste Treatment and Storage Facilities, Los Alamos National Laboratory Report LA-UR-95-2555, Los Alamos, NM, 1995.

16. A. F. Gallegos, Documentation and Utilization of the Ecological Transport Model BIOTRAN.2. Los Alamos National Laboratory Report LA-UR-97-1469, Los Alamos, NM, 1997.

17. G. J. Gonzales, A. F. Gallegos, and T. S. Foxx, Update Summary of Preliminary Risk Assessments of Threatened and Endangered Species at the Los Alamos National Laboratory, Los Alamos National Laboratory Report LA-UR-97-4732, Los Alamos, NM, 1997.

18. G. J. Gonzales, A. F. Gallegos, T. S. Foxx, P. R. Fresquez, M. A. Mullen, L. E. Pratt, and P. E. Gomez, Preliminary Risk Assessment of the Bald Eagle at the Los Alamos National Laboratory, Los Alamos National Laboratory Report LA-13399-MS, Los Alamos, NM, 1998.

19. G. J. Gonzales, A. F. Gallegos, M. A. Mullen, K. D. Bennett, and T. S. Foxx, Preliminary Risk Assessment of the Southwestern Willow Flycatcher (*Empidonax*

traillii extimus) at the Los Alamos National Laboratory, Los Alamos National Laboratory Report LA-13508-MS, Los Alamos, NM, 1998.

20. A. F. Gallegos and G. J. Gonzales, Documentation of the Ecological Risk Assessment Computer Model ECORSK.5, Los Alamos National Laboratory Report LA-13571-MS, Los Alamos, NM, 1999.

21. G. J. Gonzales, P. R. Fresquez, and M. A. Mullen, Organic Contaminant Levels in Three Fish Species Down-channel from the Los Alamos National Laboratory, Los Alamos National Laboratory Report LA-13612-MS, Los Alamos, NM, 1999.

22. LANL (Los Alamos National Laboratory), ECORISK Database (version 1.1), Los Alamos National Laboratory Environmental Restoration Project, Records Processing Facility Package ID 186, April 2000.

23. J. K. Ferenbaugh, P. R. Fresquez, M. H. Ebinger, G. J. Gonzales, and P. A. Jordan, Elk and Deer Study, Material Disposal Area G, Technical Area 54, Los Alamos National Laboratory Report LA-13596-MS, Los Alamos, NM, 1999.

24. W. N. Byer, E. E. Connor, and S. G. Gerould, Estimates of Soil Ingestion by Wildlife. *J. Wildlife Manage.*, vol. 58, no. 2, pp. 375–382, 1994.

18

Elements of Multicriteria Decision Making

**Abdollah Eskandari and
Ferenc Szidarovszky**
University of Arizona, Tucson, Arizona

Abbas Ghassemi
New Mexico State University, Las Cruces, New Mexico

1 INTRODUCTION

Our life is filled with situations when decisions have to be made. Buying a car, selecting a wife, choosing the best route to work, etc., show such situations. In engineering design we always face decisions. The consequences of our decisions are usually complex; they can by measured and evaluated only by several criteria. In environmental engineering these criteria include water and air quality, and their effects on people, livestock, plants, and wildlife, among others. Therefore all of these measures should by incorporated into the decision process, that is, we have to take multiple criteria into account. This chapter will give a brief introduction into modeling and solving such engineering decision problems.

2 MODELING DECISION PROBLEMS

Every decision-making problem is based on choice. We need to have options to select from. Options, or in other words, decision alternatives, may be the types of cars available on the market, or the different technology variants to treat a region or to build a water treatment plant. The first step in modeling decision problems is to see which *set of alternatives* we may choose from. This set can be finite or infinite. If the number of decision alternatives is finite, then the decision problem is called *discrete*. For example, if finitely many technical variants are available for an engineering project, then the decision problem is discrete. In this case we have to prepare a list of the alternatives. If the number of alternatives is infinite, then no such list is possible. In such cases decision variables have to be introduced, and the decision alternatives are identified by the different values of the decision variables. If x denotes the capacity of a wastewater treatment plant, then the different values of this variable represent the different variants for the capacity. The decision variables are usually collected in the components of a decision vector. If $x_1, x_2, \ldots, x_m$ are the decision variables, then $\mathbf{x} = (x_1, x_2, \ldots, x_m)$ is the decision vector. If the decision alternatives are represented by the different continuous values of a decision vector, then the decision problem is called *continuous*. Sometimes discrete problems with a large number of alternatives are approximated by continuous models.

The second step of the modeling process is known as the *feasibility check*. There might be many different reasons why some alternatives should be dropped from the list. For example, our budget poses a constraint on which cars could be purchased. In the case of a discrete problem, the infeasible alternatives have to be dropped from the list. The result of the feasibility check is therefore a reduced list of alternatives. In the case of a continuous problem, we are unable to check all options one by one, since there are usually too many of them or we might even have infinitely many possibilities. In such cases feasibility has to be expressed by certain constraints, which model the additional requirements that determine feasibility. As an example, assume that a function $g_o(x_1, \ldots, x_m)$ represents the cost of a project, and we cannot spend more than an amount of Q_o dollars. Then this budgetary constraint can be formulated as the inequality

$$g_o(x_1, \ldots, x_m) \leqq Q_o \tag{1}$$

In most practical problems, several similar constraints should be satisfied for feasibility. These constraints are either $\leqq$, or $\geqq$, or $=$ types. Without restricting generality, we may assume mathematically that all constraints are given by inequalities of $\geqq$ type. If an inequality is defined by a $\leq$ relation, then it must have the form

$$g_1(x_1, \ldots, x_m) \leqq Q_1 \tag{2}$$

which is equivalent to the relation

$$-g_1(x_1, \ldots, x_m) \geqq -Q_1 \tag{3}$$

which is a $\geq$ type of inequality. If a constraint is given by an equality, then it can be rewritten as a pair of $\geq$ type inequalities in the following way. Assume that

$$g_2(x_1, \ldots, x_m) = Q_2 \tag{4}$$

in a constraint of equality type. It is equivalent to the pair of inequalities

$$g_2(x_1, \ldots, x_m) \leq Q_2$$
$$g_2(x_1, \ldots, x_m) \geq Q_2 \tag{5}$$

By multiplying the first inequality by (-1) we have the pair

$$-g_2(x_1, \ldots, x_m) \geq -Q_2$$
$$g_2(x_1, \ldots, x_m) \geq Q_2 \tag{6}$$

Therefore in our further discussions we will always assume that the feasible alternative set of continuous decision problems is given by a system of $\geq$ inequalities

$$g_1(x_1, \ldots, x_m) \geq Q_1$$
$$g_2(x_1, \ldots, x_m) \geq Q_2$$
$$\vdots$$
$$g_n(x_1, \ldots, x_m) \geq Q_n \tag{7}$$

This inequality system can be formulated in a more simple and compact form by introducing the vectors

$$\mathbf{g(x)} = \begin{pmatrix} g_1(x_1, \ldots, x_m) \\ g_2(x_1, \ldots, x_m) \\ \vdots \\ g_n(x_1, \ldots, x_m) \end{pmatrix} \quad \text{and} \quad \mathbf{Q} = \begin{pmatrix} Q_1 \\ Q_2 \\ \vdots \\ Q_n \end{pmatrix} \tag{8}$$

Then system (7) can be rewritten as

$$\mathbf{g(x)} \geq \mathbf{Q} \tag{9}$$

We will use the notation X for the set of the feasible decision alternatives of continuous problems. By using the above notation we may write

$$X = \{\mathbf{x} | \mathbf{x} \in R^m, \quad \mathbf{g(x)} \geq \mathbf{Q}\} \tag{10}$$

where R^m denotes the set of all m-element real vectors. The above discussion shows that the result of the feasibility check of continuous problems is the formulation of the system of constraint (7) or (9), and the definition of the feasible set (10). It is worthwhile to mention that even discrete problems can be demonstrated in this way. If there are r alternatives to select from, then X can be defined as the discrete set $X = \{1, 2, \ldots, r\}$.

The third step of modeling decision problems is the formulation of the *evaluation criteria*. Each criterion evaluates the goodness of the different alternatives from a specific point of view. If only one decision maker is present, then the criteria reflect his or her complicated evaluation system. In the case of multiple decision makers, the evaluation systems of all decision makers are combined and summarized in the criteria. In the first case a suitable alternative is determined that gives an acceptable trade-off among the criteria, and in the second case an appropriate compromise solution is determined that is acceptable by all parties involved in the decision-making process.

Some criteria are easy to quantify. For example, cost, profit, volume, capacity, etc., have specific values for each decision alternative. Some other criteria might not be easily quantifiable. For example, the esthetic consequences of a forestry treatment technology can be judged only subjectively. Subjective judgments can be made verbally, such as very good, good, fair, bad, unacceptable, or in any other similar way. These subjective measures can be then quantified. For example, on a [0, 100] scale, the above evaluation categories can be identified by 100, 75, 50, 25, and 0, respectively. In the case of a discrete problem, a *payoff matrix* is hence constructed, where the rows correspond to the decision alternatives, the columns correspond to the evaluation criteria, and the (i, j) element a_{ij} of this matrix is the evaluation of alternative i with respect to criterion j as shown in Figure 1.

After the payoff matrix is constructed, the decision problem is considered to be mathematically well defined. In the case of a continuous problem, let $f_1, f_2, \ldots, f_s$ denote the evaluation criteria as earlier, let $\mathbf{x}$ be the decision vector, and assume that the feasibility of the decision vector is given by the system of inequalities (7) or (9). Then the decision problem is mathematically well defined by the multicriteria optimization problem

$$
\begin{aligned}
&\text{Maximize} &&f_j(x_1, \ldots, x_m) &&(j = 1, 2, \ldots, s)\\
&\text{Subject to} &&g_l(x_1, x_2, \ldots, x_m) \geq Q_l &&(l = 1, 2, \ldots, n)
\end{aligned}
\tag{11}
$$

or, using vector notation,

$$
\begin{aligned}
&\text{Maximize} &&\mathbf{f}(\mathbf{x})\\
&\text{Subject to} &&\mathbf{g}(\mathbf{x}) \geq \mathbf{Q}
\end{aligned}
\tag{12}
$$

where

Criteria Alternatives	1	2	3	$\cdots$	s
1	a_{11}	a_{12}	a_{13}	$\cdots$	a_{1s}
2	a_{21}	a_{22}	a_{23}	$\cdots$	a_{2s}
3	a_{31}	a_{32}	a_{33}	$\cdots$	a_{3s}
$\vdots$	$\vdots$	$\vdots$	$\vdots$		$\vdots$
r	a_{r1}	a_{r2}	a_{r3}	$\cdots$	a_{rs}

FIGURE 1   Illustration of a payoff matrix.

$$f(\mathbf{x}) = \begin{pmatrix} f_1(x_1, \ldots, x_m) \\ f_2(x_1, \ldots, x_m) \\ \vdots \\ f_s(x_1, \ldots, x_m) \end{pmatrix}$$

In most applications the problem formulation is itself a complicated, lengthy project, and needs a continuous interaction with all parties involved and interested in the decision problem and its solution. The process should start with identifying decision makers, or in other words, the stakeholders. Next we have to find out what they can decide on in order to find the set of decision alternatives. There are very often regulations, requirements, public opinion, etc., which influence and restrict choices. These restrictions determine the set of feasible alternatives. The formulation of the evaluation criteria is also itself a complicated process. In every engineering project we wish to accomplish certain goals, such as cleaning wastewater, providing certain products to the public, etc. At the same time, we want to avoid certain negative effects such as worsening air quality, water quality, paying too much for the project, etc. After the goals and the negative effects have been identified verbally, we have to find those quantities which can be used to characterize the goodness of the different alternatives with respect to the goals and additional consequences. These quantities should measure how the goals are satisfied if the different alternatives are selected and carried out, how certain it is that the project will be successful under the different alternatives, and how severe the additional negative consequences will be. These measurable or at least subjectively quantifiable quantities are usually selected as the evaluation criteria. Without restricting generality, we may assume that all criteria are maximized. If smaller values are better for a criterion, then by multiplying this criterion by (-1) we obtain a maximizing criterion. If a target value has to be reached in the case of a certain criterion, then the distance between the actual and target values has to be minimized, or the negative of this distance has to be maximized. In the final problem formulation in both the discrete and continuous cases, we will always assume therefore that all criteria are to be maximized.

In the decision science literature problem (11) or (12) and the one given in Figure 1 are called multiple–criteria decision problems, or multiobjective programming problems. In the next section the solution of such problems will be defined and examined, and in the later parts of this chapter we will give the fundamentals of the most popular solution algorithms. A comprehensive summary and additional details are presented in Ref. 1.

3 MULTIOBJECTIVE PROGRAMMING

As a simple illustration, consider the problem of designing a wastewater treatment plant. Assume that there are two design alternatives which cost $2,000,000 and $2,500,000, respectively. Assume that the capacities are 1500 m^3/day and 2000 m^3/day, respectively. It is also assumed that both versions result in the same output quality. Since lower cost but higher capacity is better, the first criterion has to be transformed into a maximizing criterion by multiplying it by (-1). The resulting payoff matrix is given in Table 1. Notice that this problem is a discrete multiobjective programming problem.

As a simple continuous example, consider the following problem. Assume that a combination of three alternative technologies can be used in a wastewater treatment plant. Let x_1, x_2 denote the proportion (in percent) of the application of technology variants 1 and 2; then $1 - x_1 - x_2$ is the proportion of the third technology. Therefore we have two decision variables, x_1 and x_2, and they have to satisfy the conditions $x_1 \geq 0$, $x_2 \geq 0$, and $x_1 + x_2 \leq 1$. The feasible set is illustrated in Figure 2.

TABLE 1 Payoff Matrix for Wastewater Treatment Plan

Alternatives/Criteria	−Cost (−$)	Capacity (m^3/day)
1	−2,000,000	1,500
2	−2,500,000	2,000

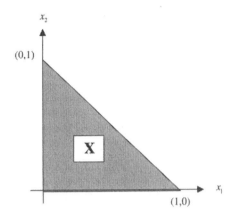

FIGURE 2   Illustration of the feasible set for the continuous example.

Assume that there are two major pollutants that have to be removed from the wastewater, and the three technology variants remove 3, 2, 1 mg/m³, respectively, of the first kind of the pollutant, and 2, 3, 1 mg/m³ of the second kind of pollutant. Then the total removed amount of the two pollutants can be obtained as $[3x_1 + 2x_2 + 1(1 - x_1 - x_2)]V$ and $[2x_1 + 3x_2 + 1(1 - x_1 - x_2)]V$, where V is the total amount of treated wastewater. Maximizing those objectives is equivalent to maximizing $2x_1 + x_2$ and $x_1 + 2x_2$. Therefore we have the following multiobjective programming problem:

$$\text{Maximize} \quad 2x_1 + x_2 \quad \text{and} \quad x_1 + 2x_2$$
$$\text{Subject to} \quad x_1 \geq 0, \quad x_2 \geq 0 \tag{13}$$
$$x_1 + x_2 \leq 1$$

Notice that (13) is a continuous problem with two decision variables and two objective functions.

In many decision problems we are also interested in knowing the set of all feasible objective values. For each $\mathbf{x} \in X$, the vector

$$\boldsymbol{f}(\mathbf{x}) = (f_1(x_1, \ldots, x_m), f_2(x_1, \ldots, x_m), \ldots, f_s(x_1, \ldots, x_m))$$

is called the *objective vector* at $\mathbf{x}$, and the set of all feasible objective vectors,

$$H = \{u | u = \boldsymbol{f}(\mathbf{x}) \text{ with some } \mathbf{x} \in X\}$$

is called the *objective space*. If there are s objectives, then $H \subset R^s$.

In our discrete problem we have two objectives, therefore the objective space is two-dimensional. Since there are only two alternatives, the objective space has only two points: $(-2,000,000, 1,500)$ and $(-2,500,000, 2,000)$. In every discrete problem, every point of the objective space corresponds to a decision alternative, but different alternatives may have the same point, if the objective function values are identical. For our discrete example, Figure 3 shows these points when the first objective is given on the horizontal axis and the second objective is identified by the vertical axis. Point $A1$ shows the simultaneous objective function values for the first alternative, and point $A2$ shows the same values for the second alternative.

The question that arises now is the choice among the two alternatives, which is the same as selecting the "better" point among $A1$ and $A2$. If cost is our only concern, then point $A1$ (that is, alternative 1) must be our choice. If capacity is our only concern, then point $A2$ with alternative 2 is our choice. The dilemma we face here is the conflict between the two objectives, since higher capacity can be obtained by selecting $A2$ but then the cost becomes worse. If both objectives are important to us, then neither of the alternatives is better than the other and hence either alternative seems to be a reasonable choice. With decision science terminology we might say that no alternative is dominated by another alternative and therefore both alternatives are *nondominated*. In the economic literature, nondominated alternatives are called *Pareto optimal*. Let us define these terms mathematically. In a discrete problem, let p and q be two alternatives, and let $\boldsymbol{f}(p)$ and $\boldsymbol{f}(q)$ denote the corresponding objective vectors. We say that alternatives p and q are *equivalent* if $\boldsymbol{f}(p) = \boldsymbol{f}(q)$. Here the equality of vectors is defined by component-wise equality. That is, two alternatives are equivalent to each other if they result in the same values in all objective functions. Similarly, we say that alternative p *dominates* alternative q, if $\boldsymbol{f}(p) \geq \boldsymbol{f}(q)$ and there is a strict inequality in at least one of the objectives. Here the notation $\boldsymbol{f}(p) \geq \boldsymbol{f}(q)$ means that each

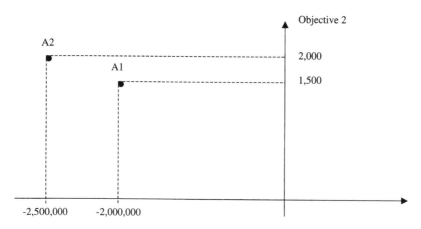

FIGURE 3 Objective space for a discrete problem.

component of vector $f(p)$ is greater than or equal to the corresponding component of $f(q)$. In our example, $f(1) = (-2,000,000, 1,500)$ and $f(2) = (-2,500,000, 2,000)$ and none of the alternatives dominates the other one. If no other alternatives dominate an alternative p, then it is called *nondominated*. In our example, therefore, both alternatives 1 and 2 are nondominated.

In the case of continuous problems, the objective space can be constructed by representing the set $H = \{\mathbf{u}|\mathbf{u} = (u_1, \ldots, u_s)|u_j = f_j(\mathbf{x})$ with some $\mathbf{x} \in X$ for $j = 1,2, \ldots, s\}$ in the s-dimensional space. In our continuous example

$$u_1 = 2x_1 + x_2 \quad \text{and} \quad u_2 = x_1 + 2x_2$$

From these equations we see that

$$x_1 = \frac{2u_2 - u_1}{3} \quad \text{and} \quad x_2 = \frac{2u_1 - u_2}{3}$$

The constraints of problem (13) have now the form

$$\frac{2u_2 - u_1}{3} \geq 0 \qquad \frac{2u_1 - u_2}{3} \geq 0$$

and

$$\frac{2u_2 - u_1}{3} + \frac{2u_1 - u_2}{3} \geq 1$$

These inequalities can be simplified as

$$u_2 \geq \frac{u_1}{2} \qquad u_2 \leq 2u_1 \qquad u_1 + u_2 \leq 3 \tag{14}$$

The set of feasible payoff vectors (u_1,u_2) satisfying these conditions are illustrated in Figure 4. The vertices A, B, C of the triangle can be obtained as the intercepts of the lines $u_2 = u_1/2$ and $u_2 = 2u_1$, $u_2 = 2u_1$ and $u_1 + u_2 = 3$, $u_2 = u_1/2$ and $u_1 + u_2 = 3$, respectively. Simple calculation shows that $A = (0, 0)$, $B = (1, 2)$ and $C = (2, 1)$. We will next show that the nondominated objective vectors form the linear segment connecting points B and C including the endpoints.

All points between A and B including point A are dominated by the objective vector B, since both coordinates of B are larger. Similarly, all points between A and C are dominated by C. Let now $\mathbf{u}$ be a point inside the triangle. The continuation of the linear segment connecting points A and $\mathbf{u}$ intercepts the linear segment between B and C in a point D. Since the slope of the line connecting A and $\mathbf{u}$ is positive (between $\frac{1}{2}$ and 2), the objective vector D dominates $\mathbf{u}$. Hence $\mathbf{u}$ is dominated. If D is any point of the linear segment connecting B and C (including the endpoints), then it is impossible to increase any objective value without worsening the value of the other objective function.

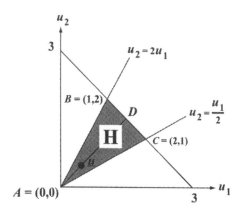

FIGURE 4 Objective space for a continuous problem.

Thus, the points of the linear segment between B and C are the nondominated objective vectors.

Any nondominated point and the corresponding decision alternative is a reasonable choice. The nondominated decision alternatives can be obtained in the following way:

$$X_o = \{x | x \in X, f(x) \in H_o\} \tag{15}$$

where H_o is the set of nondominated objective vectors. In the case of our example, the decision vector $\mathbf{x} = (x_1, x_2)$ is nondominated if and only if

$$1 \leqq 2x_1 + x_2 \leqq 2 \qquad \text{and} \qquad (2x_1 + x_2) + (x_1 + 2x_2) = 3$$

These relations can be rewritten as

$$x_1 + x_2 = 1 \qquad \text{and} \qquad 0 \leqq x_1 \leqq 1 \tag{16}$$

and the set of the nondominated decision vectors is the linear segment connecting the points $(1, 0)$ and $(0, 1)$ in Figure 2. Hence any point of this linear segment is a reasonable selection. These nondominated alternatives do not use the third technology variants.

In the case of our discrete example both alternatives are nondominated, and in the case of the continuous example there are infinitely many nondominated alternatives. Notice that the different nondominated solutions give different objective function values, contrary to single-objective optimization problems, where all alternative optimal solutions give the same optimal objective function value. In other words, in single-objective optimization problems the different optimal solutions are equivalent to each other, but in the presence of multiple objectives this is not true anymore. If we consider two different nondominated objective vectors, then the first is better in one objective and the other is better in another objective. Therefore we cannot say which objective vector is better than the other. In decision problems, however, we have to choose one alternative that could be considered as the best one overall. In order to answer the question of whether an alternative is better than another alternative overall, we need additional preference or trade-off information from the decision makers and/or from the stakeholders. Depending on the nature of this additional information, different solution algorithms are available. In the next sections the most popular methods will be outlined. In order to guarantee the existence of the solution, we will always assume that the continuous problems have continuous bounded objective functions and that the feasible set is closed.

3.1 Sequential Optimization

The application of the *sequential optimization method*, which is sometimes called the *lexicographic method*, is based on an ordinal preference order of the evaluation criteria (or objectives). This kind of information tells us which criterion is the most important, which is the second most important, and so on, to which one is the least important. Without losing generality, we may assume that the criteria are numbered in decreasing importance order; that is, the first criterion is the most important, the second is the second most important, and so on, and the sth criterion is the least important. The solution algorithm is then the following. Identify first the alternatives which give the highest value of the first objective function. If a unique alternative is found, then it is the solution of the problem. Otherwise, delete all nonoptimal alternatives, and optimize the second most important objective function on the set of the remaining alternatives. If a unique alternatives is found, it is accepted as the solution of the problem; otherwise, delete again all nonoptimal alternatives, and optimize the third objective on the remaining feasible set. Continue the process until a unique alternative is found or until the least important objective is being optimized. It can be proved that this method always results in nondominated solutions, and if there are multiple solutions, then they are equivalent to each other in the sense that they result in the same objective function values.

The above procedure can be summarized mathematically in the following way. Let X denote the (discrete or continuous) feasible alternative set and let $f_1, f_2, \ldots, f_s$ denote the objectives.

Initialization:	Set $j = 1, X_j = X$	
Optimization:	Solve problem	
	Maximize $\quad f_j(\mathbf{x})$	(17)
	Subject to $\quad \mathbf{x} \in X_j$	

If the optimal solution is unique, then it is accepted as the solution of the problem. Otherwise, go to the next step.

Adjustments: Set $j \Leftarrow j + 1$, and define

$$X_j = \{ \mathbf{x} | \mathbf{x} \in X_{j-1}, f_{j-1}(\mathbf{x}) = \text{optimal value} \},$$

and go back to optimization

In the case of a discrete problem, X has finitely many elements, $X = 1, 2, \ldots, r$, and the objective function values are represented by the a_{ij} elements of the payoff matrix (see Figure 1). The optimization step in this case can be reformulated as

$$\max_{i \in X_j} a_{ij}$$

and the adjustment step is

$$X_j = \{ i | i \in X_{j-1}, a_{i,j-1} = \text{optimal} \}$$

In the case of continuous problems the optimization step is always a single-objective optimization problem. The objective function is the jth objective of the original problem and at each optimization a new constraint has to be added to the constraints of the previous optimization step. This new constraint requires the optimality of the previous objective function. Hence the jth optimization problem can be written as

$$
\begin{aligned}
\text{Maximize} \quad & f_j(x_1, \ldots, x_m) \\
\text{subject to} \quad & g_l(x_1, \ldots, x_m) \geq Q_l \\
& f_1(x_1, \ldots, x_m) = \text{optimal value} \\
& \quad \vdots \\
& f_{j-1}(x_1, \ldots, x_m) = \text{optimal value}
\end{aligned}
\tag{18}
$$

where we assumed that the original problem is given as it was presented in the form (11). It is worthwhile to mention that if all objectives and original constraints are linear, then each optimization step requires the solution of a linear programming problem. If j is increased by 1, then an additional constraint is added to the constraints of the previous optimization problem. This makes the solution of the new linear programming problem easy based on the basic optimal solution of the previous optimization problem, since in creating a new basic solution we have to eliminate some matrix elements only from the last constraint.

Consider first the discrete problem given earlier in Table 1. If the first objective is more important than the second, then the alternative giving the largest negative cost has to be selected. In this case the first alternative is the choice. If capacity is more important than cost, then the second objective is maximized, leading to the choice of the second alternative. In both cases a unique optimal alternative is found.

Consider next the continuous problem (13). Assume first that the first objective function is the most important; then it has to be first maximized. Thus we have to solve the single-objective optimization problem

$$\text{Maximize} \quad 2x_1 + x_2$$
$$\text{Subject to} \quad x_1 \geqq 0, x_2 \geqq 0 \quad \quad \quad (19)$$
$$x_1 + x_2 \leqq 1$$

The application of the bi-variable simplex method, or elementary graphical approach, gives the unique optimal solution

$$x_1^* = 1 \quad \text{and} \quad x_2^* = 0$$

Because of uniqueness, the procedure terminates. Assume next that the second objective is more important. Then we solve the single-objective optimization problem

$$\text{Maximize} \quad x_1 + 2x_2$$
$$\text{Subject to} \quad x_1 \geqq 0, x_2 \geqq 0 \quad \quad \quad (20)$$
$$x_1 + x_2 \leqq 1$$

There is again a unique solution,

$$x_1^{**} = 0 \quad \text{and} \quad x_2^{**} = 1$$

and the algorithm terminates.

Notice that in the case of the discrete problem both nondominated alternatives could be obtained as solutions, but in the case of the continuous problem only the two endpoints of the nondominated alternative set could be obtained. The method selection has excluded all other points between (1,0) and (0,1) (see Figure 2) from the possibility to be obtained as a solution. In order to overcome this problem, the method can be modified in the following way. After maximizing the most important objective function, in later steps we do not require that all more important objectives are at their maximal levels. The optimality constraints are relaxed by requiring that in optimizing any later objective, the more important objective functions are only in the neighborhood of their optimal values. For example, we may require that their values are at least 90% of the optimal values. In the case of our continuous problem (13), assume again that the first objective is the more important and in maximizing the second objective only at least (90%) of the optimal value of the first objective is required. From the earlier discussions we know that the maximal value of the first objective is 2, which occurs at $x_1^* = 1$ and $x_2^* = 0$. Therefore in the second step we have the following problem:

Maximize $x_1 + 2x_2$
Subject to $x_1 \geq 0, x_2 \geq 0$

$$x_1 + x_2 \leq 1 \tag{21}$$

$$2x_1 + x_2 \geq 1.8$$

The feasible set of this problem is shown in Figure 5, and the optimal solution is given as $x_1^* = 0.8$ and $x_2^* = 0.2$. It can be shown that in the case of a unique solution, it is always nondominated; and if multiple solutions are present, then there is at least one nondominated alternative among the optimal solutions.

3.2 The ϵ-Constraint Method

In the application of the sequential optimization method we may have a unique optimal alternative in an early step. This was in fact the case in both the discrete and continuous examples. Since the procedure terminates, no later objective is taken into account, therefore they might have very unfavorable values. In order to avoid very "bad" values for some less important objectives, we might use the ϵ-*constraint method*. We need to know which objective is the most important, and in addition, we have to be supplied with minimal acceptable levels for all other objectives. Without losing generality, we may assume that the first objective is the most important and $f_{2*}, f_{3*}, \ldots, f_{s*}$ are the minimal acceptable levels for the other objectives. The method simply maximizes the first objective function subject to the original feasibility constraint and the additional conditions that the values of all other objectives must be at least as high as the minimal acceptable levels. It can be proven that in the case of a unique optimal alternative, the solution is nondominated; and in the case of multiple optimal solutions, there is at least one nondominated alternative among the optimal solutions.

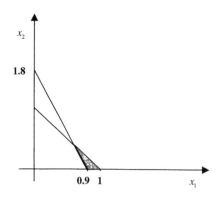

FIGURE 5 Illustration of the modified sequential optimization procedure.

In the case of the discrete problem given in Table 1, this method can be formulated as

$$\text{Maximize} \quad a_{i1}$$
$$i \in X_*$$

where

$$X_* = \{i | i \in \{1,2, \ldots, r\}, a_{ij} \geq f_{j*} \text{ for } j = 2,3, \ldots, s\} \tag{22}$$

In the case of the continuous problem (11), the ε-constraint method is the solution of the single-objective optimization problem:

$$\text{Maximize} \quad f_1(x_1, \ldots, x_m)$$
$$\text{Subject to} \quad g_1(x_1, \ldots, x_m) \geq Q_1$$
$$\vdots$$
$$g_n(x_1, \ldots, x_m) \geq Q_n \tag{23}$$
$$f_2(x_1, \ldots, x_m) \geq f_{2*}$$
$$\vdots$$
$$f_s(x_1, \ldots, x_m) \geq f_{s*}$$

Note that if all objectives and constraints of the original problem are linear, then problem (23) is a usual linear programming problem that can be solved by the simplex method.

Consider first the discrete problem of Table 1, and assume that the cost is the more important objective. Therefore the additional constraint has to be given to the capacity. If the minimal capacity requirement is 1,500 or less, then both alternatives are feasible and the less expensive alternative is selected, which is the first alternative. If the minimal capacity requirement is more than 1,500, then this additional constraint excludes the first alternative and hence the only and therefore the optimal alternative is the second one.

In the case of the continuous example (13), assume that the first objective is more important than the other, and we require that the value of the second objective is at least 1.5 (which is 75% of its maximal value shown in Figure 4). Then problem (23) can be written now as

$$\text{Maximize} \quad x_1 + 2x_2$$
$$\text{Subject to} \quad x_1 \geq 0, x_2 \geq 0$$
$$x_1 + x_2 \leq 1 \tag{24}$$
$$2x_1 + x_2 \geq 1.5$$

The feasible set of this linear programming problem is shown in Figure 6, and the optimal solution is $x_1^* = 0.5$ and $x_2^* = 0.5$. Notice that this solution gives

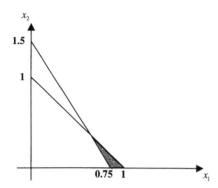

FIGURE 6 Feasible set for the ε-constraint method applied to a continuous problem.

equal proportions to the first two technologies without using the third technology at all.

3.3 The Weighting Method

In the case of the *weighting method*, it is assumed that the 100% overall importance of the objectives is divided up among the different objectives. That is, a positive weight c_j is assigned to each objective $j(j = 1,2, \ldots, s)$ such that $c_1 + c_2 + \ldots + c_s = 1$. The problem of multiobjective optimization is then reduced to optimize the weighted average of the objectives, where weights are selected as the importance factors c_j. In the case of a discrete problem (such as the one presented in Figure 1), we have to compute first the weighted averages,

$$A_i = c_1 a_{i1} + c_2 a_{i2} + \ldots + c_s a_{is}$$

for each alternative, and then the alternative i^* with the largest A_i value is selected as the solution:

$$A_{i*} = \max_{1 \leq i \leq r} \{A_i\} \tag{25}$$

In the case of a continuous problem (11), the single-objective optimization problem

$$\text{Maximize} \quad \sum_{j=1}^{s} c_j f_j(x_1, \ldots, x_m)$$

$$\text{Subject to} \quad g_1(x_1, \ldots, x_m) \geq Q_1 \tag{26}$$

$$g_n(x_1, \ldots, x_m) \geq Q_n$$

is solved, and the optimal solution is accepted as the solution of the multiobjective optimization problem. If all of the weights are possible, then the optimal solution of problem (26) is always nondominated.

Consider first the discrete problem of Table 1 and assume that the objectives (cost and capacity) are equally important: $c_1 = c_2 = 0.5$. Since

$$A_1 = 0.5(-2,000,000) + 0.5(1,500) = -999,250$$

and

$$A_2 = 0.5(-2,500,000) + 0.5(2,000) = -1,249,000$$

A_1 gives the larger value and hence the first alternative has to be selected as the solution.

Assume next that the two objectives are also equally important in the case of the continuous problem (13). Then the optimization problem (26) can be rewritten as

$$\begin{aligned}
\text{Maximize} \quad & 0.5(2x_1 + x_2) + 0.5(x_1 + 2x_2) = 1.5x_1 + 1.5x_2 \\
\text{Subject to} \quad & x_1 \geq 0, x_2 \geq 0 \qquad\qquad\qquad (27) \\
& x_1 + x_2 \leq 1
\end{aligned}$$

It is easy to see that there are infinitely many optimal solutions: $x_1^* \in [0,1]$ is arbitrary, and $x_2^* = 1 - x_1^*$. Assume next that the first objective is twice as important as the second one, then we many chose $c_1 = \frac{2}{3}$ and $c_2 = \frac{1}{3}$. Then problem (27) has to be modified is the following way:

$$\begin{aligned}
\text{Maximize} \quad & \tfrac{2}{3}(2x_1 + x_2) + \tfrac{1}{3}(x_1 + 2x_2) = \tfrac{5}{3}x_1 + \tfrac{4}{3}x_2 \\
\text{Subject to} \quad & x_1 \geq 0, x_2 \geq 0 \qquad\qquad\qquad (28) \\
& x_1 + x_2 \leq 1
\end{aligned}$$

The feasible set of this problem was shown earlier in Figure 2. The optimal solution is $x_1^* = 1$ and $x_2^* = 0$. It can be proven that this is the solution if $c_1 > c_2$, and the solution is $x_1^* = 0$ and $x_2^* = 1$ if $c_1 < c_2$. The case of $c_1 = c_2$ has been examined earlier.

The weighting method is easy to apply. If all objectives and constraints of the original problem are linear, then problem (26) becomes a linear programming problem which can be solved by the simplex method. Notice in addition that by selecting a different set of weights, only the objective function changes. Therefore the solution procedure becomes very easy if an optimal basic solution is known for a different weight set. The weighting method has a large disadvantage, however, since the solution might change if the objective functions are given in different units. As an illustration, consider again the discrete problem given in Table 1 and assume now that the cost is given in $1,000 and the capacity is presented in dm³/day. Then the payoff matrix has to be changed accordingly. The new matrix is given in Table 2.

If equal weights are assumed again, then

$$A_1 = 0.5(-2,000) + 0.5(1,500,000) = 749,000$$

and

$$A_2 = 0.5(-2,500) + 0.5(2,000,000) = 998,750$$

showing that the second alternative should be selected. This result conflicts with the selection based on the original payoff matrix. In order to overcome this difficulty, the objective functions are usually normalized. This procedure requires knowledge of the worst and best possible values of the different objectives. These values can be assessed subjectively, or can be computed by minimizing and maximizing the objectives. If f_{j*} and f_j^* denote the lowest and highest values for objective j, then the jth normalized objective is obtained by simple transformation:

$$\bar{f}_j = \frac{f_j - f_{j*}}{f_j^* - f_{j*}} \tag{29}$$

In the discrete example of Table 1 assume that the lowest cost is 0 and the largest acceptable cost is 3,000,000. Then the worst and best values for the first objective are $f_{1*} = -3,000,000$ and $f_1^* = 0$. The normalized elements of the first column of the payoff matrix become

$$\bar{a}_{11} = \frac{a_{11} - f_{1*}}{f_1^* - f_{1*}} = \frac{-2,000,000 + 3,000,000}{0 + 3,000,000} = \frac{1}{3}$$

and

$$\bar{a}_{21} = \frac{a_{21} - f_{1*}}{f_1^* - f_{1*}} = \frac{-2,500,000 + 3,000,000}{0 + 3,000,000} = \frac{1}{6}$$

TABLE 2 Modified Matrix for Wastewater Treatment Plant

Alternatives/Criteria	−Cost (−1000$)	Capacity (dm³/day)
1	−2,000	1,500,000
2	−2,500	2,000,000

Assume that the smallest acceptable capacity is 1,000 (m³/day) and the largest possibility is assessed as 3,000 (m³/day). Then $f_{2*} = 1,000$ and $f_2^* = 3,000$, so the normalized elements of the second column become

$$\bar{a}_{12} = \frac{a_{12} - f_{2*}}{f_2^* - f_{2*}} = \frac{1,500 - 1,000}{3,000 - 1,000} = \frac{1}{4}$$

and

$$\bar{a}_{22} = \frac{a_{22} - f_{2*}}{f_2^* - f_{2*}} = \frac{2,000 - 1,000}{3,000 - 1,000} = \frac{1}{2}$$

The normalized payoff table is given in Table 3. If equal weights are assumed, then

$$A_1 = 0.5(\tfrac{1}{3}) + 0.5(\tfrac{1}{4}) = \tfrac{1}{6} + \tfrac{1}{8} = \tfrac{7}{24} \approx 0.292$$

and

$$A_2 = 0.5(\tfrac{1}{6}) + 0.5(\tfrac{1}{2}) = \tfrac{1}{12} + \tfrac{1}{4} = \tfrac{1}{3} \approx 0.333$$

Since $A_1 < A_2$, the second alternative is the choice.

Consider again the continuous problem (13). Figure 4 illustrates the possible values of the objective functions, from which we see that the smallest and largest values of both objective functions are 0 and 2, respectively. Therefore we may select

$$f_{1*} = f_{2*} = 0 \quad \text{and} \quad f_1^* = f_2^* = 2$$

The normalized objective functions are the following:

$$\bar{f}_1(x_1, x_2) = \frac{(2x_1 + x_2) - 0}{2 - 0} = x_1 + \frac{1}{2}x_2$$

and

TABLE 3 Normalized Payoff Matrix for Wastewater Treatment Plan

Alternatives/Criteria	– Normalized cost	Normalized capacity
1	$\tfrac{1}{3}$	$\tfrac{1}{4}$
2	$\tfrac{1}{6}$	$\tfrac{1}{2}$

$$\bar{f}_2(x_1, x_2) = \frac{(x_1 + 2x_2) - 0}{2 - 0} = \frac{1}{2}x_1 + x_2$$

Assume again that the first objective is twice as important as the second objective, then we may chose the weights $c_1 = \frac{2}{3}$ and $c_2 = \frac{1}{3}$. Therefore the weighted average of the normalized objectives becomes

$$\frac{2}{3}(x_1 + \frac{1}{2}x_2) + \frac{1}{3}(\frac{1}{2}x_1 + x_2) = \frac{5}{6}x_1 + \frac{4}{6}x_2$$

and the optimal solution is $x_1^* = 1$ and $x_2^* = 0$.

3.4   Direction-Based Methods

Two *direction-based methods* are used in practical applications. The more frequently applied variant is based on the knowledge of a point of the objective space which is not considered good enough and therefore cannot be accepted by the decision makers (or by the stakeholders) as the solution, and a direction of improvement is specified. The point might reflect the current situation that has to be improved, or it can be selected as the "ideally worst" point, with components being the minimal values of the objective functions. Sometimes the components of this point are the result of the subjective judgement of the decision makers (or stakeholders) as their feelings about worst possibilities. If this point is constructed by the current objective function values, then it is called the *status quo point*; otherwise, it is called the *nadir*. Starting from the status quo point or from the nadir, we try to improve the objective function values in the given direction as much as possible. This idea is illustrated in Figure 7. The straight line starting from the status quo point or from the nadir pointing in the given direction has to be continued until its last intercept with the objective space.

This method is used only for solving continuous problems, since in the case of finitely many alternatives the line may not contain feasible objective vectors at all. The concept of the method can be summarized by solving the following optimization problem:

Maximize t

Subject to $\mathbf{u}_* + t\mathbf{v} \in H$ (30)

where $\mathbf{u}_*$ is the given point, $\mathbf{v}$ is the given vector defining the direction of improvement, and H is the objective space. In the case of the continuous example (13), assume that $\mathbf{u}_* = (0,0)$ and the direction of improvement is given by vector $\mathbf{v} = (3,2)$; that is, it is required that the first objective should increase 50% faster than the second objective. In this case,

$$\mathbf{u}_* + t\mathbf{v} = (0,0) + t(3,2) = (3t, 2t)$$

therefore problem (30) can be rewritten as

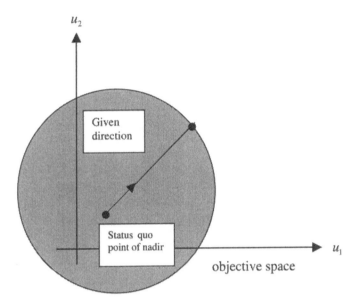

FIGURE 7 Illustration of direction-based methods starting from the status quo point or from the nadir.

$$
\begin{aligned}
\text{Maximize} \quad & t \\
\text{Subject to} \quad & t \geq 0 \\
& 3t + 2t \leq 1
\end{aligned}
$$

which has the obvious optimal solution $t^* = \frac{1}{5}$. Then the optimal alternative is $x_1^* = 3t^* = \frac{3}{5}$ and $2t^* = \frac{2}{5}$.

Another variant of this method is based on the knowledge of an "ideally best" point that can be either assessed subjectively by the decision makers (or by the stakeholders) or can be constructed by the components being the maximal values of the different objective functions, and a vector defining the direction of relaxation. Similarly to problem (30), this concept can be modeled as the following optimization problem:

$$
\begin{aligned}
\text{Minimize} \quad & t \\
\text{Subject to} \quad & \mathbf{u}^* - t\mathbf{v} \in H
\end{aligned}
\tag{31}
$$

where $\mathbf{u}^*$ is the ideally best point, and vector $\mathbf{v}$ gives the direction of relaxation.

In the case of the continuous example (13), assume that $\mathbf{u}^* = (3,3)$ and $\mathbf{v} = (2,3)$. The definition of vector $\mathbf{v}$ implies that in this case, the second objective has to be relaxed 50% faster than the first objective. In this case, problem (31) simplifies to the following:

Minimize t

Subject to $3 - 2t \geq 0, 3 - 3t \geq 0$

$(3 - 2t) + (3 - 3t) \leq 1$

since $\mathbf{u}^* - t\mathbf{v} = (3,3) - t(2,3) = (3 - 2t, 3 - 3t)$. It is easy to see that the only feasible and therefore the optimal value of t is $t^* = 1$, and the corresponding solution is $x_1^* = 1$ and $x_2^* = 0$.

There is no guarantee that the solution obtained by either variant of this method is nondominated. Such examples are shown in Figure 8, where A denotes the nadir and B is the ideal point. The directions for improvement and relaxation are shown as well. The solutions are the points S_1 and S_2; both are dominated by point C.

3.5 Distance-Based Methods

Similarly to direction-based methods, two major *distance-based methods* are applied in practical cases. The more frequently used variant is based on the knowledge of an "ideally best" point and a distance function of s-dimensional vectors. The solution that gives the closest objective vector to the ideal point is selected as the solution. Depending on the selected type of distance, different particular methods are available.

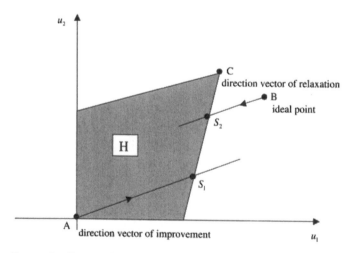

FIGURE 8 Example of dominated solutions of direction-based method.

In the case of the discrete problem of Table 1, let $f_1^*, \ldots, f_s^*$ denote the coordinates of the ideally best point. The objective vector of alternative i is given by the ith row of the payoff matrix: $\mathbf{a}_i = (a_{i1}, \ldots, a_{is})$. Similarly to the weighting method, let $c_1, \ldots, c_s$ denote the importance weights of the different objectives. Three major distance (or in other words, metric) types are selected in most practical applications:

$$d_\infty(i) = \max_j \{c_j|a_{ij} - f_j^*|\}$$

$$d_1(i) = \sum_{j=1}^{s} c_j|a_{ij} - f_j^*|$$

and

$$d_2(i) = \sqrt{\sum_{j=1}^{s} c_j\left|a_{ij} - f_j^*\right|^2}$$

The d_∞ distance is sometimes called the *Tchebyshev metric*, d_1 is known as the *first Minkowski distance*, and d_2 is referred to as the *Euclidean metric* or the *least-squares distance*. In applying the method for a discrete problem, first we have to compute the selected distances $d_p(i)$ for all alternatives $i = 1, 2, \ldots, r$ (where $p = 1, 2,$ or ∞), and the alternative i^* with the smallest distance is accepted as the solution:

$$d_p(i^*) = \min\{d_p(i)\} \tag{32}$$

In the case of the continuous problem (11), we may select one of the following distances:

$$d_\infty(\mathbf{x}) = \max_j \{c_j|f_j(\mathbf{x}) - f_j^*|\}$$

$$d_1(\mathbf{x}) = \sum_{j=1}^{s} c_j|f_j(\mathbf{x}) - f_j^*|$$

and

$$d_2(\mathbf{x}) = \sqrt{\sum_{j=1}^{s} c_j\left|f_j(\mathbf{x}) - f_j^*\right|^2}$$

The method then solves the single objective optimization problem

Minimize $\quad d_p(\mathbf{x})$

Subject to $\quad \mathbf{g}(\mathbf{x}) \geqq \mathbf{Q}$ $\qquad$ (33)

It can be proved that the selection of distances d_1 and d_2 always results in nondominated solutions. If distance d_∞ is chosen and a unique optimal solution is found, then it is also nondominated. If the optimal solution is not unique, then there is at least one nondominated solution among them.

In the case of the weighting method, we have seen that the solution might depend on the selected units of the objectives. Since the distance functions also contain weights, we face a similar situation in applying distance-based methods. Therefore we have to normalize the objective functions before using the method.

Consider first the normalized payoff matrix of the discrete problem given in Table 3. Assume that $c_1 = c_2 = \frac{1}{2}$, and the normalized ideally best values of the objectives are 1 for both objectives. If the d_∞ distance is selected, then

$$d_\infty(1) = \max\{\tfrac{1}{2}|\tfrac{1}{3} - 1|; \tfrac{1}{2}|\tfrac{1}{4} - 1|\} = \max\{\tfrac{1}{3}; \tfrac{3}{8}\} = \tfrac{3}{8}$$

and

$$d_\infty(2) = \max\{\tfrac{1}{2}|\tfrac{1}{6} - 1|; \tfrac{1}{2}|\tfrac{1}{2} - 1|\} = \max\{\tfrac{5}{12}; \tfrac{1}{4}\} = \tfrac{5}{12}$$

Since the first objective gives the smaller distance, it is considered as the better choice. If the d_1 distance is selected, then

$$d_1(1) = \tfrac{1}{2} \cdot |\tfrac{1}{3} - 1| + \tfrac{1}{2} \cdot |\tfrac{1}{4} - 1| = \tfrac{1}{2} \cdot \tfrac{2}{3} + \tfrac{1}{2} \cdot \tfrac{3}{4} = \tfrac{17}{24}$$

and

$$d_2(2) = \tfrac{1}{2} \cdot |\tfrac{1}{6} - 1| + \tfrac{1}{2} \cdot |\tfrac{1}{2} - 1| = \tfrac{1}{2} \cdot \tfrac{5}{6} + \tfrac{1}{2} \cdot \tfrac{1}{2} = \tfrac{8}{12}$$

Notice that the second alternative gives now the smaller distance, therefore the second alternative has to be selected as the solution.

In the case of the least-squares distance,

$$d_2(1) = \sqrt{\tfrac{1}{2} \cdot |\tfrac{1}{3} - 1|^2 + \tfrac{1}{2} \cdot |\tfrac{1}{4} - 1|^2} = \sqrt{\tfrac{1}{2} \cdot \tfrac{4}{9} + \tfrac{1}{2} \cdot \tfrac{9}{16}}$$
$$= \sqrt{\tfrac{145}{288}} \approx 0.710$$

and

$$d_2(2) = \sqrt{\tfrac{1}{2} \cdot |\tfrac{1}{6} - 1|^2 + \tfrac{1}{2} \cdot |\tfrac{1}{2} - 1|^2} = \sqrt{\tfrac{1}{2} \cdot \tfrac{25}{36} + \tfrac{1}{2} \cdot \tfrac{1}{4}}$$
$$= \sqrt{\tfrac{34}{72}} \approx 0.687$$

therefore the second alternative is the better one.

Consider next the technology selection continuous problem (13). In Section 3.3 we have shown that the normalized objective functions are

$$\bar{f}_1(x_1, x_2) = x_1 + \tfrac{1}{2}x_2 \quad \text{and} \quad \bar{f}_2(x_1, x_2) = \tfrac{1}{2}x_1 + x_2$$

Select again equal weights, $c_1 = c_2 = \tfrac{1}{2}$, and assume that the normalized ideal values are $f_1^* = 1$ and $f_2^* = 1$. If the d_∞ distance is used, then problem (33) can be rewritten as

$$\begin{aligned}
\text{Minimize} \quad & \max\{\tfrac{1}{2} \cdot |x_1 + \tfrac{1}{2}x_2 - 1|, \tfrac{1}{2} \cdot |\tfrac{1}{2}x_1 + x_2 - 1|\} \\
\text{Subject to} \quad & x_1 \geq 0, x_2 \geq 0 \quad\quad\quad\quad\quad\quad\quad\quad\quad\quad (34) \\
& x_1 + x_2 \leq 1
\end{aligned}$$

The choice of distance d_1 leads to the optimization problem

$$\begin{aligned}
\text{Minimize} \quad & \tfrac{1}{2} \cdot |x_1 + \tfrac{1}{2}x_2 - 1| + \tfrac{1}{2} \cdot |\tfrac{1}{2}x_1 + x_2 - 1| \\
\text{Subject to} \quad & x_1 \geq 0, x_2 \geq 0 \quad\quad\quad\quad\quad\quad\quad\quad\quad (35) \\
& x_1 + x_2 \leq 1
\end{aligned}$$

If the d_2 distance is selected, then these optimization problems are modified as follows:

$$\begin{aligned}
\text{Minimize} \quad & \tfrac{1}{2} \cdot |x_1 + \tfrac{1}{2}x_2 - 1|^2 + \tfrac{1}{2} \cdot |\tfrac{1}{2}x_1 + x_2 - 1|^2 \\
\text{Subject to} \quad & x_1 \geq 0, x_2 \geq 0 \quad\quad\quad\quad\quad\quad\quad\quad\quad (36) \\
& x_1 + x_2 \leq 1
\end{aligned}$$

where we deleted the square root in the objective function. The solutions of these problems will be computed later in this section.

We will next show that in the linear case (that is, when all objective functions and constraints are linear), problem (33) with distances d_∞ and d_1 can be solved by the simplex method, and with distance d_2 it can be solved as a convex quadratic programming problem. Assume that for $j = 1, 2, \ldots, s$,

$$f_j(\mathbf{x}) = \mathbf{f}_j^T \mathbf{x}$$

with $\mathbf{f}_j^T$ being a row vector of the coefficients of the jth objective function, and

$$g(\mathbf{x}) = \mathbf{G}\mathbf{x}$$

where $\mathbf{G}$ is a constant matrix.

With the choice of distance d_∞, problem (33) becomes

$$\begin{aligned}
\text{Minimize} \quad & \max_{j}\{c_j|\mathbf{f}_j^T\mathbf{x} - f_j^*|\} \\
\text{Subject to} \quad & \mathbf{G}\mathbf{x} \geq \mathbf{Q} \quad\quad\quad\quad\quad\quad\quad\quad\quad\quad (37)
\end{aligned}$$

Introduce the new variable F for the objective function. Then for all j,

$$c_j |f_j^T \mathbf{x} - f_j^*| \leqq F$$

which can be rewritten as the pair of inequalities

$$f_j^T \mathbf{x} - \frac{1}{c_j} F \leqq f_j^*$$

$$f_j^T \mathbf{x} + \frac{1}{c_j} F \geqq f_j^*$$

Hence problem (37) is equivalent to

Minimize F

Subject to $\mathbf{Gx} \geqq \mathbf{Q}$ (38)

$$\left. \begin{array}{l} f_j^T \mathbf{x} - \dfrac{1}{c_j} F \leqq f_j^* \\[2mm] f_j^T \mathbf{x} + \dfrac{1}{c_j} F \geqq f_j^* \end{array} \right\} (j = 1, 2, \ldots, s)$$

which is a linear programming problem and can be solved by the simplex method. In the case of problem (34), we have the following linear programming problem:

Minimize F

Subject to $x_1 \geqq 0, x_2 \geqq 0$

$x_1 + x_2 \leqq 1$

$x_1 + \frac{1}{2} x_2 - 2F \leqq 1$ (39)

$x_1 + \frac{1}{2} x_2 + 2F \geqq 1$

$\frac{1}{2} x_1 + x_2 - 2F \leqq 1$

$\frac{1}{2} x_1 + x_2 + 2F \geqq 1$

The application of the simplex method gives the unique solution

$$x_1^* = x_2^* = \tfrac{1}{2} \quad \text{and} \quad F^* = \tfrac{1}{8}$$

If distance d_1 is chosen, then problem (33) becomes

Minimize $\displaystyle\sum_{j=1}^{s} c_j |f_j^T \mathbf{x} - f_j^*|$

                                                                                 (40)

Subject to $\mathbf{Gx} \geqq \mathbf{Q}$

Introduce the new variables

$$F_j = c_j |f_j^T \mathbf{x} - f_j^*|$$

for $j = 1, 2, \ldots, s$. Then, similar to problem (38), we have again a linear programming problem,

$$\text{Minimize} \quad \sum_{j=1}^{s} F_j$$

$$\text{Subject to} \quad \mathbf{G}\mathbf{x} \geq \mathbf{Q} \tag{41}$$

$$\left. \begin{array}{l} \mathbf{f}_j^T \mathbf{x} - \dfrac{1}{c_j} F_j \leq f_j^* \\[2mm] \mathbf{f}_j^T \mathbf{x} + \dfrac{1}{c_j} F_j \geq f_j^* \end{array} \right\} \quad (j = 1, 2, \ldots, s)$$

which can be also solved by using the simplex method. In the case of problem (35), we have to solve the problem

$$\begin{array}{ll} \text{Minimize} & F_1 + F_2 \\ \text{Subject to} & x_1 \geq 0, x_2 \geq 0 \\ & x_1 + x_2 \leq 1 \\ & x_1 + \tfrac{1}{2}x_2 - 2F_1 \leq 1 \\ & x_1 + \tfrac{1}{2}x_2 + 2F_1 \geq 1 \\ & \tfrac{1}{2}x_1 + x_2 - 2F_2 \leq 1 \\ & \tfrac{1}{2}x_1 + x_2 + 2F_2 \geq 1 \end{array} \tag{42}$$

when we have infinitely many optimal solutions:

$$x_1^* \in [0, 1] \text{ is arbitrary}, \quad x_2^* = 1 - x_1^*$$

$$F_1^* = \tfrac{1}{4}(1 - x_1^*), \quad \text{and} \quad F_2^* = \tfrac{1}{4}x_1^*$$

And finally, with the selection of the d_2 distance problem (33) becomes

$$\text{Minimize} \quad \sum_{j=1}^{s} c_j |\mathbf{f}_j^T \mathbf{x} - f_j^*|^2$$

$$\text{Subject to} \quad \mathbf{G}\mathbf{x} \geq \mathbf{Q} \tag{43}$$

where we deleted the square root in the objective function. This is a convex quadratic programming problem with linear constraints and therefore it can be solved by routine methods (such as gradient-type iteration methods). In the special case of problem (36), we have to following:

$$\begin{array}{ll} \text{Minimize} & \tfrac{1}{2}(x_1 + \tfrac{1}{2}x_2 - 1)^2 + \tfrac{1}{2}(\tfrac{1}{2}x_1 + x_2 - 1)^2 \\ \text{Subject to} & x_1 \geq 0, x_2 \geq 0 \\ & x_1 + x_2 \leq 1 \end{array} \tag{44}$$

which has a unique optimal solution $x_1^* = x_2^* = \tfrac{1}{2}$.

Another variant of the method is based on the knowledge of the nadir, and the alternative with objective vector having the largest distance from the nadir is selected as the solution. The two versions of the method are illustrated in Figure 9.

In the discrete case, the problem can be written as

$$d_p(i^*) = \max_i \{d_p(i)\} \tag{45}$$

where $d_p(i)$ is the same as before. In the continuous case, problem (33) has to be modified in the following way:

$$\begin{array}{ll} \text{Maximize} & d_p(\mathbf{x}) \\ \text{Subject to} & \mathbf{g}(\mathbf{x}) \geqq \mathbf{Q} \end{array} \tag{46}$$

Consider next the normalized discrete problem given in Table 3 and assume that the normalized nadir is given as $f_{1*} = f_{2*} = 0$ and equal weights are selected: $c_1 = c_2 = \frac{1}{2}$. If the d_∞ distance is used, then

$$d_\infty(1) = \max\{\tfrac{1}{2} \cdot |\tfrac{1}{3} - 0|; \tfrac{1}{2} \cdot |\tfrac{1}{4} - 0|\} = \max\{\tfrac{1}{6}; \tfrac{1}{8}\} = \tfrac{1}{6}$$

and

$$d_\infty(2) = \max\{\tfrac{1}{2} \cdot |\tfrac{1}{6} - 0|; \tfrac{1}{2} \cdot |\tfrac{1}{2} - 0|\} = \max\{\tfrac{1}{12}; \tfrac{1}{4}\} = \tfrac{1}{4}$$

showing that the second alternative is better since it has the larger distance.

If the d_1 distance is selected, then

$$d_1(1) = \tfrac{1}{2} \cdot |\tfrac{1}{3} - 0| + \tfrac{1}{2} \cdot |\tfrac{1}{4} - 0| = \tfrac{1}{2} \cdot \tfrac{1}{3} + \tfrac{1}{2} \cdot \tfrac{1}{4} = \tfrac{7}{24}$$

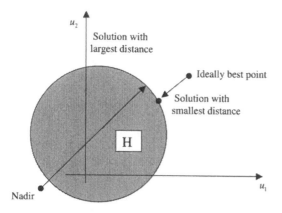

FIGURE 9 Illustration of distance-based methods.

and

$$d_1(2) = \tfrac{1}{2} \cdot |\tfrac{1}{6} - 0| + \tfrac{1}{2} \cdot |\tfrac{1}{2} - 0| = \tfrac{1}{2} \cdot \tfrac{1}{6} + \tfrac{1}{2} \cdot \tfrac{1}{2} = \tfrac{1}{3}$$

The larger distance is obtained for the second alternative, so it is selected as the solution.

Using the d_2 distance, we have

$$d_2(1) = \sqrt{\tfrac{1}{2} \cdot |\tfrac{1}{3} - 0|^2 + \tfrac{1}{2} \cdot |\tfrac{1}{4} - 0|^2} = \sqrt{\tfrac{1}{2} \cdot \tfrac{1}{9} + \tfrac{1}{2} \cdot \tfrac{1}{16}}$$

$$= \sqrt{25/288} \approx 0.295$$

and

$$d_2(2) = \sqrt{\tfrac{1}{2} \cdot |\tfrac{1}{6} - 0|^2 + \tfrac{1}{2} \cdot |\tfrac{1}{2} - 0|^2} = \sqrt{\tfrac{1}{2} \cdot \tfrac{1}{36} + \tfrac{1}{2} \cdot \tfrac{1}{4}}$$

$$= \sqrt{10/72} \approx 0.373$$

and the second alternative becomes the better choice again.

The technology selection example will be next solved by this method variant. Assume equal weights and that the normalized nadir is again $f_{1*} = f_{2*} = 0$. By applying the d_∞ distance, problem (33) reduces to

Maximize	$\max \{ \tfrac{1}{2} \cdot |x_1 + \tfrac{1}{2}x_2 - 0|, \tfrac{1}{2} \cdot |\tfrac{1}{2}x_1 + x_2 - 0| \}$	
Subject to	$x_1 \geq 0, x_2 \geq 0$	(47)
	$x_1 + x_2 \leq 1$	

Similarly to the previous case, one may verify that there are two optimal solutions: $x_1^* = 1, x_2^* = 0$ and $x_1^* = 0, x_2^* = 1$. In the case of the d_1 distance, the optimization problem can be written as

Maximize	$\tfrac{1}{2} \cdot |x_1 + \tfrac{1}{2}x_2 - 0| + \tfrac{1}{2} \cdot |\tfrac{1}{2}x_1 + x_2 - 0|$	
Subject to	$x_1 \geq 0, x_2 \geq 0$	(48)
	$x_1 + x_2 \leq 1$	

which has infinitely many optimal solutions: $x_1^* \in [0,1]$ is arbitrary and $x_2^* = 1 - x_1^*$.

Selecting the d_2 distance, problem (33) reduces to

Maximize	$\tfrac{1}{2} \cdot |x_1 + \tfrac{1}{2}x_2 - 0|^2 + \tfrac{1}{2} \cdot |\tfrac{1}{2}x_1 + x_2 - 0|^2$	
Subject to	$x_1 \geq 0, x_2 \geq 0$	(49)
	$x_1 + x_2 \leq 1$	

which has two optimal solutions: $x_1^* = 1, x_2^* = 0$ and $x_1^* = 0, x_2^* = 1$.

4 SOCIAL CHOICE

In many practical cases we have finitely many alternatives to select from, and the objective functions cannot be or are difficult to quantify. In such cases we apply methods that require ranking the alternatives with respect to each criterion. If we imagine the situation as that each criterion is associated with a decision maker, then a compromise selection can be found. This solution may not be the best for any of the decision makers, but it is a solution that can be accepted by all of them. This kind of methodology is known as *social choice*. A comprehensive summary of social choice procedures can be found in Refs. 2 and 3. This section introduces the fundamentals of this methodology.

The data, or mathematical model, consists of the sets of alternatives and criteria, the importance weights of the different criteria, and, furthermore, the rankings of the alternatives with respect to all criteria. These data can be summarized in the way shown in Figure 10. The rows correspond to the alternatives, the columns to the criteria. The last row shows the weights of the different criteria.

Each column of the matrix has the ranking of the alternatives with respect to the criterion associated to that column. That is, each column is a permutation of the numbers $1,2,3,\dots,r$. Here 1 indicates the most preferred alternative, 2 indicates the second most preferred one, and so on; r indicates the least preferred alternative.

In this section, four particular methods will be introduced for finding the best "common" choice.

4.1 Plurality Voting

Under plurality voting, for each alternative we find the sum of the weights of all criteria which give best ranking for the given alternative. Then the alternative with the highest sum is the social choice.

Alternatives \ Criteria	1	2	...	s
1	a_n	a_{12}	...	a_{1S}
2	a_{21}	a_{22}	...	a_{2S}
3	a_{31}	a_{32}	...	a_{3S}
$\vdots$	$\vdots$	$\vdots$		$\vdots$
r	a_{r1}	a_{r2}	...	a_{rs}
	w_1	w_2	...	w_s

FIGURE 10 Data for social choice.

The mathematical formulation of this method is the following. Introduce the function

$$f(x) = \begin{cases} 1 & \text{if } x = 1 \\ 0 & \text{otherwise} \end{cases}$$

and for each alternative i define

$$A_i = \sum_{j=1}^{s} w_j f(a_{ij}) \qquad (50)$$

The social choice is alternative k if

$$A_k = \max\{A_i\} \qquad (51)$$

As an illustration of this method, consider the problem given in Table 4. Simple calculations show that $A_1 = 0.4$, $A_2 = 0.3$, $A_3 = 0.1$, $A_4 = 0.2$. Since A_1 is the largest among the A_i numbers, alternative 1 is the social choice.

4.2 Borda Counts

The previous method considers the number and weight of first rankings only. The Borda count takes all rankings into account by computing the quantities

$$B_i = \sum_{j=1}^{s} w_j \, a_{ij}$$

and alternative k is the social choice if

$$B_k = \min\{B_i\}$$

As an illustration, consider again the problem given in Table 4. In this case,

$$B_1 = 2(0.3) + 4(0.2) + 1(0.4) + 4(0.1) = 2.2$$

TABLE 4 Data for Social Choice Example

Alternatives/Criteria	1	2	3	4
1	2	4	1	4
2	1	2	3	2
3	4	3	2	1
4	3	1	4	3
	0.3	0.2	0.4	0.1

$$B_2 = 1(0.3) + 2(0.2) + 3(0.4) + 2(0.1) = 2.1$$
$$B_3 = 4(0.3) + 3(0.2) + 2(0.4) + 1(0.1) = 2.7$$
$$B_4 = 3(0.3) + 1(0.2) + 4(0.4) + 3(0.1) = 3.0$$

Since B_2 is the smallest, alternative 2 is the social choice.

4.3 Hare System

The Hare method is based on successive deletion of less desirable alternatives. If there is an alternative that is selected as best by criteria with total weights at least 0.5, then it is the social choice. Otherwise, the alternative with the smallest such total weight is deleted from the table, and the table is adjusted. If more than one alternative has the smallest such total weight, then all of them are deleted. After this step the number of alternatives becomes less. Then we go back to the beginning of the procedure by looking for an alternative that is selected the best (in the new table) by criteria with total weight at least 0.5. If we can find such an alternative, then it is the social choice. Otherwise we continue the deletion.

In the case of the example of Table 4, there is no alternative that is considered the best by criteria with total weights at least 0.5, since the total weights are $A_1 = 0.4$, $A_2 = 0.3$, $A_3 = 0.1$, and $A_4 = 0.2$, as was seen in applying plurality voting. Since A_3 is the smallest A_i number, alternative 3 is deleted from the table. The modified table is shown in Table 5. The modification is

$$a_{ij} = \begin{cases} a_{ij} & \text{if } a_{ij} < a_{kj} \\ a_{ij} - 1 & \text{if } a_{ij} > a_{kj} \end{cases} \tag{52}$$

where k is the deleted alternative.

In the new table $A_1 = 0.4$, $A_2 = 0.4$, $A_4 = 0.2$. There is no A_i value that is at least 0.5, so no social choice can be made immediately. Since A_4 is the smallest A_i number, alternative 4 is deleted from the table, which is modified as shown in Table 6. Now we have $A_1 = 0.4$ and $A_2 = 0.6$, so alternative 2 is the social choice.

TABLE 5 Modified Table for Hare System Made According to the Formula

Alternatives/Criteria	1	2	3	4
1	2	3	1	3
2	1	2	2	1
4	3	1	3	2
	0.3	0.2	0.4	0.1

TABLE 6 Further Modified Table for Hare System

Alternatives/Criteria	1	2	3	4
1	2	2	1	2
2	1	1	2	1
	0.3	0.2	0.4	0.1

4.4 Pairwise Comparison

To apply the pairwise comparison method, first we have to agree on an order of pairwise comparisons. Mathematically, an order of the alternatives has to be specified in the following way. Let $i_1, i_2, \ldots, i_r$ be the order. First alternatives i_1 and i_2 are compared. The preferred alternative is then compared to i_3. The winner is then compared to i_4, and so on; the winner of the last comparison is finally accepted as the social choice.

If i_1 and i_2 are any two alternatives, then they are compared as follows. Define

$$N(i_1, i_2) = \sum_j \{w_j | a_{i_1 j} < a_{i_2 j}\} \tag{53}$$

which is the sum of the weights of all criteria which prefer alternative i_1 against i_2. Similarly,

$$N(i_2, i_1) = \sum_j \{w_j | a_{i_2 j} < a_{i_1 j}\} \tag{54}$$

and alternative i_1 is considered better than i_2 if and only if

$$N(i_1, i_2) > N(i_2, i_1) \tag{55}$$

Consider again the example of Table 4, and assume that comparisons are made in the order 1, 2, 3, 4. We first compare alternatives 1 and 2 by computing

$$N(1,2) = 0.4$$

and

$$N(2,1) = 0.3 + 0.2 + 0.1 = 0.6$$

Since $N(2,1) > N(1,2)$, we drop alternative 1 and the comparison continues with alternative 2. In comparing the winner to alternative 3, we have

$$N(2,3) = 0.3 + 0.2 = 0.5$$

and

$$N(3,2) = 0.4 + 0.1 = 0.5$$

So alternatives 2 and 3 are equivalent, and we have to continue the comparison with both of them. Simple calculations show that

$$N(2, 4) = 0.3 + 0.4 + 0.1 = 0.8$$

and

$$N(4, 2) = 0.2$$

Furthermore,

$$N(3, 4) = 0.4 + 0.1 = 0.5$$

and

$$N(4, 3) = 0.3 + 0.2 = 0.5$$

Therefore alternatives 3 and 4 are equivalent, but alternative 2 is better than 4. Hence alternative 4 is dropped and both alternatives 2 and 3 are considered as the social choice solution.

5 CONFLICT RESOLUTION

In this section we assume that there are s decision makers, and each has a well-defined objective function on the set of all alternatives. Let X denote the set of feasible alternatives and let $f_i : X \mapsto R$ denote the real-valued objective function of decision maker j. In formulating a conflict we need a reference point, the coordinates of which give the objective function values if no agreement is reached. The existence of an unfavorable reference point gives the incentive to negotiate and to reach a compromise.

In the conflict resolution literature there are many different solution concepts; they are discussed in detail in Ref. 4. In this section the most commonly used concept of Nash will be introduced.

It is assumed that the solution is given in the objective space H and it depends on only the objective space

$$H = \{\mathbf{u} | \mathbf{u} = \boldsymbol{f}(\mathbf{x}) \text{ with some } \mathbf{x} \in X\}$$

and the reference point

$$\mathbf{d} = (d_1, d_2, \dots, d_s)$$

so we may denote it as $\varphi(H, \mathbf{d})$. It is also assumed that function φ satisfies the following properties:

1. $\varphi(H,\mathbf{d}) \in H$ (that is, the solution has to be feasible).
2. $\varphi(H,\mathbf{d}) \geq \mathbf{d}$ (for all decision makers, any agreement has to be at least as good as the payoff without agreement).
3. $\varphi(H,\mathbf{d})$ is a Pareto solution in H (that is, the payoff of any decision maker can be improved only by worsening the payoff of another decision maker).
4. If $H_1 \subset H$ and $\varphi(H,\mathbf{d}) \in H_1$, then $\varphi(H,\mathbf{d}) = \varphi(H_1,\mathbf{d})$ (that is, if the feasible objective space is reduced and the solution remains feasible, then the solution does not change).
5. Define the linear transformation $L(f) = \mathbf{D}f + \mathbf{b}$ on H with a diagonal matrix $\mathbf{D}$ having positive diagonal elements and with an arbitrary s-element vector $\mathbf{b}$. Then it is required that $\varphi(L(H), L(\mathbf{d})) = L(\varphi(H,\mathbf{d}))$ (that is, scaling and shifting do not alter the solution).
6. If for a conflict $(H,\mathbf{d})$ there are two indices i and j such that $d_i = d_j$, and $(f_1, \ldots, f_s) \in H$ if and only if $(\bar{f}_1, \ldots, \bar{f}_s) \in H$ with $\bar{f}_i = f_j$, $\bar{f}_j = f_i$ and $\bar{f}_k = f_k$ for $k \notin \{i,j\}$, then at the solution, $\varphi_i(H,\mathbf{d}) = \varphi_j(\mathbf{H},\mathbf{d})$ (that is, if two decision makers are equivalent in the model, then they have to get the same payoff at the solution).

Nash (5) has proven that if H is convex, closed, and bounded, and there is at least one $f \in H$ such that $f > \mathbf{d}$, then there is a unique solution $\varphi(H,\mathbf{d}) = f^*$ which is the solution of the optimization problem

$$\begin{array}{ll} \text{Maximize} & (f_1 - d_1)(f_2 - d_2) \ldots (f_s - d_s) \\ \text{Subject to} & f \in H \\ & f \geq \mathbf{d} \end{array} \qquad (56)$$

Notice that this problem has a quasi-concave objective function, so routine methods can be used to solve the above problem (see Ref. 6).

As an illustration, consider the continuous bi-objective problem (13). We have seen that the feasible payoff set is characterized by inequalities

$$f_2 \geq \frac{f_1}{2} \qquad f_2 \leq 2f_1 \qquad f_1 + f_2 \leq 3$$

and is illustrated in Figure 4. The worst point of this set is $(0, 0)$, so it is considered as the reference point. Thus, in this case, problem (56) has the special form

$$\begin{array}{ll} \text{Maximize} & f_1 \cdot f_2 \\ \text{Subject to} & f_1, f_2 \geq 0 \\ & f_2 \geq \dfrac{f_1}{2} \\ & f_2 \leq 2f_1 \end{array}$$

$$f_1 + f_2 \leqq 3$$

Since the solution has to be Pareto optimal, it is on the linear segment connecting points (1, 2) and (2, 1). This segment has the equation

$$f_1 + f_2 = 3,$$

so $f_2 = 3 - f_1$, and the objective function is therefore

$$f_1 f_2 = f_1(3 - f_1) = -f_1^2 + 3f_1 = -(f_1 - 3/2)^2 + 9/4$$

The global maximum occurs at $f_1 = 3/2$ and $f_2 = 3 - 3/2 = 3/2$, which is the middle point of the segment. Notice that the solution of equations

$$f_1 = 2x_1 + x_2 = 3/2$$
$$f_2 = x_1 + 2x_2 = 3/2$$

gives the corresponding decision variable values:

$$x_1 = x_2 = 1/2$$

6 SUPPLEMENTARY NOTES

In this chapter the most popular multiobjective programming methods were discussed. Each method is based on a certain way of expressing preference orders. Ordinal preference order was the basis of the sequential optimization method, and the specification of the most important objective and defining minimal acceptable levels for all other objectives were needed in applying the ε-constraint method. The weighting method is the appropriate method choice if importance weights are assigned to the objectives. The knowledge of the status quo point or a nadir and the suggested direction of improvement or the specification of an ideally best point and the required direction of relaxation makes the use of direction-based methods possible. Distance-based methods are used when the directions of improvement or relaxation are replaced by distance functions. This diversity of types of preference information makes the application of this methodology complicated, but also gives users several ways and viewpoints by which to examine the decision problems in hand. Applying only one method gives the answer from only one viewpoint, and therefore the answer cannot be considered convincing. If different methods are used to solve the same problem, in most cases the results will be different. The reason for this inconsistency is the fact that even an experienced decision maker is unable to express his or her preferences accurately via a small number of quantities (such as minimal acceptable levels, weights, ideal values, etc.). If multiple decision makers (or stakeholders) are present, a certain compromise is needed in order to find a commonly acceptable preference order. Compromises via different quantities are usually different. Therefore we strongly believe that multiobjective decision making has to be a repeated interactive process with the decision makers (or stakeholders). The preference orders have to be assessed by several different ways, and the corresponding methods have to be applied simultaneously. It is very seldom the case that the same answer is found by using different methods. If this is the case, then a satisfactory solution has been obtained. Otherwise the results have to be analyzed in order to recover and understand the reasons for the discrepancies of the results. The preference orders then have to be modified and the computations redone. This process has to be repeated until a satisfactory solution is obtained. No mathematical algorithms are available to show how preference information has to be adjusted. These adjustments are usually based on subjective judgments, and intuitions.

If only rankings by the different criteria or decision makers are available, then social choice is applied. Each method is based on a certain interpretation of fairness. Different methods result in different solutions. The simultaneous application of several methods and repeated adjustments of weights can lead to a satisfactory solution.

In conflict resolution we require that the solution satisfies a system of reasonable properties. We discussed here the solution concept of Nash, but modified axiom systems usually give different solutions. The most suitable axiom system should be selected, and the corresponding solution then accepted as the final solution.

REFERENCES

1. F. Szidarovszky, M. Gershon, and L. Duckstein, *Techniques for Multi-objective Decision Making in Systems Management*. Amsterdam: Elsevier, 1986.
2. A. D. Taylor, *Mathematics and Politics*. New York: Springer-Verlag, 1995.
3. J. Bonner, *Introduction to the Theory of Social Choice*. Baltimore: Johns Hopkins University Press, 1986.
4. F. Forgo, J. Szep, and F. Szidarovszky, *Introduction to the Theory of Games*. Dordrecht: Kluwer, 1999.
5. J. Nash, The Bargaining Problem. *Econometrica*, vol. 18, pp. 155–162, 1956.
6. O. L. Mangasarian, *Nonlinear Programming*. New York: McGraw-Hill, 1962.

19

Environmental Considerations and Computer Process Design

Victor R. Vasquez
University of Nevada, Reno, Reno, Nevada

1 INTRODUCTION

The goal of this chapter is to summarize the main computational methods used in computer simulation and design of chemical processes that can be used to identify and analyze opportunities in pollution prevention and control. The material covered is by no means exhaustive. Therefore, the reader is encouraged to review the references for more specific details. The philosophy behind pollution prevention involves finding technological pathways and operations to make products in such way that the environmental impact of the process is minimum. We can see that such a framework or approach will necessarily create design and analysis of different alternatives. The total number of alternatives will depend on the type of application and objectives to be accomplished and can vary from a few to hundreds. The evaluation of the different alternatives can be greatly accelerated by the use of appropriate computational tools or procedures.

In addition to the potential of computational methods for selection and analysis of technological alternatives, there has been significant development of new concepts toward more efficient processes from an environmental protection standpoint. Among these is the concept of "green" engineering, which include ideas such as process integration to broaden the design procedures by incorporating new engineering concepts to avoid pollution at early stages of design. In other words, the idea is to avoid the generation of waste through process design. Regardless of the approach used, pollution prevention and control through process analysis and design is probably the most powerful approach to dealing with environmental issues in the chemical process industry (CPI). Increasing research efforts are reported in the literature involving the applications of concepts of process synthesis and integration toward pollution prevention and waste minimization. Development and application of techniques such as heat exchanger networks (pinch analysis), mass-exchange networks (MENs), synthesis of environmentally acceptable reaction paths coupled with mathematical optimization techniques (e.g., linear and nonlinear programming), and knowledge-based systems (expert systems) establish the current basis used to approach pollution prevention issues involving computational techniques. The methodologies and the current development of new technologies are mainly in the area of process synthesis, which is very important for new plants or expansion of existing ones. However, there is substantial lack of methodologies for pollution prevention analysis of existing plants based on the study of operational conditions without making drastic modifications in the process flowsheets. Introducing changes in process flowsheets of existing plants is a very difficult task because of the inherent reluctance of the CPI to change processes that are already producing.

Even though computer-based tools for pollution prevention can play a very significant on the identification and analysis of pollution prevention opportunities, current tools lack many capabilities for appropriate assessment of environmental impact of new and existing plants (1). Efforts from several organizations have been made during the last 10 years to address these needs and to prioritize research and development. Among the organizations involved are the American Institute of Chemical Engineering (AIChE), the U.S. Environmental Protection Agency (EPA), and the U.S. Department of Energy (DOE). Some of the needs identified are as follows (2–4):

1. Develop methodologies and tools to identify and evaluate alternative, environmentally benign reaction pathways.
2. Develop algorithms for accurately tracking trace components in process and waste streams.
3. Integrate process synthesis and process simulation.
4. Develop optimization approaches for synthesis of new processes.
5. Develop techniques to deal with the stochastic nature of variables in process modeling and simulation.
6. Identify better techniques to assess environmental costs and impacts.

It is clear that to achieve these goals, the computer tools needed have to be sophisticated, but with the growing computer industry this is becoming achievable. An important issue to be considered is how to deal with uncertainties present at different stages of process design and development. Current process simulators and computer tools used for process design, in general, lack capabilities to deal with this problem, which is very important in the evaluation of robustness and probabilistic assessment of models or technological alternatives. Appropriate risk analysis studies (economical and technical) provide a strong basis for decision making in the development of new "clean" chemical processes. Figure 1 illustrates the role of computational tools in an integrated approach to pollution prevention and design.

2 COMPUTERS AND POLLUTION PREVENTION ANALYSIS

Computer-based pollution prevention analysis plays, and will continue to play, a very significant role in the identification of potential opportunities for decreasing the environmental impact of the chemical process industry and the development of environmentally friendly technologies. The literature on pollution prevention is vast, and the descriptions of the following sections are focused on computational methods available to facilitate flowsheet analysis toward pollution prevention. From a computational standpoint, four main areas can be identified: (a) process simulation and numerical optimization, (b) process integration, (c) process synthesis and assessment of alternatives, and (d) sensitivity and uncertainty analysis of process design and simulation. Notice that these areas are strongly interconnected, and the classification is mainly for discussion purposes only.

One of the main limitations of the approaches described above is how to deal with the technical and economical uncertainty built into the processes. That is why sensitivity and uncertainty analysis of process design and simulation is becoming more important in assessing technological designs and alternatives. On the other hand, most of the methods mentioned have practical limitations in the sense that their actual application toward pollution prevention in a real plant could be cumbersome and time consuming because of the difficulty of handling complex processes realistically. However, for analysis and design of new plants, they are very valuable.

3 PROCESS SIMULATION

Process simulation tools consist of a series of computer software resources based on mathematical models used to model and analyze a given process. This section will focus on simulators related to chemical processes. In practice, any computational procedure used to model a certain operation can be called a simulator.

Several elements characterize the process simulators available in the market. A good chemical process simulator should be able to predict physical properties or use databanks to retrieve them. Under many circumstances, good

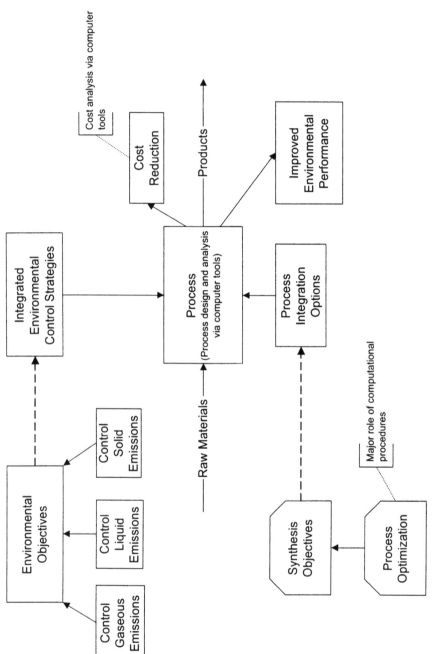

FIGURE 1 Integrated approach to pollution prevention involving computer design tools. (Adapted from Ref. 2.)

property prediction is essential. This is because experimental measurements are always more expensive than predictions, and the time required to obtained them is usually longer. For example, it is usually difficult to find experimental values of physical properties for combustion gases at high temperatures, which are essential for proper design of treatment technology of exhaust gases in combustion processes. Thus, physical properties are an elementary component of a chemical process simulator and their quality has to be assessed carefully before designing or making technical decisions. The second integral part of a process simulation tool is the unit operation models. These are subroutines or subpackages that model various unit operations such as mixers, reactors, distillation columns, heat exchangers, separators, and so forth.

Also, a process simulator must have a software component (usually called the executive) that administrates the different modules and interacts with the user. Chemical process simulators are very common in process modeling of industrial processes design to transform raw materials into products and by-products. A classical example of the application of chemical process simulators is in the petroleum refining industry. Every time that the composition of the crude oil changes due to changes in the source, process simulators are used to evaluate the new operation conditions required to keep the quality of the products within specifications or to estimate the new product flows caused by the change in composition of the raw materials.

Process simulators also differ on how the models inside the package interact and how the different systems of equations involved are solved or administrated. There are two main architectures for process modeling tools, modular process modeling and equation-oriented process modeling. A simplified diagram showing the basic architecture of a modular process tool is presented in Figure 2. The process modeling executive takes responsibility for interacting with the modules that describe the unit operations involved in the process. Notice that in this case, each unit operation module is responsible for calling subroutines or computational packages to obtain or estimate the physical properties required for the calculations. Additionally, all the equations involved in a given unit operation are solved within the unit operation module or by calling external subroutines or packages. In the modular approach, the executive plays a key role in managing and interacting with the different unit operations modules (5). The executive administrates the inputs and outputs of the different modules involved in the process. Usually, the output of a unit operation module is the input of another unit operation module; therefore, the solution process becomes sequential. The computational strategies used in the modular approach are closely related to the actual flowsheet of the process (see Figure 2).

An example taken from Westerberg et al. (5) illustrates the sequential modular approach. Figure 3 shows a simple flowsheet for the conversion of propylene to 2-hexene by the reaction

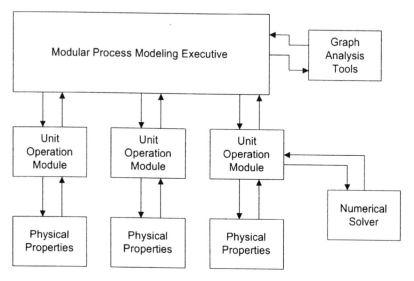

FIGURE 2 Typical architecture for modular process modeling computer tools. (Adapted from Ref. 7.)

$$2 \text{ propylene } [CH_3CH\!\!=\!\!CH_2] \rightarrow \text{2-hexene } [CH_2\!\!=\!\!CH(CH_2)3CH_3] \quad (1)$$

The propylene feed is at 150°C and 20 bar, and it contains 2 mol% propane. The process simulator will have a flowsheeting system to build the model interactively in the computer by placing icons representing the unit operations and then connecting through the different streams involved in the process. The graphical output on the screen will be something like the sketch in Figure 3. The next step for the modular executive will be to specify the components to be used in the simulation. Each stream will have different species, and the physical property methods used for the different mixtures or pure components also have to be chosen. In modern chemical process simulators, the options can be numerous and some skills are required to choose among them. This is particularly important when dealing with thermodynamic models. Some process simulators have expert systems (e.g., ChemCAD) to help the user choose the thermodynamic property package. For simple physical systems, ideal methods can be chosen to begin with and then the calculations can be tuned using more sophisticated techniques. The example, Figure 3 shows recycles; therefore, initial guesses for some of the variables or conditions have to be given before starting the computations. Commonly, the software package will present several possibilities for conditions to be guessed. The choices made may influence significantly the convergence of the simulation. Sometimes it is very important to have good initial guesses for certain conditions.

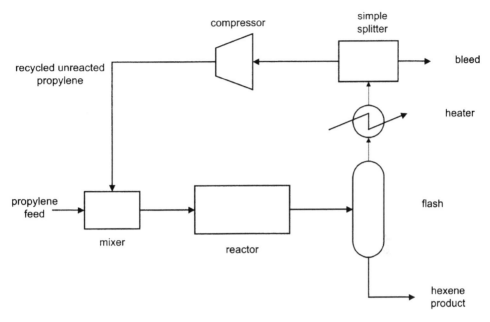

FIGURE 3 Simplified process flowsheet for the production of 2-hexene from 2-propylene. (Adapted from Ref. 5.)

Next, the models to be used in the unit operations calculations have to be selected. Here, the user is faced with different options depending on the unit operation. The correct choice is very important for the degree of accuracy required in the calculations. If only rough estimates are required, the choices will be to use simple or ideal calculation methods. For more accurate calculations, rigorous models have to be chosen. The user has to keep in mind that the more complex the model, the more time is required to perform the calculations. Sometimes convergence problems may arise with the use of complex models. For example, for the flash unit in Figure 3, we might have three unit operation models available. The first might be a component splitter, for which the user specifies the percentage of components entering that has to exit in the top vapor stream and in the bottom liquid stream. The second might be a constant relative volatility flash unit, for which the user specifies the relative volatilities for the species or requests the process simulator to estimate them using some thermodynamic method such as Raoult's law at a given temperature and pressure. A third option might be a more sophisticated or rigorous method which uses a physical property library to compute nonideal equilibrium K values and nonideal mixture enthalpies.

The same process is repeated for all the units. Then the input streams have to be specified together with the various conditions required, such as temperature, composition, pressure, and state of aggregation (i.e., gas, liquid, solid). The user may also have to choose specific numerical packages or methods to solve the equations in the unit operation modules. Once all the conditions have been specified for the streams and the unit operations, the process simulator solves the model. Because it is a sequential modular process simulator, it solves first the mixer, then the reactor, the flash, the heat exchanger, the splitter, and finally the compressor. It compares the output of the compressor with the one guessed, and if they do not agree, the recycle is re-guessed and the calculation repeated until it converges within the accuracy specified by the user.

Figure 4 shows a typical architecture for an equation-oriented process simulator. We can see that the structure does not differ that much with respect to the sequential modular (see Figure 2). The fundamental difference is that the unit operation modules no longer have the responsibility for solving the equations. Instead, the equation-oriented process modeling executive collects all the equations to assemble a single system of equations (usually large), which is solved with one or more numerical solvers. One of the main problems of this approach is robustness. In other words, solving the system of equations, particularly if they are highly nonlinear, can be difficult, but if convergence is not a problem this approach is faster than using a sequential modular. Equation-oriented systems are suitable for synthesis and optimization because the simulations are usually faster.

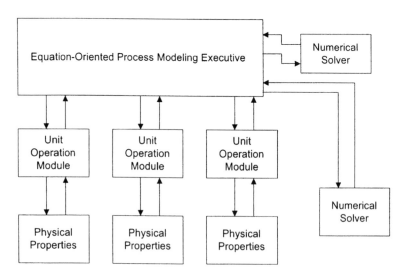

FIGURE 4 Typical architecture for equation-oriented process modeling tools. (Adapted from Ref 7.)

For example, when dealing with recycles, sequential modular systems usually have to do many iterations before they reach a solution, a problem that is avoided in an equation-oriented simulator. However, it is substantially easier to set up a flowsheet in a sequential modular than in an equation-oriented simulator, because the former is more logical for the user. This is one of the reasons why most commercial process simulators are sequential (see Table 1). In other words, they are user-friendlier than equation-oriented ones. Another disadvantage of equation-oriented systems is that it is not always obvious how to set up the global or total system of equations.

Besides the two main classifications described above, process simulators can be classified as steady state versus dynamic or unsteady-state. For many processes (particularly batch), it is very important to determine how physical conditions and operation characteristics change over time. Examples are simulation of heat transfer processes or reaction kinetics. Usually, modeling of unsteady-

TABLE 1 Partial List of Commercial Process Simulators[a]

Simulation package	Description
ASPEN PLUS	Sequential modular chemical process simulator produced by Aspen Technologies, Inc. (see www.aspentech.com for details).
ASCEND	Equation-oriented process simulator developed at Carnegie-Mellon University (see www.cs.cmu.edu/~ascend for details).
ChemCAD	Sequential modular chemical process simulator (see www.c-com.net/~chemstat/ for details).
FLOWTRAN	Sequential modular simulator for fluid flow in pipe networks (see www.FlowTran.com.au for details).
HYSIM	Sequential modular chemical process simulator (see www.hyprotech.com for details).
gPROMS	Equation-oriented general process modeling tool developed by Process Systems Enterprise Ltd. (see www.psenterprise.com/gPROMS/ for details)
PRO II	Sequential modular chemical process simulator (see www.simsci.com for details).
RIP	Sequential modular stochastic process simulator oriented to environmental applications (see www.golder.com/rip/ for details).
SPEEDUP	Equation-oriented dynamic process simulator (see www.aspentech.com).

[a]For a more comprehensive list, see www.interduct.tudelft.nl/Pitools/tools.html.

state process is more difficult than modeling of steady-state calculations; therefore unsteady-state process simulators are more difficult to find in the market and are usually designed for specific applications. Table 1 shows a partial list of chemical process simulators. The list is not exhaustive, but information about them can be easily obtained through the World Wide Web. Also, it is suggested to review Ref. 6. Another important characteristic of modern chemical process simulators is the capability to perform cost analysis including cost estimation of equipment and their operation, energy, raw materials, etc. These options greatly facilitate the economical evaluation of technical alternatives.

Another issue that modern engineering practice faces is information transfer among different computational platforms or tools. For example, Figure 5

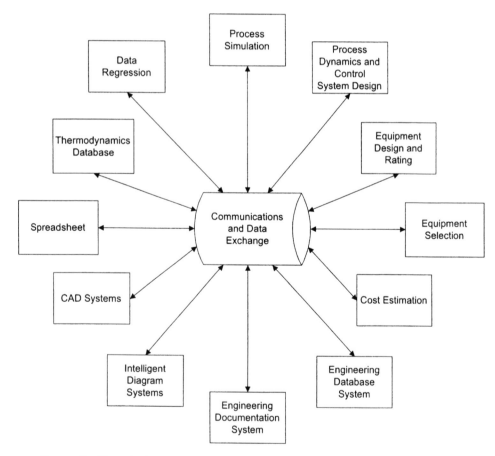

Figure 5 Chemical engineering information transfer practices in process design and simulation. (Adapted from Ref 8.)

illustrates current practice of chemical engineering information transfer. Each of the boxes is based on computer applications, and all of them have to interact. How these boxes interact has been the subject of discussion in many companies and organizations (see Refs. 7 and 8). Some initiatives such as the AIChE Process Data eXchange Institute (pdXi) (8) and the project CAPE-OPEN (Computer Aided Process Engineering. Open Simulation Environment) (7), involving companies, academic institutions, and process engineering software vendors, were created to address compatibility issues and data exchange among the different computer tools available for engineering design and simulation. It is expected that in the near future most engineering computer tools will be able to exchange data with standard process modeling environments (PMEs). These projects will substantially enhance the role of computer simulation in designing cleaner processes from an environmental standpoint.

There are other issues, however, that most computer process simulators do not address properly. For instance, the development of cleaner processes involves necessarily more factors and variables than traditional chemical process design, so evaluation of robustness and probabilistic modeling becomes very important. As mentioned earlier, most of the commercial process simulators do not have capabilities for probabilistic modeling of processes. They are, however, capable of performing some basic sensitivity analysis by performing successive incremental evaluations of certain variables such as flows or the number of stages in a distillation column.

4 OPTIMIZATION AND MATHEMATICAL PROGRAMMING

Process simulation coupled with numerical optimization techniques is commonly used for process optimization, production programming, scheduling, and alternative analysis in process synthesis. Process simulators have been useds extensively for material and energy balances and for unit operations modeling but, as mentioned earlier, they lack sufficient capabilities to study pollution prevention options (2,9–11). Numerical optimization or mathematical programming consists basically of identifying an n-dimensional vector of variables that maximizes (or minimizes) an *objective function* subject to a series of equality and inequality constraints (12–17). Mathematically, the problem can be stated as follows:

$$
\begin{array}{lll}
\text{Optimize} & Z = z(x,y) & \\
\text{Subject to} & h(x,y) = a & \quad (2) \\
& g(x,y) = b &
\end{array}
$$

where Z is the objective function, x is a vector of design variables, and y can be a vector of discrete parameters to further constrain the problem or to study different alternatives.

Within this framework, any pollution prevention problem can be studied as long as the setup and computations are feasible from a practical standpoint. Mathematical programming techniques are classified mainly as *linear programming* (LP) and *nonlinear programming* (NLP). Also, depending on the nature of the optimization, further classifications are found. For instance, formulations with real continuous variables mixed with integer variables are called *mixed-integer programming* (MIP), with two flavors: *mixed-integer linear programming* (MILP) and *mixed-integer nonlinear programming* (MINLP). The principles and theory of these approaches can be found in several references (18–21). Development of methods to find global solutions of MINLP problems is still a very active research field (22–26). Notice that, in principle, these methods or techniques can be used to set up objective functions that involve process synthesis or finding of optimal configurations to minimize the environmental impact. This approach is referred as *structural optimization* (2). The principles and theory of optimization techniques are covered in many references (see, for example, Refs. 18–21). Also, several commercial software packages are available. Edgar and Himmelblau (18) present an early survey of computer packages for optimization. Optimization tools can also be part of integrated modeling systems. Examples of these are GAMS (27) and LINDO (28). A great deal of information about optimization methods and packages can be found on the World Wide Web. For instance, the NEOS Website at www-neos.mcs.anl.gov provides a guide to optimization software, containing around 130 entries. Case studies that demonstrate the use of the algorithms are also presented (for more details and other Websites, see Ref. 28).

5 PROCESS INTEGRATION

Process integration involves mainly the interconnection and use of existing equipment in more effective ways. Two main classifications are found: (a) pinch analysis, and (b) knowledge-based approaches. In pinch analysis (29–32), the heat flows through industrial processes are systematically analyzed based on fundamental thermodynamics in order to determine an optimal heat exchanger network from an energy-savings standpoint. A trade-off between energy consumption and capital investment has to be defined. The use of pinch technology results in energy savings and also leads to waste minimization and emission reduction (2). Commercial software is available for pinch technology, including ADVENT from Aspen Technology, Inc., and SuperTarget from Linnhoff March, which evaluate the trade-off between energy savings and the associated capital costs. Using similar ideas from pinch analysis, El-Halwagi and Manousiouthakis (33) developed the notion of *mass-exchange networks* (MENs). Basically, MENs consist of synthesizing a network of mass exchangers given a number of waste-(rich) streams and a number of lean streams (MSAs). The network can transfer species from the waste streams to the MSAs at a minimum cost. The main disadvantage of this approach is the fact that mass-exchanging technology is substantially more complex and unique to the application than the technology for transferring energy (heat exchangers). However, significant efforts have been made to continue the development and application of mass-exchange networks, with promising results. (34–40).

On the other hand, knowledge-based systems include *expert systems* and *artificial intelligence* applications. They are based on expert knowledge, usually generated through experience in various processes (41–43). The use of heuristics and numerical optimization approaches play a significant role in the development of the process integration methods mentioned.

6 PROCESS SYNTHESIS AND ASSESSMENT OF ALTERNATIVES

Methodologies in the area of process synthesis and assessment of chemical processes also are very important for analysis of pollution prevention alternatives. These methods require extensive amounts of information to generate a series of technological options and then, through the use of assessment methodologies, determine the most appropriate from both technical and economical standpoints. Many times the assessment methodologies are linked directly to numerical optimization techniques (optimization and mathematical programming). Basically, any methodology developed for process synthesis and assessment can potentially be applied to pollution prevention analysis by setting appropriate objective functions or goals [see, for example, Refs. 44–46. For instance, Linninger et al. (47) describe methods for synthesis and assessment of batch processes for pollution prevention. Their work is focused toward the pharmaceutical and specialty chemicals industry, which use primarily batch processing. Some of the concepts presented include ideal ones such as zero avoidable pollution (ZAP) and more pragmatic ones such as minimum avoidable pollution (MAP), which represents a trade-off between cost and environmental impact. Environmental acceptable reactions (EARs) is another computational-based methodology used to identify environmentally safe overall chemical reactions (48). The task is formulated as a mixed-integer nonlinear optimization problem that examines overall reactions occurring to produce a given product. Other ideas include synthesis of reactor networks (49,50), targeting waste minimization. Their success has been limited mainly due to the mathematical complexity of the problem (highly nonlinear differential models) and substantial uncertainty in the definition of pollution-control cost functions.

7 SENSITIVITY AND STOCHASTIC ANALYSIS OF PROCESS DESIGN AND SIMULATION

The values of design parameters and decision variables used in any process design project are not 100% certain: there is always uncertainty associated with them. The problem of uncertainty increases particularly in environmental applications, due to the increase in the number of variables and parameters that have to be monitored or used in the calculations. For example, every time an experimental measurement is repeated, a different value is obtained even though, in theory, the conditions are kept constant (experimental error). Thus, from a modeling and design standpoint, we would like to know all the possible values that a calculation may have due to uncertainties in the process variables. The collection of all these potential values is called a distribution and is characterized in terms of probability. Before defining other terms required to perform uncertainty or probabilistic analysis in process design and simulation, let us illustrate further the role of uncertainty.

Figure 6 shows a simplified sketch of a stochastic modeling problem. In this figure, we have a series of conditions or parameters that we use for design and analysis, but we are not certain of their values requiring other type of representations. In this case, probability distributions are used to represent their values. Now we can see that the problem is not deterministic anymore. Therefore, it is expected that the results of the modeling process will be probability distributions. The main goal of stochastic modeling is to determine the probability distributions of the results and use them to analyze the robustness of the design or calculations. Probability distributions are more valuable than deterministic calculations because they tell the designer the certainty that the results will be within the specifications or objectives. Also, this information can be used to select safety factors more rationally. Traditionally, safety factors in design are chosen based on experience, by applying a percentage to the deterministic answer of the problem and usually leading to expensive designs (significant overdesign) and poor evaluation of robustness.

Stochastic modeling analysis can be carried out in several ways, depending on the nature of the problem being studied. Computer-based approaches usually follow four basic steps:

1. Specification of probability distributions for the stochastic variables involved
2. Sampling process
3. Simulation
4. Statistical analysis of the results

7.1 Specification of Probability Distributions for Stochastic Variables

In order to specify probability distributions for stochastic variables, reasonable information has to be found about all the parameters and variables used in the

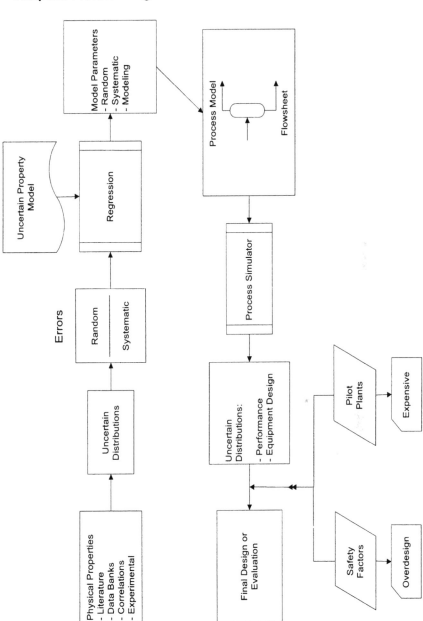

FIGURE 6 Role of uncertainty (stochastic variables) in process design and simulation.

evaluation or design. For instance, for experimental data, probabilistic information can be found from the uncertainties of the apparatus used to obtain the
measurements. Usually, the manufacturer will report what type of probabilistic
distribution can be used to characterize the errors (e.g., Gaussian). Various
probability distributions can be used to represent stochastic variables, depending
on the amount of information available. Figure 7 shows some classical probability

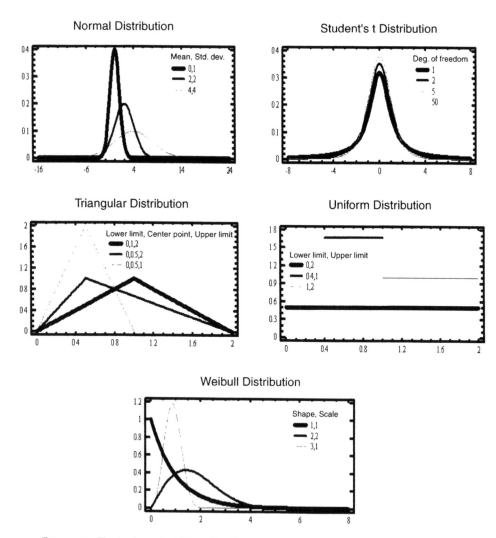

FIGURE 7 Typical probability distributions used in stochastic analysis and
modeling of chemical processes.

distributions. Among them, the Gaussian or normal and uniform are very common in many processes. The parameters usually used to characterize probability distributions are the mean, variance, skewness, and kurtosis (51).

7.2 Sampling Process

From Figure 6, we can easily conclude that it is not practical to evaluate all possible values of the probability distributions that characterize the design variables. Thus, a suitable sampling procedure is required to take representative values of the probability distributions in such way that representative probability distributions are obtained for the output variables. This process involves two main steps: (a) defining the sample size and (b) choosing an appropriate sampling technique.

7.2.1 Sample Size

For simulation purposes, the sample size is mainly an economic decision, assuming that at least a reasonable minimum sample size is guaranteed from a statistical viewpoint. Basically, for computer-based simulation of chemical processes, we are interested in either the precision of the mean of the output variables or the precision of the estimate of the median and other fractiles. The latter involves the precision of the estimated cumulative distributions.

Let us assume that we have available a random sample of n output values generated for a given sampling technique such as Monte Carlo simulation:

$$(y_1, y_2, y_3, \ldots, y_n) \tag{3}$$

Applying the classical definitions for the mean and standard deviation of y:

$$\bar{y} = \sum_{i=1}^{n} \frac{y_i}{n} \tag{4}$$

and

$$s^2 = \sum_{i=1}^{n} \frac{(y_i - \bar{y})^2}{n - 1} \tag{5}$$

The confidence interval for y with a confidence level α can be defined as (60)

$$\left(\bar{y} - c \frac{s}{\sqrt{n}}, \bar{y} + c \frac{s}{\sqrt{n}} \right) \tag{6}$$

where c is the deviation for the unit normal enclosing probability α.

In this case we want precision around the mean value, so we have to specify the precision desired. Let us say that we want a total width w in the interval as a maximum value. Therefore,

$$2c\,\frac{s}{\sqrt{n}} \le w \tag{7}$$

Rearranging,

$$n \ge \left(\frac{2cs}{w}\right)^2 \tag{8}$$

The equation above gives an estimate of the samples needed for a given w and α. In order to use it, notice that an estimation of the standard deviation is required, which can be obtained through simulation.

The other criterion mentioned is the precision of the estimated cumulative distributions. It is possible to obtain a confidence interval for Y_p, the pth fractile of Y. We need to reorder the n elements of the sample y as follows:

$$y_1 \le y_2 \le \dots y_n \tag{9}$$

In this case, y_i is an estimate of fractile Y_p, where $p = i/n$.

The α confidence level for Y_p is given by (y_i, y_j), where y_i is the lower point and y_j is the upper one, and

$$i = \lfloor np - c\sqrt{np(1 - p)} \rfloor$$
$$i = \lceil np - c\sqrt{np(1 - p)} \rceil \tag{10}$$

where $\lceil x \rceil$ means rounding x down and $\lfloor x \rfloor$ means rounding it up.

Equating the two expressions for i and j, the following expression for n is obtained:

$$n = p(1 - p)\left(\frac{c}{\Delta p}\right)^2 \tag{11}$$

The term Δp corresponds to the width of the confidence interval centered on fractile p.

For more details about this topic, see Morgan and Henrion (52). Authors such as Vasquez and Whiting (53), Reed and Whiting (54), and Whiting and Tong (55) have reported the use of this approach successfully in uncertainty studies related to chemical process design and simulation. According to these authors, 100 simulations give a 90% confidence region of 0.85–0.95 about the 0.90 percentile.

7.2.2 Sampling Techniques

Several sampling methods are available in the literature for selecting samples from probability distributions [Kalagnanam and Diwekar (56), Morgan and Henrion (52)], such as using combinatorial scenarios, probability tree approaches, and Monte Carlo methods. Several of them are well known in the literature for the last 25 years: (a) Monte Carlo sampling (MCS), (b) Latin hypercube sampling (LHS), (c) median Latin hypercube sampling (MLHS), and (d) shifted Hammersley sampling (SHS), which is one of the most recent reported in the literature (56).

In crude Monte Carlo sampling, for a sample of size n, a value is drawn at random from the distribution n times. If the model evaluation is not expensive, MCS has the advantage of avoiding the discretization of continuous distributions and is very to implement for discrete distribution as well.

Most of the time, the primary objective is to obtain accurate quantile estimations for the output of the model. In this case, *stratified sampling* is more convenient. A classical stratified sampling technique is Latin hypercube sampling, where the distribution is divided into intervals of equal probability and one sample is taken at random from within each interval [Iman and Shortencarier (57)]. One important advantage of this technique is that it produces a faster approximation of the output probability distribution than would the random Monte Carlo sampling technique. This is because, in principle, stratified sampling always performs a more uniform coverage of the full range of the probability distributions. Figure 8 illustrates the basic principle of stratified sampling. Notice that the sampling is performed using the cumulative frequency distribution (cfd) and not the probability density function (pdf). This facilitates the sampling

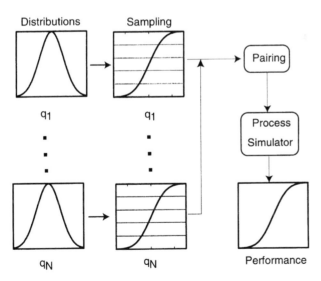

FIGURE 8 Stratified sampling procedure. Random numbers uniformly distributed between 0 and 1 are used on the y axis of the cumulative probability distributions to read the samples on the x axis.

process substantially, by using uniformly distributed random numbers on the y axis and then using the cfd curve to obtain the samples.

In LHS, the order of sampling is random, but the pairing of the selected values is done either randomly or using a technique that enforces correlation between the variables. For several correlated variables, the correlation can be established using the correlation matrix. The goal is to perform the pairing process in such way that the final sample has a correlation matrix nearly equal to the one specified. To do this, the method introduced by Iman and Conover (58) is widely used. This method rearranges the input variable sets so that they have nearly the same rank correlation matrix as the specified rank correlation matrix. The values of the input variables are not modified; only their pairing is modified. The details are provided in Section 7.3.

A modification of the LHS technique is to choose the median of each of the n probability intervals. This method is called median LHS or midpoint LHS. It has the advantage that the approximation of the output probability distributions is even faster than with the LHS. With this method the mean and variance of the sample can often be almost exact, and if the model is linear, the mean of the output will converge more rapidly (52). In general, median LHS is considered better than crude LHS, but it is subject to problems when dealing with models that present periodicity characteristics, producing misleading results in the output (52). For example, Lee et al. (59) reported the use of median LHS in thermal design studies.

Kalagnanam and Diwekar (56) developed a sampling technique called shifted Hammersley sampling (SHS). This technique is similar in principle to median LHS, with the difference that the sequence of points for sampling from the distribution is generated by a Hammersley sequence introduced by Wozniakowski (60). This sequence has been shown to have a minimum discrepancy for the class of real continuous functions (on average) (60). When there are many stochastic inputs and the model is highly nonlinear, all these stratified sampling techniques including SHS are not much better than crude Monte Carlo sampling (52). In these cases, more sophisticated sampling techniques are recommended (61,62).

7.3 Correlated Samples

It is well known that the assumption of independence among input variables is not appropriate in many cases. When performing sampling, the correlation structures among the input variables have to be taken into account in some way. Most of the approaches to this problem consider linear combinations of independent random variables to obtain a given correlation structure. Now, when dealing with stratified sampling, the way of keeping the correlation patterns is not straightforward in order to maintain the integrity of the sampling.

Iman and Conover (58) proposed a distribution-free approach to inducing rank correlation among input variables. Their method has been widely used to keep correlations structures when using stratified sampling (2,56).

Iman and Conover use the fact that a sample matrix R of n rows of k independent random variables can be transformed into a sample matrix R^* from a correlated multivariate distribution with correlation matrix C via multiplication of R by the upper-triangular matrix S^T:

$$R^* = RS^T \qquad (12)$$

where

$$C = ST S^T \qquad (13)$$

where S^T is the transpose of S and T is the correlation matrix of R.

When performing a Monte Carlo simulation, we first generate n sets of k independent variables, where n is the number of simulations to be performed. It is often useful (or necessary) to retain all of the original nk values of the variables, but to transform the pairing so that the variables are correlated. For example, we use Latin hypercube sampling to stratify our input variables according to their respective distributions. We would prefer to retain this one-dimensional stratification while inducing multidimensional correlation.

To allow the retention of the original variable values, Iman and Conover developed a technique that matches the rank correlation matrix rather than the correlation matrix. This is not only desirable for stratified sampling techniques but also essential if the individual distributions for the variables are non-Gaussian.

First, Iman and Conover create a generic $n \times k$ matrix that represents n sample sets of k independent Gaussian variables. They used "van der Waerden scores" to sample the Gaussian distributions, but any stratified sampling technique could be used. [The van der Waerden scores are similar to LHS except that the former uses the median of the n intervals as the samples, rather than taking a stochastic sample within the interval as in LHS. Also, the upper and lower $(2n + 2)^{-1}$ of the Gaussian distribution is not sampled by van der Waerden scores.]

The desired rank correlation matrix is C^R. For a Gaussian distribution, the rank correlation matrix is equal to the correlation matrix. Therefore, Iman and Conover transform the generic Gaussian sample matrix with Eqs. (12) and (13), using C^R, which is equal to C. The resultant sample matrix has a correlation matrix and a rank correlation matrix that are equal to the desired rank correlation matrix C^R. The final step is to use the transformed sample matrix as a template to reorder the values in our original LHS sample matrix so that the ranks (1 to n) for each of the k variables are the same as those for the respective variable in the transformed generic matrix. Thus, for a given sample size $n \times k$ and desired rank correlation matrix C^R, all the steps of the procedure are identical (regardless of the distributions of the variables) until the final template matching of the ranks.

This procedure can be used with any sampling technique that starts with independence between the variables. The individual variable distributions (which need not be Gaussian) are not changed. Also, the procedure does not induce a desired correlation matrix on the sample; rather, it induces a desired rank correlation matrix. Also, it does not capture the complete interdependence of the variables in the multivariate distribution; it only matches the rank correlation matrix. For highly nonlinear models, the sampling technique proposed by Vasquez et al. (69) is recommended.

7.4 Simulation

Once the samples are generated, the input values are passed through the model. For chemical engineering purposes, the model can represent a thermodynamic equation plus procedures that involve the simulation of one or several unit operations. All the most popular chemical process simulators (see Table 1), such as FLOWTRAN, ChemCAD, ASPEN, PROVISION, and HYSIM, work in a deterministic way, implying that the samples have to be passed through the software manually or using batch mode.

After performing the simulations, the output information has to be processed to construct the cumulative probability distributions. It is important to mention that we have to take into account that there are several sources of uncertainty in the simulators, such as modeling error and round-off error, that can affect the output cumulative distributions. These effects are very difficult to estimate, but it is important to be aware of them in the process of decision making.

7.5 Statistical Analysis

7.5.1 Cumulative Distributions

The cumulative probability distributions are basic tools for the understanding and quantification of the output of stochastic models. They give a direct measure of the probability of the occurrence of a given event. The definition of a cumulative distribution can be found in any statistics book (63,64); summarizing, suppose that X is a random variable that takes values on ___[1]. The cumulative distribution function of X is defined as

$$F(x) = \Pr\{X \leq x\} \qquad (14)$$

If $F(x)$ is differentiable, the function

$$f(x) = F'(x) \qquad (15)$$

is called the density function of X. Then, for any real numbers a and b, we have that the probability of X being between a and b is given by

$$\Pr\{a < X \leq b\} = F(b) - F(a) = \int_a^b f(x)\, dx \qquad (16)$$

Figure 9 shows performance results for the extraction of acetic acid from an aqueous solution using diisopropyl ether as solvent. Three sets of parameters for the NRTL equation were used to simulate the extraction [for details, see Vasquez and Whiting (53)]. For example, analyzing Figure 9, we can see that there is almost 30% probability of getting a percentage extracted less than the value obtained with the optimum parameters for Set 1. In other words, there is approximately 70% chance of getting an extraction greater than that obtained with the optimum parameters. Thus, using the optimal parameter set (in this case, the deterministic calculation) could lead to considerable overdesign.

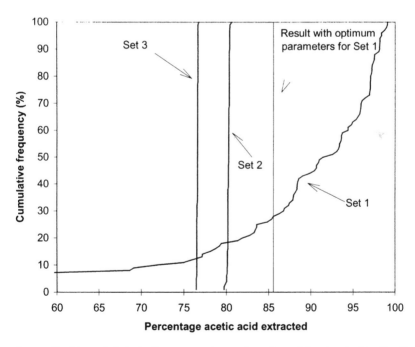

FIGURE 9 Uncertainty in the percentage of acetic acid extracted in the liquid–liquid extraction from an aqueous solution using diisopropyl ether as solvent. The three sets corresponds to different values of the binary interaction parameters of the NRTL equation regressed from different experimental data sets. For details, see Vasquez and Whiting (53).

7.5.2   Sensitivity Analysis

Additional statistical measures are used to identify the importance of the input variables in relation to the output variable probability distribution. These measures are the standardized regression coefficient (SRC), partial correlation coefficient (PCC), standardized rank correlation coefficient (SRRC), and partial rank correlation coefficient (PRCC). Rank correlation coefficients are the same as SRCs and PCCs, but are computed on the ranks of the original observations. A larger absolute value of any of these coefficients means a stronger effect by that input variable on the output variable.

Standardized Regression Coefficients.   The SRC provides a direct measure of the relative importance of the input variables, as long as the linear relation used to compute the SRC provides a reasonable relation between the output and input variables. To compute the SRC, a multilinear regression has to be performed to get the best fit for the following model:

$$\frac{(Y - \bar{Y})}{s_Y} = \sum_i \text{SRC}_i \frac{X_i - \bar{X}_i}{s_{X_i}} \tag{17}$$

In the above equation, Y is the output variable and X_i are the input variables, $\bar{Y}$ is the mean for input and output variables, and and are the standard deviations for the output and input variables, respectively.

Partial Correlation Coefficients.   The PCC is a measure of the linear relationship between the output and input variables, and it explains the unique relation between two variables that cannot be explained in terms of the relations of these two variables with any other variables (65). In other words, the PCC provides a way of identifying which additional variables have to be added to a linear model with Y as a function of the input variables with the strongest linear relationships.

In order to compute the PCC, a measure of the adequacy of the regression model is required. The lack of fit from the analysis of variance can be used for this purpose, so the adequacy of the model is measured as

$$R_y^2 = \frac{\Sigma_i(\hat{Y}_i - \bar{Y})^2}{\Sigma_i(Y_i - \bar{Y})^2} \tag{18}$$

R_y^2 varies between 0 and 1 and is called the *coefficient of determination*. It provides a measure of the percent of the variation in Y explained by regression on the X's.

The *partial correlation coefficient* PCC_i is computed using the following equation:

$$PCC_i = SRC_i \sqrt{\frac{1 - R_{X_i}^2}{1 - R_y^2}}$$

(19)

where $R_{X_i}^2$ is the coefficient of determination of the multilinear regression of X_i as a function of Y and all the other X's, and R_y^2 is the coefficient of determination of the multilinear regression of Y as a function of all of the X's.

More details about these regression coefficients can be found in Iman et al. (65), Reed et al. (55), Reed and Whiting (54), Diwekar and Rubin (11), and Morgan and Henrion (52).

Rank Transformation. When nonlinear regressions are involved, rank correlation coefficients are preferred. The only difference between the SRC and the PCC is that the variable ranks are used in place of the actual values for the variables. When computed in this way they are called standardized rank correlation coefficients (SRRC) and partial rank correlation coefficients (PRCC), respectively (65). In this work, both approaches are used for comparison and analysis.

To compute the ranks, the smallest value of each variable is assigned the rank 1, the next smallest value is assigned rank 2, and so on up to the largest value, which is assigned the rank n, where n is the number of observations.

7.6 Computer Tools

The computer tools available in the market for stochastic modeling of chemical processes are very limited. There are some software packages for risk analysis in general-purpose applications, but these are not specific to the chemical industry. For instance, the package @RISK, from Palisade Corporation (66), is a risk analysis and simulation add-in for Microsoft Excel or Lotus 1-2-3. A similar package, but directed more toward environmental applications, is RIP Integrated Probabilistic Simulation for Environmental Systems, from Golder Associates (67). This is self-standing software, but it is very limited in terms of unit operation modules.

7.7 Example: Modeling the Effect of Uncertainties on the Performance of a Liquid/Liquid Extraction Operation

The purpose of this example is to show the significance of stochastic modeling using probabilistic approaches. We chose a single unit operation (liquid–liquid extraction) to illustrate some of the ideas discussed. In order to model a liquid–liquid extraction operation, a suitable thermodynamic model is required. In this case, we used the NRTL equation (68) to calculate the liquid–liquid equilibria of the system 1,1,2-trichloroethane (1)–acetone (2)–water (3) system at 25°C and 1 atm. The NRTL equation requires at least six binary interaction parameters, which can be regressed using liquid–liquid equilibria experimental measurements. The experimental data used to regress the binary interaction parameters and their values are reported by Vasquez and Whiting (53).

The example selected is reported by Schweitzer (69), and for this case the acetone concentration of 45 kg/h of feed is 50% (mass) acetone–50% water and the solution is to be reduced in its acetone concentration with 13.5 kg/h of 1,1,2-trichloroethane as solvent in a countercurrent multiple-contact operation at 25°C in five stages. The output variable for this case is the percentage of acetone extracted. The extraction was simulated using the ChemCAD chemical process simulator.

One hundred simulations were performed using this set of parameters. The results are shown in Figure 10. The shape of the output distribution is an "S shape," and it is quite wide, from 75.38% to 94.43% in the percentage of acetone extracted. Also, the value using the optimum parameters is 86.40%. Looking at the figure, we can see that, almost 42% of the time, we get a percentage extracted less than the extraction reached with the optimum parameters. In other words, we

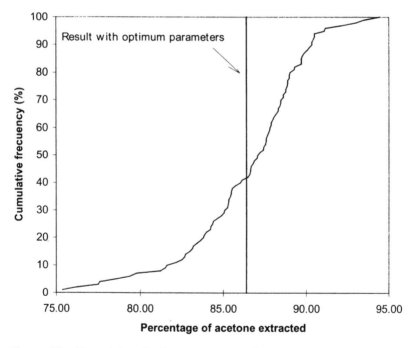

FIGURE 10 Uncertainty in the percentage of acetone extracted in the liquid–liquid extraction an aqueous solution using 1,1,2-trichloroethane as solvent. For details, see Vasquez and Whiting (53).

TABLE 2 Liquid–Liquid Extractor Regression Results[a]

Input variable	PCC	SRC	PRCC	SRRC
b_{12}	−0.795883	−1.8013	−0.562165	−1.18981
b_{21}	−0.752577	−1.82991	−0.549007	−1.48629
b_{13}	−0.274897	−0.17939	−0.073065	−0.05707
b_{31}	−0.002288	−0.00143	−0.060129	−0.05175
b_{23}	0.853524	2.566886	0.719530	2.11771
b_{32}	0.824773	2.394583	0.670185	2.03531

[a]Output variable: percentage acetone extracted. System: 1,1,2-trichloroethane (1)–acetone (2)–Water (3).

have over 58% probability of getting an extraction greater than calculated using the optimum parameters or deterministic value.

The SRC, PCC, SRRC, and PRCC computed are shown in Table 2. Observing the data, we see that the parameters b_{13} and b_{31} are the least influential. These parameters correspond to the pair 1,1,2-trichloroethane + water. If we want to reduce the effect of uncertainty in the output variable distribution, we have to reduce the variance associated with the parameters b_{12}, b_{21}, and b_{23}. Even though this is a very simple example, it clearly illustrates the usefulness of stochastic modeling. With a performance cumulative frequency curve such as the one presented in Figure 10, the design engineer can rationalize better the safety factors selection among other applications. This example also demonstrates the hidden underdesign and overdesign bias that can result from using the optimum parameter set from a regression (deterministic values). Notice that a hidden overdesign bias is detected.

8 CONCLUSIONS

From a chemical process standpoint, pollution prevention is nothing but good engineering practice. However, environmental control problems are very complex in nature. Therefore, the design tasks become complex, creating a need for more advance design tools. The future role of computer applications toward the design of clean processes will be very significant. The capability of studying many technical options in a reasonable period of time is going to boost and improve current design methodologies. Tasks such as life-cycle analysis, multiobjective optimization, and environmental impact assessment will become more and more computerized. We believe that the role of uncertainty and therefore stochastic modeling is very important in the determination of robustness of process modeling and simulation. Also, with the development of new computer architectures, the study of many stochastic scenarios related to environmental protection will be feasible. For instance, scenarios such as contaminants leaks, runaway reactions, trace pollutants, and so forth, will create more certainty in the design of clean processes.

REFERENCES

1. D. Petrides, V. Aelion, S. Mallick, and K. Abeliotis, Pollution Prevention through Process Simulation. In H. Freeman (ed.), *Industrial Pollution Prevention Handbook*, pp. 409–432. New York: McGraw-Hill, 1995.
2. P. Chaudhuri and U. Diwekar, Pollution Prevention Design. In R. Meyers (ed.), *Encyclopedia of Environmental Analysis and Remediation*, pp. 3772–3791. New York: Wiley, 1998.
3. J. M. Persichetti and S. M. Desai, Recent Advances in Process Simulation for Environmental Considerations. AIChE Spring Natl. Meeting, Atlanta, GA, 1994.
4. J. Eisenhauer and S. McQueen, Environmental Considerations in Process Design and Simulation. CWRT-AIChE, ISBN 0-8169-0614-9, 1993.
5. A. W. Westerberg, H. P. Hutchinson, R. L. Motard, and P. Winter, *Process Flowsheeting*. Cambridge, UK: Cambridge University Press, 1979.
6. W. Marquardt, Trends in Computer-Aided Process Modeling. *Comput. Chem. Eng.*, vol. 20, no. 6/7, pp. 591–609, 1996.
7. B. L. Braunschweig, C. C. Pantelides, H. I. Britt, and S. Sama, Open Software Architectures for Process Modeling: Current Status and Future Perspectives. FOCAPD'99, Breckenridge, CO, 1999.
8. T. L. Teague and J. T. Baldwin, The Emerging Discipline of Chemical Engineering Info Transfer. FOCAPD'99, Breckenridge, CO, 1999.
9. U. M. Diwekar, A Process Analysis Approach to Pollution Prevention. In M. M. El-Halwagi and D. P. Petrides (eds.), *Pollution Prevention via Process and Product Modifications*, AIChE Symp. Ser., vol. 9, pp. 3772–3791. New York: American Institute of Chemical Engineering, 1994.
10. P. D. Chaudhuri and U. M. Diwekar, Process Synthesis under Uncertainty: A Penalty Function Approach. *AIChE J*, vol. 42, no. 3, pp. 742–748, 1996.
11. U. M. Diwekar and E. S. Rubin, Stochastic Modeling of Chemical Processes. *Comput. Chem. Eng.*, vol. 15, no. 2, pp. 105–114, 1991.
12. M. M. El-Halwagi, Introduction to Numerical Optimization Approaches to Pollution Prevention. In A. P. Rositer (ed.), *Waste Minimization through Process Design*, pp. 199–208. New York: McGraw-Hill, 1995.
13. I. Grossamnn, Mixed-Integer Programming Approach for the Synthesis of Integrated Process Flowsheets. *Comput. Chem. Eng.*, vol. 9, no. 5, p. 463, 1985.
14. I. Grossmann, Mixed-Integer Nonlinear Programming Techniques for the Synthesis of Engineering Systems. *Res. Eng. Des.*, vol. 1, p. 205, 1990.
15. G. R. Kocis and I. Grossmann, A Modeling and Decomposition Strategy for MINLP Optimization of Process Flowsheets. *Comput. Chem. Eng.*, vol. 13, p. 797, 1989.
16. I. P. Androulakis, C. D. Maranas, and C. A. Floudas, αBB: A Global Optimization Method for General Constrained Nonconvex Problems. *J. Global Optimization*, vol. 7, pp. 337–363, 1995.

17. C. S. Adjiman, I. P. Androulakis, C. D. Maranas, and C. A. Floudas, A Global Optimization Method, αBB, for Process Design. *Comput. Chem. Eng. Suppl.*, vol. 20, p. S419, 1996.

18. T. F. Edgar and D. M. Himmelblau, *Optimization of Chemical Processes.* New York: McGraw-Hill, 1988.

19. S. F. Hiller and G. J. Lieberman, *Introduction to Operations Research.* New York: McGraw-Hill, 1986.

20. G. V. Reklaitis, A. Ravindran, and K. M. Ragsdell, *Engineering Optimization.* New York: Wiley, 1983.

21. G. S. G. Beveridge and R. Schechter, *Optimization: Theory and Practice.* New York: McGraw-Hill, 1970.

22. R. Vaidyanathan and M. M. El-Halwagi, Global Optimization of Nonconvex Nonlinear Programs via Interval Analysis. *Comput. Chem. Eng.*, vol. 18, no. 10, pp. 889–897, 1994.

23. C. D. Maranas and C. Floudas, Finding All Solutions of Nonlinearly Constrained Systems of Equations. *J. Global Optimization*, vol. 7, p. 143, 1995.

24. C. S. Adjiman, S. Dallwig, C. A. Floudas, and A. Neumaier, A Global Optimization Method, αBB, for General Twice-Differentiable NLPs-I. Theoretical Advances. *Comput. Chem. Eng.*, vol. 22, no. 9, pp. 1137–1158, 1998.

25. C. S. Adjiman, I. P. Androulakis, and C. A. Floudas, A Global Optimization Method, αBB, for General Twice-Differentiable NLPs-II. Implementation and Computational Results. *Comput. Chem. Eng.*, vol. 22, no. 9, pp. 1159–1179, 1998.

26. C. S. Adjiman and C. A. Floudas, Rigorous Convex Underestimators for General Twice-Differentiable Problems. *J. Global Optimization*, vol. 9, p. 23, 1996.

27. A. Brooke, D. Kendrick, and A. Meeraus, GAMS Release 2.25: A User's Guide. Danvers, MA: International Thomson, 1992.

28. S. J. Wright, Algorithms and Software for Linear and Nonlinear Programming. FOCAPD'99, Breckenridge, CO, 1999.

29. B. Linnhoff, D. R. Mason, and L. Wardle, I. Understanding Heat Exchanger Networks. *Comput. Chem. Eng.*, vol. 3, pp. 295–302, 1979.

30. B. Linnhoff and E. Hindmarsh, The Pinch Design Method of Heat Exchanger Networks. *Chem. Eng. Sci.*, vol. 38, no. 5, pp. 745–763, 1983.

31. B. Linnhoff and J. R. Flower, Synthesis of Heat Exchanger Networks: Part I: Systematic Generation of Energy Optimal Networks. *AIChE J.*, vol. 24, no. 4, pp. 633–642, 1978.

32. B. Linnhoff and J. R. Flower, Synthesis of Heat Exchanger Networks: Part II: Evolutionary Generation of Networks with Various Criteria of Optimality. *AIChE J.*, vol. 24, no. 4, pp. 642–654, 1978.

33. M. M. El-Halwagi and V. Manousiouthakis, Synthesis of Mass Exchange Networks. *AIChE J.*, vol. 35, no. 8, pp. 1233–1244, 1989.

34. M. M. El-Halwagi and H. D. Spriggs, Solve Design Puzzles with Mass Integration. *Chem. Eng. Prog.*, vol. 94, no. 8, p. 25, 1998.

35. M. M. El-Halwagi, A. A. Hamad, and G. W. Garrison, Synthesis of Waste Interception and Allocation Networks. *AIChE J.*, vol. 42, no. 11, pp. 3087–3101, 1996.

36. C. Stanley and M. M. El-Halwagi, Synthesis of Mass-Exchange Networks Using

Linear Programming Techniques. In A. Rositer (ed.), *Waste Minimization through Process Design*, pp. 209–224. New York: McGraw-Hill, 1995.

37. M. Zhu and M. M. El-Halwagi, Synthesis of Flexible Mass-Exchange Networks. *Chem. Eng. Commun.*, vol. 138, pp. 193–211, 1995.

38. G. W. Garrison, W. B. Cooley, and M. M. El-Halwagi, Synthesis of Mass-Exchange Networks with Multiple Target Mass-Separating Agents. *Dev. Chem. Eng. Mineral Proc.*, vol. 3, no. 1, pp. 31–49, 1995.

39. B. K. Srinivas and M. M. El-Halwagi, Synthesis of Reactive Mass-Exchange Networks with General Nonlinear Equilibrium Functions. *AIChE J.*, vol. 40, no. 3, pp. 463–472, 1994.

40. B. K. Srinivas and M. M. El-Halwagi, Synthesis of Combined Heat and Reactive Mass-Exchange Networks. *Chem. Eng. Sci.*, vol. 49, no. 13, pp. 2059–2074, 1994.

41. K. E. Nelson, Use These Ideas to Cut Waste. *Hydrocarbon Processing*, vol. 69, no. 3, pp. 93–98, 1990.

42. J. M. Douglas, Process Synthesis for Waste Minimization. *Ind. Eng. Chem. Res.*, vol. 31, pp. 238–243, 1992.

43. A. P. Rossiter, H. D. Spriggs, and H. Klee, Apply Process Integration to Waste Minimization. *Chem. Eng. Prog.*, vol. 89, no. 1, pp. 30–36, 1993.

44. M. M. Daichent and I. E. Grossmann, Integration of Hierarchical Decomposition and Mathematical Programming for the Synthesis of Process Flowsheets. *Comput. Chem. Eng.*, vol. 22, pp. 147–175, 1998.

45. H. Yeomans and I. E. Grossmann, A Systematic Modeling Framework of Superstructure Optimization in Process Synthesis. *Comput. Chem. Eng.*, vol. 23, no. 6, pp. 709, 1999.

46. B. T. Safrit and A. W. Westerberg, Synthesis of Azeotropic Batch Distillation Separation Systems. *Ind. Eng. Chem. Res.*, vol. 36, no. 5, pp. 1841–1854, 1997.

47. A. A. Linninger, S. A. Ali, E. Stephanopoulos, C. Han, and G. Stephanopoulos, Synthesis and Assessment of Batch Processes for Pollution Prevention. In M. M. El-Halwagi and D. P. Petrides (eds.), *Pollution Prevention via Process and Product Modifications*, AIChE Symp. Ser., vol. 90, pp. 46–58. New York: American Institute of Chemical Engineers, 1994.

48. E. W. Crabtree and M. M. El-Halwagi, Synthesis of Environmentally Acceptable Reactions. In M. M. El-Halwagi and D. P. Petrides (eds.), *Pollution Prevention via Process and Product Modifications*, AIChE Symp. Ser., vol. 90, pp. 117–127. New York: American Institute of Chemical Engineers, 1994.

49. A. Lakshmanan and L. T. Biegler, Reactor Network Targeting for Waste Minimization. In M. M. El-Halwagi and D. P. Petrides (eds.), *Pollution Prevention via Process and Product Modifications*, AIChE Symp. Ser., vol. 90, pp. 128–138. New York: American Institute of Chemical Engineers, 1994.

50. D. Hildebrandt and L. T. Biegler, Synthesis of Reactor Networks. In *Proc. VI FOCAPD Conf.*, Snowmass, CO, 1994.

51. S. Dowdy and S. Wearden, *Statistics for Research*, 2nd ed. New York: Wiley, 1985.

52. M. G. Morgan and M. Henrion, *Uncertainty: A Guide to Dealing with Uncertainty in Quantitative Risk and Policy Analysis*. New York: Cambridge University Press, 1990.

53. V. Vasquez and W. Whiting, Uncertainty of Predicted Process Performance Due to

Variations in Thermodynamic Model Parameter Estimation from Different Experimental Data Sets. *Fluid Phase Equilibria*, vol. 142, pp. 115–130, 1998.

54. M. E. Reed and W. B. Whiting, Sensitivity and Uncertainty of Process Designs to Thermodynamic Model Parameters: A Monte Carlo Approach. *Chem. Eng. Commun.*, vol. 124, pp. 39–48, 1993.

55. W. B. Whiting, T. Tong, and M. E. Reed, Effect of Uncertainties in Thermodynamic Data and Model Parameters on Calculated Process Performance. *Ind. Eng. Chem. Res.*, vol. 32, pp. 1367–1371, 1993.

56. J. Kalagnanam and U. M. Diwekar, Efficient Sampling Technique for Optimization under Uncertainty. *AIChE J.*, vol. 43, no. 2, 1997.

57. R. L. Iman and M. J. Shortencarier, A FORTRAN 77 Program and User's Guide for the Generation of Latin Hypercube and Random Samples for Use with Computer Models, Report NUREG/CR-3624. Springfield, VA: National Technical Information Service, 1984.

58. R. L. Iman and W. J. Conover, A Distribution-Free Approach to Inducing Rank Correlation among Input Variables. *Commun. Statist.-Simul. Comput.*, vol. 11, no. 3, pp. 311–334, 1982.

59. S. Lee, H. Kim, S. Park, and S. Chang, Development of Statistical Core Thermal Design Methodology Using a Modified Latin Hypercube Sampling Method. *Nuclear Technol.*, vol. 94, p. 407, 1990.

60. H. Wozniakowski, Average Case Complexity of Multivariate Integration. *Bull. Am. Math. Soc.*, vol. 24, no. 1, p. 185, 1991.

61. V. R. Vasquez and W. B. Whiting, Uncertainty and Sensitivity Analysis of Thermodynamic Models Using Equal Probability Sampling (EPS). *Comput. Chem. Eng.*, vol. 23, pp. 1825–1838, 2000.

62. V. Vasquez, M. Meerschaert, and W. Whiting, A Sampling Technique for Correlated Parameters in Nonlinear Models Based on Equal Probability Sampling (EPS). Submitted, 1998.

63. M. Meerschaert, *Mathematical Modeling*, 2nd ed. San Diego, CA: Academic Press, 1999.

64. R. V. Hogg and E. A. Tanis, Probability and Statistical Inference, 4th ed. New York: Macmillan, 1993.

65. R. L. Iman, M. J. Shortencarier, and J. D. Jhonson, A FORTRAN 77 Program and User's Guide for the Calculation of Partial Correlation and Standardized Regression Coefficients, Report NUREG/CR-4122. Springfield, VA: National Technical Information Service, 1985.

66. Palisade Corporation, Guide to Using @RISK: Advance Risk Analysis for Spreadsheets. Newfield: Palisade Corporation, 1997.

67. Golder Associates, RIP Integrated Probabilistic Simulator for Environmental Systems: Theory Manual and User's Guide, 1999 (see www.golder.com/rip/ for more information).

68. J. P. Novak, J. Matous, and J. Pick, *Liquid-Liquid Equilibria*. New York: Elsevier, 1987.

69. P. A. Schweitzer, *Handbook of Separation Techniques for Chemical Engineers*. New York: McGraw-Hill, 1979.

20

Minimization and Use of Coal Combustion By-Products (CCBs): Concepts and Applications

Harold W. Walker, Panuwat Taerakul, Tarunjit Singh Butalia, and William E. Wolfe
The Ohio State University, Columbus, Ohio

Warren A. Dick
The Ohio State University, Wooster, Ohio

1 INTRODUCTION AND BACKGROUND

During coal-fired electric power production, four main types of coal combustion by-products (CCBs) are produced: fly ash, bottom ash, boiler slag, and flue gas desulfurization (FGD) material (1,2). In 1998, 97.7 million metric tons of CCBs were produced in the United States (see Figure 1). Fly ash was generated in the largest quantity (57.1 million metric tons), with FGD material the second most abundant CCB (22.7 million metric tons). Roughly 15.1 million metric tons of bottom ash were generated and 2.7 million metric tons of boiler slag were produced. Although the majority of CCBs produced currently enters landfills and surface impoundments, there is great potential for the effective and environmentally sound utilization of these materials.

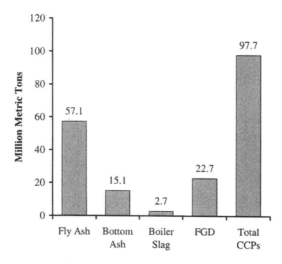

FIGURE 1 CCB production in million metric tons in the United States in 1998 (1).

Currently, the amount of CCBs entering landfills and surface impound-ments is greater than half of the total municipal solid waste (MSW) disposed of in the United States (see Table 1). Of the 97.7 million metric tons of CCBs generated in 1998, 69.4 million metric tons of CCBs (or 70%) were disposed of in landfills or surface impoundments (1). In 1997, the most current year for which data are available, the total MSW disposed of in landfills was 119.6 million tons (3). The amount of CCBs disposed each year is greater than the amount of paper (37.4 million metric tons), plastic (15.5 million metric tons), wood (8.4 million metric tons), and glass (6.9 million metric tons) discarded.

TABLE 1 Amount of CCBs Disposed of in Landfills in the United States in 1998 Compared to Disposal of Municipal Solid Waste (MSW)[a]

Material	Metric tons $\times 10^6$	Reference
Total MSW	119.6	(3)
CCBs	69.4	(1)
Paper	37.4	(3)
Plastic	15.5	(3)
Wood	8.4	(3)
Glass	6.9	(3)

[a]Data for disposal of MSW are for 1997, the most current year for which data are available.

TABLE 2 Amount of CCBs Produced in the United States in 1998 Compared to Traditional Nonfuel Mineral Commodities

Commodity	Metric tons $\times 10^6$	Reference
Crushed stone	1,500[a]	(4)
Sand and gravel	1,020[a]	(4)
CCBs	97.7	(1)
Cement	85.5[a]	(4)
Iron ore	62[a]	(4)

[a]Estimated.

Recently, the American Coal Ash Association (ACAA) proposed that CCBs be considered a product, and therefore they recommend that these materials be referred to as coal combustion products (CCPs). Considered as a commodity, CCBs are ranked as the third largest nonfuel mineral commodity produced in the United States (1,4). As shown in Table 2, the amount of CCBs generated every year exceeds the amount of Portland cement generated in the United States, is significantly greater than the production of iron ore, and falls behind the production of crushed stone, sand, and gravel.

The purpose of this chapter is to review the current state of the art in technology for minimizing CCB generation, maximizing CCB use, and reducing the disposal of CCBs in landfills and surface impoundments. This chapter will first present a review of important federal regulations influencing the generation and utilization of CCBs in the United States. Next, the physical, chemical, and engineering properties of CCBs will be discussed, and the operational factors affecting CCB generation will be presented. The chapter will conclude with a discussion regarding strategies for minimizing CCB production and maximizing the utilization of CCBs. Potential barriers to utilization and minimization in the future will also be discussed.

2 FEDERAL REGULATIONS INFLUENCING CCB GENERATION AND USE

Governmental regulations of emissions from electric power plants combined with efforts to improve air quality have had a profound effect on the amount and type of CCBs produced in the United States over the past 25 years. The Clean Air Act of 1967 was the first legislation to establish the authority of the federal government to promulgate air quality criteria (5). It set the groundwork for future "technology-forcing legislation," i.e., legislation that sets standards unattainable utilizing existing technology. This regulatory approach required industry and utilities to develop new technologies to meet promulgated standards.

The Clean Air Act Amendments of 1970 established Natural Ambient Air Quality Standards (NAAQS) and set specific pollutant removal requirements (New Source Performance Standards or NSPS) for both stationary and mobile sources (5). NSPS, which are applicable to coal-fired utilities, were written in part 60, subpart D, Da, Db, and Dc of 40 *CFR* (*Code of Federal Regulation*) (6). NSPS in 40 *CFR*, part 60, subpart D, set air pollutant levels for coal-fired steam generators with heat input rates over 73 megawatts (MW), constructed or substantially modified after August 17, 1971. Amendments to the Clean Air Act in 1990 added new provisions to reduce the formation of acid rain by decreasing sulfur and nitrogen oxide emissions. Key to these provisions was the requirement to reduce annual SO_2 emissions by 10 million tons below 1980 levels, and to reduce NO_x emissions by 2 million tons below 1980 levels. To achieve these emission reductions, the Clean Air Act Amendments of 1990 promulgated NO_x and SO_2 performance standards and set up an innovative emission trading system for SO_2 reduction. In phase I of the SO_2-reduction program, the legislation required 110 identified utilities to reduce SO_2 emissions to 2.5 lb/mmBTU by January 1995. Phase II mandated further reductions in emissions to 1.2 lb/mmBTU for all utilities generating at least 25 MW of electricity. It is estimated that phase II requirements will affect 2128 utilities in the United States (7). The NO_x reduction program was also separated into two phases. In phase I, Group 1 boilers (dry-bottom wall and tangentially fired boilers) were required to meet NO_x performance standards by January 1996 (8). Phase II set lower NO_x emission limits for Group 1 boilers and established initial NO_x emission limitations for Group 2 boilers (cell burner technology, cyclone boilers, wet bottom boilers, and other types of coal-fired boilers) (7).

To meet these federal regulations, coal-fired utilities have switched to alternative fossil fuels or installed air pollution control technologies such as electrostatic precipitators, baghouses, and wet or dry SO_2 scrubbing systems. Currently, CCBs generated as a result of air pollution control processes are regulated under subtitle D of the Resource Conservation and Recovery Act (RCRA), which pertains to nonhazardous solid wastes (9). In 1988, and then again in 1999, the U.S. Environmental Protection Agency (EPA) issued a Report to Congress examining the environmental impacts associated with CCB use and disposal (10,11). Reports in both 1988 and 1999 concluded that CCBs were nonhazardous and nontoxic materials. In early 2000, based on its own findings in the Report to Congress as well as input from environmental groups, the EPA maintained its previous ruling that CCBs will continue to be regulated under subtitle D of the RCRA. As a result, the use and/or disposal of CCBs is regulated at the state level. For example, regulations in the state of Ohio consider fly ash, bottom ash, boiler slag, and FGD generated from coal or other fuel combustion sources to be exempt from regulation as hazardous waste (12).

3 PHYSICAL, CHEMICAL, AND ENGINEERING PROPERTIES OF CCBS

Information regarding the physical, chemical, and engineering properties of CCBs is required before these materials can be safely and effectively utilized. The physical and engineering properties, in particular, are important parameters affecting the behavior of CCBs in various engineering applications. Information regarding the chemical composition is important for addressing potential environmental impacts associated with CCB utilization and disposal. Chemical data are also useful for explaining physical properties when pozzolanic or cementitious reactions take place.

As mentioned above, the four main types of CCBs are fly ash, bottom ash, boiler slag, and FGD material. Fly ash is a powdery material removed from electrostatic precipitation (ESP) or baghouse operations, while bottom ash is a granular material removed from the bottom of dry-bottom boilers. Boiler slag is a granular material that settles to the bottom of wet-bottom and cyclone boilers. It forms when the operating temperature in the boiler exceeds the ash fusion temperature. Boiler slag exists in a molten state until it is drained from the boiler. The majority of FGD material is a mixture of fly ash and dewatered scrubber sludge. Scrubber sludge is produced when flue gases are exposed to an aqueous solution of lime or limestone. The wet scrubber sludge is dewatered and stabilized with fly ash and extra lime. Alternatively, the scrubber sludge can be oxidized to calcium sulfate ($CaSO_4$) to produce synthetic FGD gypsum. Dry FGD processes are widely used, in which limestone is injected directly into the boiler or flue gas stream. Dry FGD by-products are removed from the flue gas by electrostatic precipitation or baghouse operations.

3.1 Physical and Engineering Properties of CCBs

A number of the physical and engineering properties of fly ash, bottom ash, boiler slag, and FGD material are summarized in Table 3 (10,11,13–16,18,19). Fly ash is usually spherical, with a diameter ranging from 1 to 100 μm. Fly ash has the appearance of a gray cohesive silt and has low permeability when compacted. Bottom ash and boiler slag are granular in shape, with sizes ranging from 0.1 to 10.0 mm. Boiler slag has a glassy appearance. Bottom ash has a permeability higher than fly ash, while boiler slag has a permeability similar to that of course sand. Fly ash, bottom ash, and boiler slag have dry densities that range between 40 and 100 lb/ft^3 (10,11,15,16). Fly ash has lower shear strength than both bottom ash and boiler slag.

TABLE 3 Summary of Physical Characteristics and Engineering Properties of Fly Ash, Bottom Ash, Boiler Slag, and FGD Material (10,11,13–16,18,19)

Physical characteristics	Fly ash	Bottom ash/ boiler slag	FGD material	
			Wet	Dry
Particle size (mm)	0.001–0.1	0.1–10.0	0.001–0.05	0.002–0.074
Compressibility (%)	1.8	1.4		
Dry density (lb/ft^3)	40–90	40–100	56–106	64–87
Permeability (cm/s)	10^{-6}–10^{-4}	10^{-3}–10^{-1}	10^{-6}–10^{-4}	10^{-7}–10^{-6}
Shear strength				
Cohesion (psi)	0–170	0		
Angle of internal friction (deg)	24–45	24–45		
Unconfined compressive strength (psi)			0–1600	41–2250

The physical characteristics of FGD material depend on the type of FGD system used: wet or dry (see Table 3). Wet FGD systems generate by-products with diameters ranging from 0.001 to 0.05 mm. Dry FGD systems produce by-products with diameters ranging from 0.002 to 0.074 mm. FGD material generally has low permeability, ranging from 10^{-4} to 10^{-7} cm/s. The unconfined compressive strength is affected by the moisture content of FGD, and the percentages of fly ash and lime. For example, wet FGD scrubber sludge is similar to toothpaste in consistency and has little unconfined compressive strength. However, the strength of wet FGD is greatly improved when FGD sludge is stabilizing by mixing with lime and fly ash.

3.2 Chemical Properties of CCBs

The chemical characteristics of fly ash, bottom ash, and boiler slag depend greatly on the type of coal used and the operating conditions of the boiler (10,11). Over 95% of fly ash consists of oxides of silicon, aluminum, iron, and calcium, with the remaining 5% consisting of various trace elements (10,11). The chemical composition of fly ash is affected by the operating temperature of the boiler, because the operating temperature influences the volatility of certain elements. For example, sulfur may be completely volatilized at high temperature and removed during lime scrubbing, thus reducing the amount in the fly ash, bottom ash, and boiler slag (10,11).

Table 4 shows the trace-element content of fly ash, bottom ash, boiler slag, and FGD material (10,11,20). The elemental composition of fly ash from two

TABLE 4 Trace Element Composition of Fly Ash, Bottom Ash, Boiler Slag (10), and FGD Material (20)

| Element (mg/kg) | Fly ash | | | | Bottom ash/boiler slag | | Dry FGD material | |
| | Mechanical | | ESP/baghouse | | | | | |
	Range	Median	Range	Median	Range	Median	Range	Median
Arsenic	3.3–160	25.2	2.3–279	56.7	0.50–168	4.45	44.1–186	86.5
Boron	205–714	258	10–1300	371	41.9–513	161	145–418	318
Barium	52–1152	872	110–5400	991	300–5789	1600	100–300	235
Cadmium	0.40–14.3	4.27	0.10–18.0	1.60	0.1–4.7	0.86	1.7–4.9	2.9
Cobalt	6.22–76.9	48.3	4.90–79.0	35.9	7.1–60.4	24	8.9–45.6	26.7
Chromium	83.3–305	172	3.6–437	136	3.4–350	120	16.9–76.6	43.2
Copper	42.0–326	130	33.0–349	116	3.7–250	68.1	30.8–251	80.8
Fluorine	2.50–83.3	41.8	0.4–320	29.0	2.5–104	50.0	—	—
Mercury	0.008–3.0	0.073	0.005–2.5	0.10	0.005–4.2	0.023	—	—
Manganese	123–430	191	24.5–750	250	56.7–769	297	127–207	167
Lead	5.2–101	13.0	3.10–252	66.5	0.4–90.6	7.1	11.3–59.2	36.9
Selenium	0.13–11.8	5.52	0.6–19.0	9.97	0.08–14	0.601	3.6–15.2	10.0
Silver	0.08–4.0	0.70	0.04–8.0	0.501	0.1–0.51	0.20	—	—
Strontium	396–2430	931	30–3855	775	170–1800	800	308–565	432
Vanadium	100–377	251	11.9–570	248	12.0–377	141	—	—
Zinc	56.7–215	155	14–2300	210	4.0–798	99.6	108–208	141

types of collection methods is shown. Mechanical collection methods generally collect larger particles from the flue gas, while finer ash particles are collected by electrostatic precipitators (ESPs) or baghouses. However, similar ranges of most trace elements are found in both types of collection methods. Some exceptions to this are arsenic, boron, lead, and selenium, which may be found at slightly higher fractions in fly ash collected by ESPs or baghouses. Cadmium and fluorine may be present at higher levels in ash collected by mechanical methods.

The chemical characteristics of FGD by-products depend on the type of absorbent used and the sulfur content of the coal. In the United States, approximately 90% of FGD systems use lime or limestone as a sorbent (17). In lime-based FGD processes, the absorbent reacts with sulfur in the flue gas and forms a calcium compound, either calcium sulfite or calcium sulfate, or a calcium sulfite–sulfate mixture (10,11). In systems that use dual-alkali scrubber technology, sodium hydroxide, sodium sulfite, or lime is used as absorbent solution. These types of systems generate calcium sulfite and sodium salts (10,11). In spray-drying scrubber systems, sodium sulfate and sodium sulfite are produced with sodium-based reagents. When fly ash is added to FGD, the quantity and characteristics of the fly ash will also affect FGD chemical characteristics.

The most significant components in FGD include calcium and sulfur, with lesser amounts of silica, aluminum, iron, and magnesium if fly ash is added. The elemental composition of dry FGD materials has been determined based on data from a variety of dry-scrubber technologies, including spray dryer systems, duct injection, lime injection multistage burner (LIMB) processes, and a number of fluidized bed combustion (FBC) processes (i.e., bed-ash process and cyclone ash process) (18,19). The calcium content of dry FGD material varies in the range from 10% to 30% depending on the particular scrubber technology. The sulfur content of dry FGD material typically varies between 4% and 11%. The silicon content of dry FGD may range from 2% to 11%, while the aluminum content can vary from 1% to 7%. Table 4 shows the trace-element content of dry FGD materials (20). Although detectable amounts of arsenic, cadmium, chromium, copper, lead, molybdenum, nickel, selenium, and zinc are present in dry FGD materials, levels of these constituents are typically lower than EPA land application guidelines for sewage sludge.

For many CCB applications, it is important to understand the leaching behavior of these materials. The EPA's Toxicity Characteristic Leaching Procedure (TCLP) is a commonly used method for characterizing the leaching potential of organics, metals, and other inorganic constituents from CCB matrices (21). Table 5 shows the results of TCLP analyses of dry FGD materials and ash produced from various air pollution control technologies. Typically, very low levels of organic materials are found in CCBs, and therefore, TCLP tests focus on examining the leaching behavior of inorganic constituents. The TCLP values for FGD shown in Table 5 were determined for a variety of dry scrubber

TABLE 5 Range of Values Observed for TCLP Analysis of
Dry FGD Materials (19,20) and Ash (14)

Chemical constituent (mg/liter)	FGD	Ash
pH	9.58–12.01	—
TDS	11,840–13,790	—
Ag	<0.024	0.0–0.05
Al	0.12–0.20	—
As	<0.005	0.026–0.4
B	0.543–2.17	0.5–92
Ba	<0.002	0.30–2.0
Be	0.141–0.348	<0.0001–0.015
Ca	1,380–3,860	—
Cd	<0.003	0.0–0.3
Co	<0.014–0.026	0.0–0.22
Cr	<0.005–0.028	0.023–1.4
Cu	<0.013	0.0–0.43
Fe	<0.029	0.0–10.0
Hg	<0.0002	0.0–0.003
K	1.3–22.1	—
Li	0.04–0.18	—
Mg	<0.04–1,360	—
Mn	<0.001	0.0–1.9
Mo	0.025–0.088	0.19–0.23
Na	1.32–9.82	—
Ni	<0.01	0.0–0.12
P	<0.12	—
Pb	<0.001–0.017	0.0–0.15
S	132–979	—
Sb	<0.24	0.03–0.28
Se	<0.001–0.005	0.011–0.869
Si	0.10–0.33	—
Sr	0.83–3.38	—
V	<0.019–0.024	—
Zn	<0.006	0.045–3.21
Cl^-	19.6–67.8	—
SO_3^{2-}	<1.0–43.2	—
SO_4^{2-}	236–2,800	—

technologies. TCLP leachate typically meets most primary and secondary drinking water standards. Levels of silver, arsenic, barium, cadmium, copper, iron, mercury, manganese, nickel, phosphorus, antimony, and zinc in leachate are typically below the limit of detection. For all FGD materials shown in Table 5, high pH values are observed, thus making FGD an attractive product for applications requiring alkaline materials. Typically, with the exception of sulfur and calcium, higher levels of most inorganic elements are found for TCLP tests carried out with ash than for FGD. It should be noted that the acidic conditions and high liquid-to-solids ratio of the TCLP test are perhaps more favorable for leaching than conditions typically observed in field applications.

4 FACTORS AFFECTING CCB GENERATION

The physical and chemical properties of CCBs and the quantity of CCBs produced will depend on the mechanical design and operation of the combustion process, the type of air pollution control equipment utilized, as well as the characteristics of the coal used in the combustion process (11). In order to minimize CCB generation, it is important to understand how these factors affect the type and amount of solid by-product produced. In all cases, however, efficient energy production and low-pollutant air emissions must be maintained.

4.1 Boiler Technology

The boiler used in an electric power plant is a closed vessel that is heated from the combustion of coal to produce hot water or stream. There are four major types of boiler technologies in current commercial application: pulverized coal (PC) boilers, stokers, cyclones, and fluidized bed combustion systems. Figure 2 shows the approximate distribution of ash and slag produced by different kinds of boiler technology.

The most widely used boiler technology is the PC boiler. The coal used in PC boilers is finely ground prior to combustion. The large effective surface area of finely ground coal used in PC boilers increases combustion efficiency. The greater efficiency of combustion reduces the total volume of ash by-products produced. There are two types of pulverized coal boilers; wet-bottom and dry-bottom boilers. The larger-sized ash that falls to the bottom in a dry-bottom process remains dry and becomes bottom ash. For the wet-bottom process, ash is removed as a flowing slag. Large ash particles fall to the bottom of the furnace and flow out of the furnace in a molten state which later solidifies as slag (10,11). As seen in Figure 2, dry-bottom PC boilers produce 80% fly ash and 20% bottom ash. PC boilers with a wet-bottom design produce 50% fly ash and 50% slag. The predominance of fly ash in these two types of boilers is primarily a result of the small particle size of ground coal used in the combustion process.

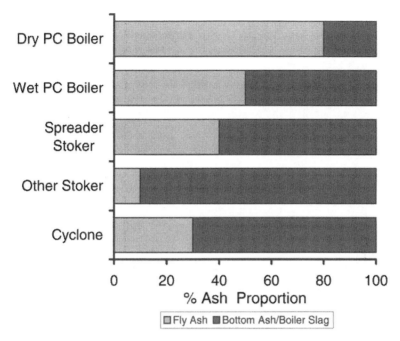

FIGURE 2 Approximate ash distribution as a function of boiler technology (10,11,22).

Stoker boiler technology is typically used in smaller utility plants. A stoker boiler is classified based on the location of the stoker, the method of coal feeding, and the method of grating coal in the furnace. Spreader stokers are the most widely used of all stoker technologies (10,11). The bottom ash generated by spreader stokers ranges from free-flowing ash to fused slag, while the bottom ash created from other types of stokers is normally slag (10,11,22). Figure 2 shows that spreader stokers produce 35–60% fly ash and 40–65% bottom ash and slag. Other types of stokers produce about 10% fly ash and 90% bottom ash and slag.

Cyclone boilers are used for coal combustion and are designed to circulate air to enhance the combustion of fine coal particles in suspension. This design helps to reduce erosion and fouling problems in the boiler. Larger ash particles stick to the molten layer of slag and flow out. Combustion occurs in a horizontal cylindrical vessel attached to the boiler. This kind of design facilitates the flow of molten slag and also reduces the cost of particulate collection (10,11). Most of the by-product from the cyclone design is in the form of slag. Figure 2 shows that cyclones produce 30% fly ash and 70% slag.

Other technologies are also used for coal combustion. These alternative boilers may also aid in controlling air emission. Fluidized bed combustion (FBC) is a boiler technology that can be used with a variety of fuels (22). This technology has a high combustion efficiency at low operating temperatures (11). The fluidized bed combustion system consists of a blower that injects preheated air into the fluidization vessel, and a bed material that can be sand or a reactive solid. Injection of air into the vessel fluidizes the bed material and aids in combustion. The amount of CCBs produced from FBC is based on the type of FBC. Two types of FBC include bubbling fluidized bed systems and circulating fluidized bed systems.

Bubbling fluidized bed systems have gas velocities of 5–12 ft/s. Gas flow passes through the bed and causes the bed material to "bubble." In bubbling FBC systems, the particle size of bottom ash in the bed is usually larger and is packed denser ($\sim$45 lb/ft^3) than in circulating FBC systems (11,22). CCBs generated from bubbling FBC systems include ash, sand, and other inert bed material. Lime or limestone may be added directly to the bed to aid in sulfur emission control (11,22). As a result, by-products from FBC boilers may also contain unreacted lime, calcium sulfate, and/or calcium sulfite.

Circulating fluidized bed combustion systems have higher gas velocities of about 30 ft/s. In circulating systems, some of the bed material is recovered from the gas phase and reinjected into the fluidized bed vessel. Bottom ash in the bed of circulating FBC systems is usually finer and more densely packed ($\sim$35 lb/ft^3) than in bubbling FBC systems (11,22). Ash generated from circulating FBC systems consists mainly of fly ash, with lesser amounts of bottom ash (22).

4.2 Air Pollution Control Technology

The type of technology used for controlling pollutants released to the atmosphere during coal combustion influences the generation and characteristics of CCBs. There are two main categories of air pollution control technologies that generate CCBs during coal combustion: particulate control and gaseous emission control technologies.

Particulate control technologies during coal combustion capture fly ash from the flue gases before they are released to the atmosphere. The processes most often used for particulate control are electrostatic precipitation, fabric filtration, scrubbers, and mechanical collectors. The electrostatic precipitator is the most common process used for capturing fine ash particles in coal-fired utilities (11,22). ESPs capture ash by applying an electrical charge to the ash particles. The charged particles are subsequently attracted to oppositely charged collector surfaces in an intense electrical field. Following collection, the particles are sent to a hopper. This technology is appropriate for capturing fly ash from coal with high sulfur content. In fact, sulfur oxides in the flue gas may increase the efficiency of particle capture in the ESP (10,11,22). The capability of ESP to capture fly ash in the flue gas is more than 99% when this process is properly operated and maintained (10,11).

A fabric filter unit, or baghouse, is an appropriate technology for particulate control in combustion processes that use coal with low sulfur content. This technology operates by forcing the flue gas through a fine mesh filter. Fly ash is trapped and builds up on the filter surface. The ash on the filter forms a cake, which is then periodically removed. The efficiency of the filter increases as ash forms a thick layer on the filter surface. However, thick cake formation also leads to greater head losses in the process. Fabric filters can remove over 99% of fly ash from the flue gas in coal-fired utilities (10,11).

Scrubbers can also be used for particulate control and operate by applying water to contact the fly ash in the flue gas in a spray tower. This technology also can remove over 99% of large ash particles, but less than 50% for particles with size smaller than 1 or 2 μm (11,22). Mechanical collectors are instruments used for removing primarily large ash particles. They operate by forcing the ash particles against a collector wall, where the dry ash by-product is collected. The efficiency is lower than 90% for small particles.

Desulfurization technology is used for capturing gaseous sulfur oxides from flue gas in coal-fired utilities. The use of desulfurization technology results in the generation of FGD material. There are two major types of FGD systems: nonrecovery and recovery systems. In the United States, 95% of FGD systems are nonrecovery systems (11,22). Nonrecovery systems produce by-product material, mainly calcium sulfate or sulfite, that has to be disposed or used. The nonrecovery FGD process can be separated into two types, wet and dry systems. Wet systems operate by contacting the flue gas with a slurry of water and sorbent. Examples of wet scrubber systems include direct lime, direct limestone, alkaline fly ash, and dual-alkali. As mentioned earlier, approximately 90% of FGD systems in the United States use lime or limestone as a sorbent (17). Typically, the calcium sulfite/sulfate sludge produced in wet systems is dewatered and mixed with fly ash and lime to produce "stabilized" FGD. Examples of dry nonrecovery FGD systems include spray drying and dry sorbent injection (11). Wet FGD systems produce more FGD per pound of coal than that of dry systems because of the use of water in the process. Recovery systems produce materials that can be used again in the FGD process, because the sorbent can be recycled. Recovery FGD processes also have wet and dry systems. Examples of recovery FGD systems include Wellman-Lord and magnesium oxide systems and aluminum sorbent and activated-carbon sorbent systems (11).

4.3 Types of Coal

Different types of coal have different heating values and also different ash contents. The highest-ranked coal with respect to heating value is anthracite, while the lowest-ranked coal is lignite (22). The generation of ash and slag from the combustion process is affected by the ash content, which is determined by the rank and geographic source of the coal. Certain anthracite coals in the United States generate about 30% ash. Bituminous coal ash content ranges from 6% to 12%, and subbituminous and lignite coals have ash contents ranging from 6% to 19% (10,11). The rank of coal is affected by the specific region of coal mining, mine, seam, and production method (11,22). The use of low-ash-content coal reduces the management cost for removing particulate matter. In the United States, the average ash content of coal used decreased from 13.5% in 1975 to 9.22% in 1996 (11,22,23). In addition, coal can be cleaned before the combustion process to reduce the quantity of CCBs. The cleaning process as a pretreatment for coal can reduce the ash content by 50–70% (22).

The generation of FGD depends on the sulfur content of the coal. The sulfur content varies from region to region. Some coal produced in Iowa has a sulfur content as high as 8% by weight, while coal from Wyoming may have an average sulfur content of less than 1% by weight (22). Coal cleaning processes that reduce the ash content of coal can reduce sulfur emissions by removing pyrites and other metal sulfides from the coal.

Technologies for precombustion coal desulfurization are characterized as physical, chemical, or biological. Currently, physical cleaning processes are the most widely used. Physical processes use density differences to separate out pyrites from the coal (24). Chemical processes use a chemical agent to desulfurize coal. Chemical desulfurization processes that use chemical reagents such as ferric salts, chlorine, and ozone are currently not cost effective, due to high chemical recovery costs. In addition, chemical processes are also energy intensive due to high operating pressures (600–1000 psi) and temperatures (100–500°C) (25–28). Recently, new chemical agents have been developed that may provide a cheaper approach to the chemical cleaning of coal (29).

In biological processes, microorganisms are used to remove sulfur from coal. Biological processes can be operated at room temperature and atmospheric pressure, and therefore have lower costs than some physical and chemical coal cleaning techniques. In addition, biological pretreatment of coal does not reduce the BTU value of the coal, but instead, may increase coal energy content due to the remaining biomass (25–27,30). Biological coal cleaning processes can remove both inorganic and organic sulfur. *Thiobacillus ferrooxidans* can oxidize inorganic pyrite (FeS_2) in coal, while microorganisms such as *Rhodococcus rhodochrous, Sulfologus brierleyi,* and *Sulfolobus acidocaldarius* remove organic sulfur compounds (31–35). Although precombustion cleaning can reduce flue gas desulfurization requirements, additional by-products may be produced and energy may be required.

5 STRATEGIES FOR MINIMIZATION OF CCBS

The established hierarchy for minimization of waste materials in any process consists of the following: reduction > recycle/reuse > treatment > disposal. In this hierarchy, strategies for reduction and recycle/reuse are favored over "end-of-pipe" treatment and disposal options for waste material.

5.1 Reduction at Source

Strategies for the reduction of CCBs at the source include process modifications, feedstock improvements, improvements in efficiency of equipment, better management practices, and recycling of material within or between processes. Possible process modifications may include changes to boiler operating conditions, selection of wet versus dry scrubbing technologies, or addition of precombustion coal cleaning, to name a few. Changes in feedstock may also aid in CCB minimization. Recent advances are providing more efficient technologies for removal of pollutants from flue gas in coal combustion facilities. For example, new sorbents with high efficiency for SO_2 capture do a better job of removing SO_2 from flue gas while at the same time reducing the amount of solid by-products produced during the process. Better management of the coal combustion process may also lead to improved CCB minimization. Internal audits provide opportunities to optimize process operations, thus maximizing energy production and minimizing CCB production. For example, one factor limiting use of CCBs is the variability in CCB properties. Better management practices may reduce this variability, lower total volume of CCBs produced, and enhance utilization.

5.2 Use of Coal Combustion By-products

A number of applications of CCBs have been developed and demonstrated in order to reduce the amount of CCBs disposed of in landfills. The first column in Table 6 shows demonstrated applications of CCBs. In 1998, the American Coal Ash Association (ACAA) reported that 28 million metric tons of CCBs were used in the United States. This represented 30% of all CCBs generated in that year. One demonstrated application is the use of CCBs in cement, concrete, and grout. CCBs may also be used in flowable and structural fill applications. Demonstrated applications of FGD material include use in the production of wallboard or in mine land reclamation.

The use of CCBs reduces the need for landfill space and also reduces the utilization of natural resources. Table 6 presents some of the environmental benefits associated with CCB use. Approximately 10 million metric tons of fly ash and bottom ash were used as a replacement for cement in nonfill applications. This was the largest application of fly ash and represented roughly half of all the

TABLE 6 Major CCB Applications and Environmental Benefits of CCB Use

CCB utilization	Reduction in emission or natural resource utilization per year	
	Current utilization rate	100% CCB utilization[a]
Cement/concrete/grout	10×10^6 tons cement	32×10^6 tons cement
	10×10^6 tons CO_2[b]	32×10^6 tons CO_2
Flowable fill	0.36×10^6 tons CO_2	1.08×10^6 tons CO_2
Structural fill	3.7×10^6 tons Soil[c]	11.2×10^6 tons soil
Wallboard gypsum	1.6×10^6 tons gypsum	16.5×10^6 tons gypsum
Mining applications	0.6×10^6 tons clay	2.2×10^6 tons clay
Total CCB reduction in landfills	17×10^6 m^3	59×10^6 m^3

[a]The environmental benefit for 100% CCB utilization is calculated assuming the percent of material used for a particular application (e.g., cement/concrete/grout) is independent of overall CCB utilization.
[b]Assumes 1 lb CO_2/1 lb cement not used, and 1 lb of cement replaced/1 lb fly ash used.
[c]Assumes 1 lb soil/1 lb of fly ash used.

fly ash used in 1998. It has been estimated that every ton of cement replaced with fly ash eliminates the emission of approximately 1 ton of CO_2 to the atmosphere (36). Based on this figure, the use of fly ash in cement applications reduced the release of CO_2 by 10 million tons in 1998. Also, in 1998, 0.36 million metric tons of fly ash were used for flowable fill. This application also replaced cement and represented approximately 1.8% of all fly ash used. The use of fly ash in flowable fills reduced the emission of CO_2 by 0.36 million metric tons in 1998. If all of the fly ash generated were used at current utilization percentages (50% for cement replacement and 1.8% of flowable fill), CO_2 emissions would be reduced by roughly 32 million metric tons (31 million tons as a result of cement replacement and 1 million ton due to the use of fly ash in flowable fill). Structural fill is an application in which CCBs, and in particular fly ash, are used to replace natural soil. In 1998, the use of fly ash and bottom ash for structural fill saved 3.6 million metric tons of soil and could save up to 11.2 million metric tons of soil if 100% of all CCBs were used at the current utilization rate for structural fill (13.2%).

In 1998, 2.2 million metric tons of FGD were used, which represented only 8% of all FGD produced. Of this amount, 1.6 million metric tons of FGD were used in the production of wallboard. This represented the single largest use of FGD and 73% of all FGD used in that year. As a result, approximately 1.6 million metric tons of natural gypsum were saved by using FGD gypsum in the wallboard industry. FGD could replace up to 18 million tons of gypsum if all FGD in the

United States were reused, 73% of which for wallboard manufacturing. This amount of FGD wallboard could supply a good fraction of the 1.6 million new houses built in the United States every year. About 0.6 million tons of clay used for mining applications were saved in 1998 by replacement with CCBs, and up to 2.2 million tons could be saved if all CCBs were used (6.8% for mining applications). The use of CCBs in 1998 reduced landfill space consumption by about 17 million m^3. If all CCBs were used, landfill space consumption would be reduced by 59 million m^3 each year.

5.2.1 Cement/Concrete/Grout Application

Using fly ash in cement and concrete increases the strength, workability, and resistance to alkali–silica reactivity and sulfate, and reduces permeability, bleeding, and heat of hydration (13). The Federal Highway Administration (FHWA) and ACAA have reported that fly ash-enhanced concrete has lower strength than pure Portland cement in early periods, but provides higher strength in the long term (13). In 1998, 10.2 million metric tons of CCBs were used in concrete applications. The amount of fly ash used with cement varies from 15% to 20% of the total weight. Typically, 1–1.5 kg of fly ash is used for every 1 kg of cement replaced in concrete applications (13). Lowering the cement content reduces the emission of CO_2 caused by the calcination of limestone and fuel burning during cement production.

Future efforts to reduce nitrogen oxide emissions from coal combustion facilities may negatively impact the utilization of fly ash in concrete. To control emission of NO_x, many coal-fired utilities may utilize new low-NO_x emission technologies. Some of these technologies operate at lower temperatures than traditional boilers. Operation at lower temperature may result in an increase in carbon and ammonia content of fly ash (13). Carbon content, which is typically measured by determining the weight loss on ignition (LOI), affects concrete strength development and so is restricted by industry standards. In the United State, the specification of LOI of fly ash is between 3% and 5% for use in ready-mix concrete (37,38). The use of low-NO_x technology may increase the LOI above 5% and thus reduce the utilization of fly ash in concrete applications (39). For example, in 1998, the ACAA reported that 19 out of 20 coal-fired utilities in Ohio may be affected by the low-NO_x emission technologies, and 46% of all coal-fired utilities in the United States may be required to control NO_x (39).

5.2.2 Flowable Fill

Flowable fill is another application in which fly ash can be used in place of cement. Flowable fill is defined as a self-leveling, self-compacting cementitious material that is in a flowable condition at the time of placement and has a compressive strength of 1200 psi or less at 28 days (40). Flowable fill is also known as control density fill (CDF), controlled low-strength material (CLSM), unshrinkable fill, flowable mortar, plastic-soil cement slurry, K-Krete, and/or Flash Fill (40). Flowable fill may contain a mixture of fly ash, bottom ash, water, and Portland cement. This application is especially suitable for filling in void spaces that are difficult to reach.

The use of fly ash and bottom ash in flowable fill applications is increasing. Recently, a technical guidance manual for flowable fill applications has been written: the ACI229 Committee Report on Controlled Low Strength Material (40). However, the current use of CCBs in this application is still relatively small. As mentioned above, only 360,000 metric tons of fly ash and bottom ash were used in flowable fill applications in the United States, which represented only 1.8% of all fly ash and 0.3% of all bottom ash used.

5.2.3 Embankment/Structural Fill

Embankment and structural fill is currently the second largest application of CCBs. CCBs are used to replace conventional soil in structural fill applications. CCBs have many advantages over natural soil for use as structural fill, including lower unit weight, high shear strength-to-unit weight ratio, and good availability in bulk (13). These benefits lead to significant savings in material costs when CCBs are used. For example, in 1993, FGD material from pressurized fluidized bed combustors (PFBCs) was used as embankment to repair part of Ohio State Route (SR) 541 near Coshocton, Ohio (41–43). The cost of this project was $77,000, while the estimated cost of using conventional materials would have been between $105,000 and $120,000. This represented a savings of between 26% and 36%, before counting the environmental benefits associated with saving existing natural resources.

The use of CCBs for structural fill has increased in recent years. American Society for Testing and Materials (ASTM) Standard E1861 defines the appropriate guidelines for the use of CCBs for structural fill (36,44). In 1998, approximately 4 million tons (12.7%) of CCBs were used as structural fill. Fly ash and bottom ash were used the most, 2.5 and 1.1 million metric tons respectively. Although only 18,000 metric tons of FGD were used in 1998, this material has demonstrated excellent strength over a wide range of moisture contents compared to natural soils (41–43). FGD could be an excellent alternative for use in embankment and structural fill applications in the future (41–43).

5.2.4 Stabilized Base/Subbase

Mixing fly ash with lime and aggregate can produce a good-quality road base and subbase. This material is also called lime–fly ash aggregate (LFA) or pozzolanic-stabilized mixture (PSM) base. The fly ash content in the material for road base typically varies from 12% to 14%. Lime content also varies from 3% to 5% (13). The proportion of lime can be replaced with Portland cement or cement kiln dust. The advantages of using LFA or PSM for road base and subbase applications include increased strength and durability of the mixture, lower cost, autogenous (self-generating) healing, and less energy consumption (13). Use of LFA reduces the energy to produce cement. In addition, it does not require heat, as an asphalt base does.

In 1998, 3.6 million tons of CCBs were used in road base and subbase applications. The CCBs used most commonly for these applications were fly ash and bottom ash (1.4 million tons of fly ash and 1.6 million tons of bottom ash). For example, 10 municipal and commercial projects in and around the City of Toledo used approximately 1 million tons of LFA from 1970 to 1985 (45). In 1998, Hunt et al. (46) developed an economic analysis comparing the use of LFA and other pavement materials. It was found that LFA base pavement was 20% cheaper than aggregate base pavement and 15% cheaper than bituminous base pavement (46).

5.2.5 Mining Applications

Use of CCBs in mining applications can aid in the abatement of acid mine drainage (AMD), reduce subsidence, and reduce off-site sedimentation control. Acid mine drainage is an environmental problem caused by water drainage from abandoned mines and coal refuse piles. Water in abandoned mines reacts with pyrite and other metal sulfides in the presence of oxygen and produces acidity (47). Fly ash and/or FGD can be used to minimize the exposure of pyrite to water and oxygen. Also, the alkalinity in FGD can neutralize AMD already generated. In 1998, 2 million metric tons of CCBs in the United States were used for mining applications. This represented 6.8% of all CCBs used. Most of material used in this application was fly ash (1.9 million tons).

At mine sites, refuse waste materials such as soil, rock, slate, and coal are commonly found and can pose serious environmental problems. These refuse waste piles, commonly called "gob piles," contain pyrite and produce acidity. FGD material may be used as a liner to construct ponds to collect runoff from gob piles. The low permeability of FGD may also be used as a cap to minimize the amount of water entering the gob pile. For example, the Rock Run reclamation site at New Straitsville, Ohio, had approximately 14 acres of gob piles prior to reclamation utilizing CCBs (39). The drainage from these gob piles to Rock Run had a pH of 2.27. At this site, 2 ft of stabilized FGD from the Conesville coal-fired power plant in Ohio was used to cover the gob piles. Utilization of FGD as a cap material resulted in improvements in water quality at the site. Estimated cost saving from using FGD for gob pile reclamation, instead of clay, ranged from $8,350 to $12,600 (39).

Another potential application of FGD is in the reclamation of abandoned surface mines (39,48). An example of such an application was carried out at the Fleming site located in Franklin Township of Tuscarawas County, Ohio (49,50). The Fleming site was an abandoned clay and coal mine. In the past, flooding downstream of the Fleming site resulted in offsite soil sedimentation at an estimated rate of 450 tons/acre/year. In 1994, several AMD treatment approaches were developed at this site utilizing limestone, a dry FGD material from an PFBC plant, and a 2.5:1 mixture of FGD and yard waste. All treatments resulted in neutralization of mine drainage. Trace-metal analysis showed water quality improved and, in fact, met all drinking water standards.

5.2.6 Wallboard Manufacture

Instead of being stabilized with fly ash and sent to landfill, FGD material can be used as a material to manufacture wallboard. The calcium sulfite in FGD can be oxidized to calcium sulfate and dewatered to produce synthetic FGD gypsum (39,51). In 1998, 1.6 million tons of FGD were used in wallboard manufacturing (1). This represented 72.9% of all FGD used. There will be a $20 million investment in oxidation and dewatering equipment to produce synthetic FGD gypsum at the Zimmer plant (located in Moscow, Ohio) by Cinergy, American Electric Power, and Dayton Power and Light (52–54). The synthetic FGD gypsum produced at the Zimmer plant will supply a wallboard plant at Silver Grove, Kentucky. Upon completion, the Silver Grove wallboard plant will have the highest wallboard production capacity in the world, 900 million ft^2 per year.

5.2.7 Agricultural Applications

There are many advantages to using CCBs in agricultural applications, and addition of CCBs can improve the properties of soil and increase plant growth. For example, the lime in CCBs can raise the pH of acidic soil to neutral levels, and can increase yields of alfalfa above those observed using agricultural limestone (18). Trace elements in FGD can be utilized by plants and may aid in plant growth (24). In 1998, 920,000 metric tons of CCBs were used in agricultural applications. It has been estimated that 25% of the agricultural lime used in Ohio could be replaced by FGD (39). Assuming that every ton of agricultural lime is equivalent to 1.67 tons of FGD, the potential use of FGD for replacing agricultural lime in Ohio alone would be 365,000 tons per year (55).

    CCBs can also be used in animal production facilities as a base for livestock feeding and hay storage (56). The moisture from mud can deteriorate the quality of hay bales and decrease animal yields. Utilizing CCBs as a base material can reduce muddy conditions. In 1997, 24 livestock feeding and hay storage pads (ranging size from 1,500 to 15,000 ft^2) were built in eastern and southern Ohio. Over 150 FGD pads were constructed in 12 counties in Ohio in 1998 (56). An economic analysis of the construction of FGD pads performed for Gallia County, Ohio, in 1997 showed that FGD pads were 26% cheaper than aggregate pads and 65% cheaper than concrete pads (57).

    FGD can also be used for constructing low-permeability liners for water-holding ponds and manure storage. Normally, the construction of manure facilities in the United States utilizes compacted clay. The successful utilization of FGD requires that the material provide sufficiently low permeability and not degrade groundwater quality. A study has been conducted to evaluate the permeability and the water quality of the leachate from an FGD liner (58–60). It was found that the FGD liner has a permeability as low as 10^{-7} cm/s. Moreover, trace-element concentrations in the leachate were generally lower than the drinking water standards (60). It has been estimated that replacing clay or geomembranes with FGD material for pond liners could save construction costs by as much as \$2–\$3 per square foot (58–60).

5.3 Treatment and Disposal

Options for treatment and disposal are important considerations for evaluating the ultimate fate CCBs. For example, in wet scrubbing technologies a calcium sulfite slurry is produced that is dewatered prior to disposal. The dewatering step reduces the total volume of material going to final disposal, but it also generates a liquid waste stream. The dewatered scrubber sludge is typically mixed with equal amounts of fly ash in order to improve the handling characteristics of the material. The exact proportion of fly ash to dewatered scrubber sludge has a significant impact on the performance of these materials during disposal and/or use.

6 LIFE CYCLE ASSESSMENT (LCA) MODEL FOR MINIMIZATION OF CCBS

A life cycle assessment model (61) can be used to determine optimum strategies for minimizing the environmental impacts associated with coal combustion processes. The benefit of this approach is that it does not trade gains made in minimizing the amount of CCBs with other environmental impacts. In a life cycle assessment model, the coal combustion process can be broken down into the following elements: resource extraction, electricity generation, electricity transmission, and electricity use. In developing an LCA model, each element of the coal combustion process must be evaluated with respect to the following variables: materials choice, energy use, solid residues, liquid residues, and gaseous residues. Once this has been carried out, an LCA matrix can be developed which includes a numerical score (from 1 to 4) for each variable and element. The rank of the entire process can then be determined as

$$R = \sum_i \sum_j M_{ij} \tag{1}$$

where R is the rank of the process and M_{ij} is the numerical score summed over each variable (i) and element (j). It should be noted that in certain cases it might not be possible or appropriate to develop a single score defining a process. In such cases, ISO 14040 provides criteria by which a comparison can be made between two or more potential options using the LCA framework (62).

The LCA model provides a framework for analyzing different process alternatives and determining the most environmentally sound option. For example, it was mentioned above that one strategy to minimize CCB generation is to utilize a low-sulfur, low-ash coal source. However, utilizing low-sulfur, low-ash coal may not be the most environmentally sound alternative in all cases. If a preferable coal source is located far from the coal-fired utility, significant environmental impacts may be associated with the transportation of these materials.

Recently, a comparison between the use of fly ash and natural soil for structural fill was conducted using the LCA framework (63). This study considered a number of factors, such as natural resource use, energy consumption, air emissions, and solid waste. It was concluded that the use of fly ash as structural fill resulted in less use of natural raw materials, significant reductions in solid waste disposal, and was more energy efficient than the use of natural resources. In this particular study, however, the impact of fly ash on water quality was not assessed quantitatively, due to variations in soil and fly-ash leaching behavior.

7   BARRIERS TO CCB UTILIZATION

Currently, the cost of disposing and managing CCBs in landfills is relatively low, and this reduces the incentive to utilize CCBs. For example, landfill costs for coal-fired utilities in Ohio range from $2 to $40 per ton (18,39). A lack of standards for using CCBs is also a barrier for CCB utilization. For example, the use of CCBs in structural fill has typically fluctuated in the past. Increasing use of CCBs for structural fill is expected in the future as a result of ASTM Standard E1862 (44). New, large-volume, cost-effective applications of CCBs are needed.

Because CCBs are secondary products of energy production, few controls on their generation are in place. As a result, the physical, chemical, and engineering properties of CCBs may vary. Significant variations can be observed in CCB properties within a given plant, as well as variations between plants. Future emission standards for coal-fired power plants will also affect the utilization of CCBs. As mentioned above, NO_x control technologies influence the carbon content of fly ash. The increased carbon content of fly ash generated from low-NO_x boilers could reduce the potential for use in cement applications, the biggest current CCB market.

Although extensive testing has shown that CCBs are nontoxic and nonhazardous, public concern regarding the environmental impacts of CCBs may limit utilization. Continued research on the environmental impacts associated with CCB use and public education are needed. As is the case for any product or process, efforts should be focused primarily on reducing CCB generation, while maintaining efficient energy production and air pollution controls.

8 CONCLUSIONS

Fly ash, bottom ash, boiler slag and FGD material are by-products from the combustion of coal and are considered to be solid wastes from a federal regulatory perspective. Currently, 100 million tons of CCBs are produced in the United States every year. Approximately 70 million tons of CCBs are disposed of in landfills and surface impoundments.

The two primary strategies for minimizing CCBs include reduction at source and effective utilization. A life cycle assessment model should be used to determine the most environmentally beneficial approach for minimizing CCB generation and disposal. A number of applications have been developed for using CCBs, including the use of fly ash as a substitute for cement in concrete and grout applications, the use of fly ash in flowable and structural fill, the use of calcium sulfate-rich FGD scrubber sludge as a replacement for natural gypsum in wallboard manufacturing, and a variety of mine reclamation applications. The utilization of CCBs reduces the consumption of natural resources, reduces emissions of greenhouse gases to the atmosphere, and reduces the need for new landfill construction. Potential barriers to CCB use include the low cost of landfilling, the lack of available large-volume or high-value applications, and variations in CCB material properties.

ACKNOWLEDGMENTS

The authors would like to thank the Ohio Coal Development Office (OCDO) for its support of much of the research cited herein. We would also like to thank the reviewers for their helpful comments.

REFERENCES

1. B. R. Stewart, S. S. Tyson, and G. J. Deinhart, American Coal Ash Association Programs to Advance the Management and Use of Coal Combustion Products (CCPs). 1999 Int. Ash Utilization Symp., October 18–20,1999.

2. American Coal Ash Association, Coal Combustion Byproduct (CCB) Production & Use: 1966–1994. Report of the Coal Burning Electric Utilities in the United States, Alexandria, VA, May 1996.

3. U.S. Environmental Protection Agency, Characterization of Municipal Solid Waste in the United States 1998 Update, July 1999.

4. U.S. Geological Survey, Mineral Commodity Summaries, January 1999.

5. C. D. Cooper and F. C. Alley, *Air Pollution Control: A Design Approach*, 2nd ed., Waveland Press, 1994.

6. *Code of Federal Regulations*, Title 40, Part 60, Standards of Performance for New Stationary Sources. Washington, DC: Office of Federal Register, July 1, 1999.
7. U.S. Environmental Protection Agency, Profile of the Fossil Fuel Electric Power Generation Industry, EPA /310-R-97-007, September 1997.
8. *Code of Federal Regulations*, Title 40, Part 76, Acid Rain Nitrogen Oxides Emission Reduction Program. Washington, D.C.: Office of Federal Register, July 1, 1999.
9. *United States Code*, Title 42, Section 6921 (b)(3)(A)(i), Identification and Listing of Hazardous Waste. Washington DC: U.S. Government Printing Office, 1995.
10. U.S. Environmental Protection Agency, Wastes from the Combustion of Coal by Electric Utility Power Plants, Report to Congress, EPA/530-SW-88-002, February 1988.
11. U.S. Environmental Protection Agency, Wastes from the Combustion of Fossil Fuels, Volumes 1 and 2—Methods, Findings, and Recommendations, Report to Congress, EPA 530-S-99-010, March 1999.
12. *Ohio Administrative Code*, Chapter 3745-51-04 (B) (4), Exclusion: Identification and Listing of Hazardous Waste, June 25, 1998.
13. American Coal Ash Association, Fly Ash Facts for Highway Engineers, FHWA-SA-94-081. Washington, DC: Federal Highway Administration, August 1995.
14. U.S. Environmental Protection Agency, The Class V Underground Injection Control Study, Volume 10, Mining, Sand, or Other Backfill Wells, EPA/816-R-99-014, September 30, 1999.
15. L. K. Moulton, Bottom Ash and Boiler Slag. *Proc. Third Int. Ash Utilization Symp.*, Information Circular No.8640. Washington, DC: U.S. Bureau of Mines, 1973.
16. U.S. Department of Transportation. Coal Fly Ash—User Guideline—Embankment or Fill, http://www.tfhre.gov///////hnr20/recycle/waste/cfa54.htm, 2000.
17. R. Kalyoncu, Coal Combustion Products. U.S Geological Survey, Mineral Information, 1997.
18. R. Stehouwer, W. Dick, J. Bigham, R. Forster, F. Hitzhusen, E. McCoy, S. Traina, W. Wolfe, R. Haefner, and G. Rowe, Land Application Uses for Dry FGD By-products. Phase 2. EPRI Rep. TR-109652. Palo Alto, CA: Electrical Power Research Institute, 1998.
19. W. A. Dick, J. M. Bigham, R. Forster, F. Hitzhusen, R. Lal, R. Stehouwer, S. Traina, W. Wolfe, R. Haefner, and G. Rowe, Land Application Uses of Dry FGD By-product. Phase 3. EPRI Rep. TR-112916. Palo Alto, CA: Electrical Power Research Institute, 1999.
20. W. A. Dick, J. M. Bigham, R. Forster, F. Hitzhusen, R. Lal, R. Stehouwer, S. Traina, W. Wolfe, R. Haefner, and G. Rowe, Land Application Uses for Dry Flue Gas Desulfurization By-products, Executive Summary. U.S. Geological Survey, January 1999.
21. U.S. Environmental Protection Agency, Test Methods for Evaluating Solid Waste Physical/Chemical Methods, EPA SW-846, 3rd ed. Washington, DC: U.S. Government Printing Office, 1986.
22. S. C. Stultz and J. B. Kitto (eds.), *Steam: Its Generation and Use*. Barberton, OH: Babcock & Wilcox, 1992.
23. Energy Information Administration, Cost and Quality of Fuels for Electric Utility Plants, 1996 Tables. U.S. Department of Energy, May 1996.

24. R. F. Korcak, Agricultural Uses of Coal Combustion Byproducts. In R. J. Wright, W. D. Kemper, P. D. Milner, J. F. Power, and R. F. Korcak (eds.), Agricultural Uses of Municipal, Animal, and Industrial Byproducts. U.S. Department of Agriculture, January 1998.

25. F. Kargi, Microbial Coal Desulfurization. *Enzyme Microb. Technol.*, vol. 4, pp. 13–19, 1982.

26. F. Kargi, Microbial Desulfurization of Coal. In A. Mizrahi and A. L. van Wezel (eds.), *Advances in Biotechnological Process*, vol.3, pp. 241–272. New York: Alan R. Liss, 1984.

27. T. Omori, L. Monna, Y. Saiki, and T. Kodama, Desulfurization of Dibenzothiophene by *Corynebacterium* sp. Stain SY1. *Appl. Environ. Microbiol.*, vol. 58, pp. 911–915, 1992.

28. T. D. Wheelock, *Coal Desulfurization (Chemical and Physical Methods)*, ACS Symp. Ser. Washington, DC: American Chemical Society, 1977.

29. K. K. Ho, G. V. Smith, R. D. Gaston, R. Song, J. Cheng, F. Shi, and K. L. Gholson, Desulfurization of Coal with Hydroperoxides of Vegetable Oils, Tech. Rep. December 1, 1994 through February 28, 1995. U.S. Department of Energy, March 1996.

30. P. Bos and J. G. Kuenen, Microbial Treatment of Coal. In H. L. Ehrlich and C. L. Brierly (eds.), *Microbial Mineral Recovery*, pp.343–377. New York: McGraw-Hill, 1990.

31. S. A. Denome, E. S. Olson, and K. D. Young, Identification and Cloning Genes Involved in Specific Desulfurization of Dibenzothiopene by *Rhodococus* sp. Strain IGTS8. *Appl. Environ. Microbiol.*, vol. 59, pp. 2837–2843, 1993.

32. J. R. Gallagher, E. S. Olson, and D. C. Stanley, Microbial Desulfurization of Dibenzothiophene, A Sulfur-Specific Pathway. *FEMS*, vol. 107, pp. 31–36, 1993.

33. Y. Izumi, T. Ohshiri, H. Ogino, Y. Hine, and M. Shimao, Selective Desulfurization of Dibenzothiophene by *Rhodococcus erythropolis* D-1. *Appl. Environ. Microbiol.*, vol. 60, pp. 223–226, 1994.

34. B, Lei and S. C. Tu, Gene Overexpression, Purification, and Identification of a Desulfurization Enzyme from *Rhodococcus* sp Strain IGTS8 as a Sulfide/Sulfoxide Monoxygenase. *J. Bacteriol.*, vol. 178, pp. 5699–5705, 1996.

35. C. S. Piddington, B. R. Kovacevich, and J. Rambosek, Sequence and Molecular Characterization of a DNA Region Encoding the Dibenzothiophene Desulfurization Operon of *Rhodococcus* sp. Strain IGTS8. *Appl. Environ. Microbiol.*, vol. 61, pp. 468–475, 1995.

36. B. R. Stewart and R. S. Kalyoncu, Materials Flow in the Production and Use of Coal Combustion Products. 1999 Int. Ash Utilization Symp., October 18–20, 1999.

37. Pittsburgh Mineral & Environmental Technology, Coal Ash Utilization Technologies—Introduction, 30 December 1999, http://www.pmetinc.com/coal_ash/intro.html.

38. American Society for Testing and Materials (ASTM), ASTM C618-92a, Standard Specification for Fly Ash and Raw or Calcined Natural Pozzolan for Use as Mineral Admixture in Portland Cement Concrete. ASTM Designation C618-92a-1994. West Conshohocken, PA: ASTM, 1994.

39. T. S. Butalia and W. E. Wolfe, Market Opportunities for Utilization of Ohio Flue Gas

Desulfurization (FGD) and Other Coal Combustion Products (CCPs). Columbus, OH: The Ohio State University, May, 2000.

40. American Concrete Institute, Controlled Low Strength Materials (CLSM), American Concrete Institute Committee 229, Rep. 229R-94. Detroit, MI: American Concrete Institute, July 1994.

41. S. M. Nodjomian, Clean Coal Technology By-products Used in a Highway Embankment Stabilization Demonstration Project, M.S. thesis, The Ohio State University, Columbus, OH, 1994.

42. S. M. Nodjomian and W. E. Wolfe, Field Demonstration Projects Using Clean Coal Technology By-products. Second Annual Great Lakes Geotechnical/Geoenvironmental Conf., West Lafayette, IN, May 1994.

43. S. H. Kim, S. Nodjomian, and W. E. Wolfe, Field Demonstration Project Using Clean Coal Technology By-products (CCBs). American Coal Ash Association and Electric Power Research Institute, EPRI TR-104657, vol. 1, p. 16(1–15), Orlando, FL, January 15–19, 1995.

44. American Society for Testing and Materials (ASTM), ASTM E1861 Standard Guide for Use of Coal Combustion By-products in Structural Fills, ASTM Designation E1861-97. West Conshohocken, PA: ASTM, 1997.

45. Transportation Research Board, Synthesis of Highway Practice 199: Recycling and Use of Waste Materials and By-products in Highway Construction. Washington, DC: National Academy Press, 1994.

46. R. G. Hunt, L. E. Seitter, J. C. Collins, R. H. Miller, and B. S. Brindley, Final Report: Data Collection and Analyses Pertinent to EPA's Development of Guidelines for Procurement of Highway Construction Products Containing Recovered Materials, Volume K: Issues and Technical Summary, EPA Contract 68-01-6014, July 6, 1981.

47. A. G. Kim, B. S. Heisey, R. L. P. Kleinmann, and M. Deul, Acid Mine Drainage: Control and Abatement Research, BuMines IC 8905, 1982.

48. T. S. Butalia, W. E. Wolfe, and W. A. Dick, Developments in Utilization of CCPs in Ohio. *Proc. 13th Int. Symp. on Use and Management of Coal Combustion Products (CCPs)*, Orlando, FL, January 11–15, 1999.

49. W. Dick, R. Stehouwer, J. Bigham, and R. Lal, Dry Flue Gas Desulfurization By-products as Amendments for Reclamation of Acid Minespoil. *Proc. Int. Land Reclamation and Mine Drainage Conf.*, Pittsburgh, PA, April 24–29, 1994, pp. 129–138.

50. R. C. Stehouwer and W. Dick, Soil and Water Quality Impacts of a Clean Coal Combustion By-product Used for Abandoned Mined Land Reclamation. *Proc. 12th Int. Symp. on Coal Combustion By-product (CCB) Management and Use*, American Coal Ash Association and Electric Power Research Institute, vol. 1, p. 7(1–12).

51. J. H. Beeghly, K. J. Smith, and M. Babu, The Dewatering and Agglomeration of FGD Gypsum into Micropellets for Use as a Soil Amendment. *Proc. 1997 Int. Ash Utilization Symp.*, Lexington, KY, October 20–22, 1997.

52. Zimmer to Provide Raw Material, *The Cincinnati Enquirer*, January 28, 1999.

53. WH Zimmer, Generating Station to Invest $20 Mil to Make High Quality Synthetic Gypsum to Be Sold to Lafarge Gypsum Div for a $90 Mil Wallboard Plant, *Power Eng.*, vol. 103, no. 5, p. 12, 1999.

54. Lafarge Corp Plans to Build a US $90 Mil, 900 Mil Square Feet/Year Greenfield

Gypsum Wallboard Plant in Silver Grove, KY. *Industrial Specialties News*, vol. 13, no. 3, February 8, 1999.

55. R. Stehouwer, W. Dick, J. Bigham, L. Forster, F. Hitzhusen, E. McCoy, S. Traina, and W. E. Wolfe, Land Application Uses for Dry FGD By-products, Rep. TR-105264, Res. Proj. 2769-02. Palo Alto, CA: Electrical Power Research Institute, 1995.

56. T. S. Butalia, P. Dyer, R. Stowell, and W. E. Wolfe, Construction of Livestock Feeding and Hay Bale Storage Pads Using FGD Material, The Ohio State Extension Fact Sheet, AEX-332-99. Columbus, OH: The Ohio State University, 1999.

57. Gallia Country NRCS, Heavy Use Livestock Pads Constructed of Stabilized FGD By-product, Report Prepared by Gallia County NRCS and Radian Corporation, Federal Energy Technology Center, Pittsburgh, PA, 1997.

58. T. S. Butalia, S. Mafi, and W. E. Wolfe, Design of Full Scale Demonstration Lagoon Using Clean Coal Technology By-products. 13th Int. Conf. on Solid Waste Technology and Management, Philadelphia, PA, November 16–19, 1997.

59. W. E. Wolfe and T. S. Butalia, Use of FGD as an Impervious Liner, 23rd Int. Tech. Conf. on Coal Utilization and Fuel Systems, Clearwater, FL, March 9–13, 1998.

60. W. E. Wolfe, T. S. Butalia, and C. Fortner, Preliminary Performance Assessment of an FGD-Lined Pond Facility. *Proc. 13th Int. Symp. on Use and Management of Coal Combustion Products (CCPs)*, Orlando, FL, January 11–15, 1999.

61. T. E. Graedel and B. R. Allenby, *Design for Environment.* Prentice-Hall, Englewood CLiffs, NJ, 1996.

62. Draft ISO/DIS 14040, Environmental Management, Life Cycle Assessment, Principles and Framework, 1996.

63. American Coal Ash Association, Comparison of Coal Combustion Products (CCPs) Used for Structural Fill Material vs. Disposal in a Landfill Using the Life Cycle Analysis Framework. Alexandria, VA: ACAA, September 1997.

21

Engineered Wetlands for Metal Mining-Impacted Water Treatment

Herold J. Gerbrandt
Montana Tech of the University of Montana, Butte, Montana

1 INTRODUCTION

Mining is a major point source of water pollution in America and in the world. In mining zones containing pyritic ore (sulfur-bearing ore), acid mine drainage (AMD) contributes degraded water quality in general and potentially toxic metals in specific to receiving streams, lakes, and oceans. This chapter explains the production of AMD, summarizes common solutions to the problem, and presents engineered wetlands as a possible new control of this very serious pollutant. Waste minimization is not addressed here, although many possibilities for waste minimization exist within the mining industry.

2 BACKGROUND

U.S. mining has supplied national and world demand for gold, iron, copper, silver, molybdenum, and a host of other metals for two centuries. Butte, Montana's "Richest Hill on Earth" supplied copper to electrify America in the 1920s through the 1940s. U.S. metals production helped turn the tide of World War II with an overwhelming armory of guns, planes, tanks, and ships. In more recent times, demand for jewelry and electronic parts has also been met by the national mining industry.

The success of this extractive enterprise has come at a cost to the environment. While surface impact is often limited to isolated areas, groundwater and surface water contamination has been more extensive and far more difficult to address. Sulfuric ore bodies, when brought to the surface, have the potential to produce acidic waters when exposed to water and oxygen (air). Acidic waters (pH values between 2 and 5 Standard Units) can then mobilize metals by dissolving them and transporting them downstream or vertically down to the water table.

This chapter outlines the problem of mine-impacted surface and groundwater, then discusses conventional approaches to reduce environmental impact. Lastly, results of a demonstration-scale engineered wetlands to clean metals-laden waters are presented as a natural alternative to conventional methods. Throughout the chapter, illustrations of various topics are drawn for the nation's largest Superfund site, the Butte Area/Clark Fork River Superfund Project. The site stretches from the hills above Butte, Montana, 120 miles down the Clark Fork River to the Milltown Reservoir just outside of Missoula, Montana. The Butte area has a rich history of mining beginning in the 1860s, as well as a legacy of environmental degradation resulting from the mining and smelting in the vicinity.

Major sources of acidic and metals-laden waters include the following.

Mine adits. These horizontal tunnels access ore bodies or veins within mountains. Once the underground workings are abandoned, the adits become unwanted drains for the tunnels, shafts, drifts, and stopes that make up the mine works.

Waste dumps. Made up of overburden soil and rock as well as ore too low in grade to process, waste dumps can become sources of acid rock drainage (ARD) if not isolated from the environment.

Tailings. Metals ore is milled down into fine particles so that the metals can be removed by physical/chemical processes. The remaining fine-grade material is discarded. Modern tailings disposal places this material in lined empoundments. Historic as well as present-day operations such as Freeport-McMoRan's Grasberg mine (1) often discharged tailings into nearby waterways to be washed downstream.

Pit lakes. Large open pit mines have, to date, not been required to backfill after the economical ore has been removed. If the pit intercepts the groundwater table, the groundwater will begin filling the pit when dewatering pumps are turned off. Groundwater flowing into the pit will carry dissolved metals into the lake. When the lake surface rises to the natural groundwater level, acidic water carrying dissolved metals may reverse directions and began flowing out of the lake into the aquifer (2).

In the cases of adits and pit lakes, groundwater flowing through the unmined ore body surrounding the mine workings can generate acid and dissolve metals. Waste dumps and tailings become ARD producers when rainfall and snowmelt move through the material. The resulting surface flow or infiltration carries metals and acidity to down-gradient receptors (streams, lakes, aquifers).

3 CHEMISTRY OF METALS-IMPACTED WATERS

Iron sulfide ores marcasite and pyrite (both FeS_2) are prevalent above and below coal seams as well as in metal deposits. They differ by crystal structure. Pyrite is the predominant of the two, and is the major source of acid mine drainage. Stoichiometric reactions are

$$FeS_2 + \tfrac{7}{2}O_2 + H_2O = Fe^{2+} + 2SO_4^{2-} + 2H^+ \tag{1}$$

Once the reaction is started and the ferrous iron is generated, this reaction looses significance compared to the following three.

$$Fe^{2+} + \tfrac{1}{4}O_2 + H^+ = Fe^{3+} + \tfrac{1}{2}H_2O \tag{2}$$

The oxygenation of ferrous iron in the chemistry lab is very slow, approximately 1000 days.

$$Fe^{3+} + 3H_2O = Fe(OH)_3(s) + 3H^+ \tag{3}$$

Ferric iron is hydrolyzed and 3 moles of acid (H^+) are produced.

$$FeS_2 + 14Fe^{3+} + 8H_2O = 15Fe^{2+} + 2SO_4^{2-} + 16H^+ \tag{4}$$

This reaction rate is rapid (20–1000 min), and 16 moles of acid are produced. It can be seen that the ferrous iron produced in Eq. (4) will feed the reaction represented by Eq. (2). In the presence of excess pyrite, reactions (2), (3), and (4) are now self-sustaining, yielding large quantities of acid in steps 3 and 4.

How is the "acid" produced? The hydrogen ions generated in Eqs. (3) and (4) combine with sulfate or phosphate ions to form their respective acids:

$$2H^+ + SO_4^{2-} = H_2SO_4 \quad \text{sulfuric acid} \tag{5}$$
$$3H^+ + PO_4^{2-} = H_3PO_4 \quad \text{phosphoric acid} \tag{6}$$

Sulfate ions, and often phosphate ions, are present in excess in pyritic ore.

In a hypothetical chemistry lab, acid generation should not be a problem, as Eq. (2), the oxygenation of ferrous iron, has an extremely slow rate of reaction. In nature, microorganisms speed up the rate of reaction by mediating the oxidation of ferrous iron. This is performed principally by *Thiobacillus* and *Ferrobacillus ferrooxidans*, as well as other sulfide-reducing bacteria (SRB). pH is also important, as the reactions are slow unless pH < 4. Thus, potential ARD material may lie nonreactive for years until the Eq. (1) reaction lowers the pH to around 4. Then, the generation of acid is fast. ARD from adits, waste dumps, tailings, or pit lakes may have pH values less than 2.

The final result of the acidification of water in contact with ARD materials is the dissolution of metals in the same materials or in the downstream environment. It has long been known that most metals will dissolve at extremely low or extremely high pH. A typical *metals concentration-versus-pH* plot shows a V-shaped zone of solids precipitate, with the bottom of the V near neutral (see Figure 1). Outside the V-shaped zone, metals tend to be in solution as dissolved

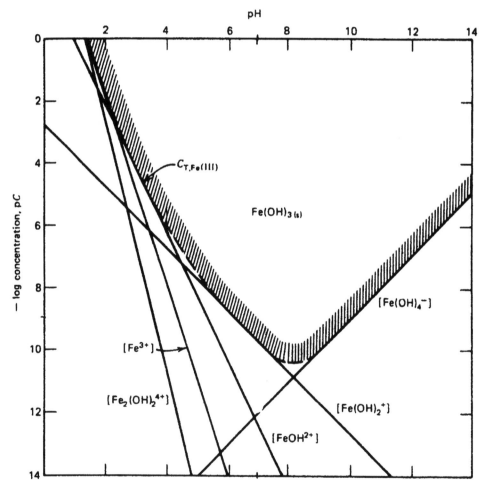

FIGURE 1 Zone of precipitation of $Fe(OH)_{3(s)}$ in contact with hydroxoiron (III) complexes at 25°C (3).

ions. Figure 1 shows equilibrium concentrations of hydroxo iron(III) complexes in a solution in contact with freshly precipitated $Fe(OH)_{3(solid)}$ (3). Most metals show a similar trend. It can be seen, then, that at pH ranges between 2 and 4, many metals will dissolve and be carried down-gradient is surface water and infiltration.

4   IMPACTS ON THE ENVIRONMENT

The acidic water leaving the mine site may prove toxic to aquatic flora and fauna in the receiving streams and lakes. Most aquatic life flourishes at near-neutral pH, and may die or fail to flourish when pH values drop significantly. A lower pH may also cause reactions in plant nutrients, changing them to forms in which they are unavailable to the aquatic floral community (4).

The low pH values additionally can cause large concentrations of metals to enter the aquatic environment: iron, aluminum, manganese, copper, cadmium, zinc, nickel, and lead, to name the typical major constituents. In higher concentrations, these metals can themselves prove toxic to aquatic life. Additionally, precipitation of metals (principally iron oxides) in downstream channels and lakes will affect the aquatic life found on the beds and banks of streams and lakes.

5   TRADITIONAL SOLUTIONS

Traditional solutions can be grouped into two categories: source control and downstream treatment (5). Source control seeks to prevent or stop ARD production at the source—the waste dumps, adits, and tailings.

5.1   Source Control

For dumps and tailings, source control usually consists of isolating the waste materials from water or oxygen or both, covering them with either soil or water. This is often successful when waste materials are contained within impoundments or engineered dumps. When waste rock or tailings are spread over large areas, it is difficult to locate, excavate, transport, and isolate these ARD-producing wastes.

Adit discharge has proven even more difficult to address at the source. Conventional approaches include plugging up the adit and attempting to prevent percolation through the mine workings drained by the adit. Plugging an adit often results in the creation of ARD springs in the vicinity of the plugged adit, as the hydraulic pressures inside the mountain create new passages to drain the mine works. Schemes to prevent percolation through the mine workings include attempting to grout fissures which carry rainfall percolation down to the mine works. Large recharge areas and complex geology often foil such attempts (6).

Source control for large pit lakes is impossible using current technology, as the lake bank surface area is too large to seal, and the inflowing groundwater too difficult to control. Backfilling of small pits attempts to limit contact between groundwater and oxygen, and may or may not prove successful. Approaches for large pits include institutional controls (fencing, restriction of pumping) if the lake will not affect the groundwater. If the pit lake water will significantly affect the groundwater, then conventional technology calls for perpetual pumping with surface treatment before discharge (7).

5.2 Downstream Treatment

Most downstream treatment approaches for surface flow ARD use neutralization followed by precipitation of metal compounds. Neutralization may be achieved by adding lime: limestone ($CaCO_3$), slaked lime [$Ca(OH)_2$], or baked lime (CaO). Limestone is often used to line channels through which the ARD is directed. It is inexpensive and dissolves slowly; however, iron precipitation in the form of FeO_2 or $FeOH$ often coats the limestone particles, decreasing their effectiveness. Slaked lime and baked lime are most often added to ARD streams as a slurry (lime mixed with water). Lime addition to the Clark Fork River near Warm Springs, Montana, is a good example. The Clark Fork is a potential trout stream heavily impacted by metal mining in its watershed. Both waste rock and tailings contribute acidity and dissolved metals to surface and groundwater, which in turn deliver these pollutants to the Clark Fork. Flow rates average 40 cubic feet per second (cfs) in spring and 8 cfs during the rest of the year (8). To raise the pH and drop out metals, a lime slurry is injected into the stream, followed by oxidation using a small waterfall. The stream flow then passes through a series of settling ponds, where metal hydroxides and oxides precipitate out. The water leaves the ponds at a pH 4 range of 7.5–9.0 and with typical metals concentrations several orders of magnitude lower than those in the upstream water.

While lime precipitation is successful at reducing metals concentrations down to levels that meet discharge regulations, several factors argue for a better solution. First, the cost of lime is significant, and increases with the purity and fineness of the lime. The Warm Springs lime treatment station spent approximately $89/ton in 1999 for slaked lime, which resulted in an annual cost of approximately $720,000 (9). Since treatment for surface runoff may continue for hundreds of years and pit lake treatment for perpetuity, a less costly alternative with minimal operations is desirable. Second, lime precipitation creates a large volume of $CaOH$, $CaCO_3$, and CaO sludges. These sludges have various waters of hydration attached, and may contain large amounts of water when pumped out of settling chambers. Additionally, these sludges are unstable and may redissolve under changing water conditions.

An innovative and still experimental approach for treating mine adit ARD is the U.S. Bureau of Mine's (USBM) "In-Mine Treatment of Acidic Drainage Using Anaerobic Bio-reactors" (10). The USBM built a water treatment bioreactor inside an abandoned mine adit and filled the reactor with limestone and compost. The mine adit discharged around 4 liters per minute with a pH of 3, 500 ppm iron, and 30 ppm aluminum. Mine drainage was directed to the reactor, where, in an anaerobic environment, sulfur-reducing bacteria (SRB) are encouraged to consume organic materials such as sugar and generate hydrogen sulfide (H_2S). The H_2S reacts with metal ions to produce metal sulfide precipitates. The metal sulfide precipitates are retained within the reactor, and have a much smaller volume and are more stable than hydroxide or oxide precipitates. Alkalinity, generated by dissolution of the limestone, raises the pH and precipitates aluminum as hydroxide [$Al(OH)_3$]. The bioreactor reportedly raised the pH to 6 and reduced the iron and aluminum concentrations by 50% and 99%, respectively. The treatment system was located inside the mine and could thus be operated year-around. The cost of the sugar was expected to be $0.67 per million liters of water.

6 ENGINEERED WETLANDS TO TREAT MINE-IMPACTED WATER

The USBM bioreactor described in the previous section, while using natural mechanisms to precipitate metal sulfides, is still operationally demanding, requiring addition of sugar on a semiweekly basis. Additionally, it is limited to warm climates or protected areas where freezing will not inhibit the operation. Also, the discharge treated was approximately a gallon per minute, on the small side of adit discharges and much less than storm discharges of surface water from mining-impacted areas. Thus, a natural technology with larger discharge capacity and reduced operational requirements is desired to address the problem of mining-impacted waters.

In the past two decades, natural and created wetlands have been used successfully to clean a variety of contaminants from water, including domestic sewage and industrial waste streams. It is logical to investigate such wetlands for removal of metals from water as well. Dominant reactions and treatments occurring within wetlands are represented by the following five processes (11).

Photosynthesis uses light energy and CO_2 to form organic matter and O_2 and reduce the partial pressure of CO_2. The production of organic matter allows other zones of the system to become anaerobic. Production of O_2 increases the oxidation potential, most probably at microenvironment sites. The change in the partial pressure of CO_2 shifts the water chemistry, usually toward higher pH.

Aerobic respiration involves the formation of energy for cell use by the consumption of organic matter and oxygen. The main benefit of aerobic respiration is the consumption of oxygen. Through aerobic respiration, the deeper portions of the free water surfaces have decreased oxygen concentrations, as do the treatment walls. The decrease of oxygen allows other organisms to exist, which use other electron acceptors for anaerobic respiration.

Anaerobic respiration involves the formation of energy for cell use by the consumption of organic matter and electron acceptors such as ferric iron, sulfate, nitrate, and CO_2. The main benefits are the reduction of sulfate and nitrate and the formation of metal sulfide complexes. Metal sulfide complexes are more stable than metal oxides/hydroxides, which are more likely to form in aerobic environments.

Settling and filtering involves the removal of suspended solids from the water. Plantings along the banks and berms of the cells will increase settling and filtering. It is anticipated that solids that are settled and filtered out will reside in and help to create anaerobic zones and transitional zones between aerobic and anaerobic.

Oxidation reactions in zones of higher oxidation potential (near the water surface) are expected to form metal oxides/hydroxides, which will settle in the open water system.

These complex natural processes are expected to combine to adjust pH values and remove excess metals from the influent water.

7 CASE STUDY: ARCO DEMONSTRATION WETLANDS

In 1996, the Atlantic Richfield Corporation (ARCO) began a series of demonstration projects to treat mining-impacted surface and groundwater in Butte, Montana. ARCO had been named in 1983 as one of the Potentially Responsible Parties (PRPs) for the Clark Fork Basin Superfund sites listed in the very first National Priorities List (12). Water components included surface water running through or over widespread waste rock and tailings as well as precipitation percolating through waste rock and tailings and reaching the groundwater. While some adit discharge may add to the storm water, adit flow is considered insignificant. The site includes a pit lake with severely degraded water quality. No pit lake water was treated in the demonstration projects.

Three demonstration-scale wetlands were constructed by ARCO to test the hypothesis that natural and chemical processes in engineered wetlands could remove metals from mining-impacted surface and groundwater. The three wetlands are identified below, with salient features of each presented.

Wetlands Demonstration Project 1 (WDP1). Seven cells were constructed, including an initial storage pond, four anaerobic subsurface cells, and two aerobic surface water cells. The volumes of the four anaerobic cells measured 30,000 ft³, 20,000 ft³, 12,000 ft³, and 15,000 ft³ for Cells 1–4, respectively, with depths varying from 2.5 to 6 ft. Hydraulic retention times varied from 3.7 days to 9.4 days at 5 gallons per minute (gpm). Three different flow paths were possible using the seven cells. The anaerobic cells' substrate was a 20% limestone/80% river cobble mix. Compost was added to three of the four anaerobic cells. Cattails were planted in the anaerobic cells as a renewable source of organic carbon (13). All of the cells except the storage pond were lined with HDPE geomembrane. Inflow water came from the Metro Storm Drain, a small channel collecting storm water at the foot of the "Richest Hill on Earth." Inflow characteristics are presented in Table 1.

Wetlands Demonstration Project 2, Butte Reduction Works (BRW). This wetland consisted of three open ponds separated by two porous treatment walls (13), the total surface area measuring slightly less than an acre with a depth of 4–5 ft. Water flowed sequentially through the five cells. The treatment walls were constructed of river cobble. The second treatment wall contained compost as well. None of the cells was lined. This wetland treated Missoula Gulch surface water, which is also impacted by mine

TABLE 1 Dissolved Metals and Major Anions (14)

Parameter	Units	Missoula Gulch	LAO groundwater	Metro Storm Drain
PH	mg/liter	7.51	3.58	7.05
Aluminum	mg/liter	<0.03	2.52	0.04
Arsenic	mg/liter	<0.04	<0.04	0.05
Cadmium	mg/liter	0.019	0.115	0.007
Calcium	mg/liter	84.8	134	214
Copper	mg/liter	0.096	14.6	0.067
Iron	mg/liter	<0.021	0.087	<0.021
Lead	mg/liter	<0.04	0.39	<0.04
Magnesium	mg/liter			
Manganese	mg/liter	1.38	17.6	13
Phosphorus	mg/liter	<0.1	<0.1	<0.1
Sodium	mg/liter		37.3	
Zinc	mg/liter	2.22	18	15.1
Chloride	mg/liter	35	0.58	69
Nitrate	mg/liter	3.34	0.15	0.87
Phosphate	mg/liter	0.09		0.11
Sulfate	mg/liter	184	594	650

[a]Combined POTW: POTW water, LAO groundwater, and LMG water.
[b]Combined w/o POTW: Metro Storm Drain (MSD) water, LAO groundwater, and LMG water.

wastes distributed throughout the watershed. Missoula Gulch water characteristics can be found in Table 1.

Wetlands Demonstration Project 2, Colorado Tailings (CT). The wetland design was similar to BRW, consisting of three sequential open ponds separated by two cobble treatment walls. The five cells are unlined, and are approximately 425 ft long by 185 ft wide, with a depth of around 4.5 ft. The wetlands floor penetrated the groundwater, so the bottom 6–18 in. can be considered groundwater flow, while the top 2–4 ft of water is surface flow. CT included lime addition at the inlet to raise the pH, remove excess metals, and pretreat the water. Influent water was pumped from a collection trench intercepting groundwater, known as Lower Area One (LAO) groundwater. This water is the most degraded of the three. Characteristics are presented in Table 1.

7.1 Summary of Performance on Each Wetland

It was hoped that the ARCO wetlands would precipitate metal sulfides via anaerobic respiration. Theoretically, in an anaerobic environment, sulfur-reducing bacteria (SRB) generate hydrogen sulfide (H_2S). The H_2S reacts with metal ions to produce metal sulfide precipitates. As stated previously, metal sulfide precipitates are more stable and of a much smaller volume than metal oxides or hydroxides. Anaerobic zones are produced in the subsurface cells in WDP1 and in the treatment walls of BRW and CT, as well as at the bottom of surface cells where sediments and decaying organic matter form an anaerobic layer. However, any of the other mechanisms—photosynthesis, oxidation, aerobic respiration, settling, and filtration—may play the major role as well. The metals of interest for the ARCO wetlands are copper, manganese, and zinc. Other metals are present at concentrations low enough to meet anticipated discharge standards. A summary of the performance of each of the wetlands is presented below.

7.2 WDP1 Performance

Cell 3, the only upflow cell of the four, failed hydraulically within the first three months. The remaining three horizontal flow cells were monitored continuously from the spring of 1996 until December of 1998. Contaminants of major concern were copper, cadmium, and zinc. Manganese may also be a problematic constituent, depending on site-specific cleanup levels yet to be determined. Table 2 presents typical summer and winter performance data from Cell 2. It can be seen that copper and cadmium concentrations were reduced more than 97% on both sampling days. Zinc concentrations were reduced by more than 99% on the summer day, but only 94% reduction was achieved for unfiltered Zn on the winter sampling date. The decrease in Zn removal efficiency is probably due to both a reduction in hydraulic retention time because of freezing of the upper

TABLE 2 Concentration of Selected Parameters in Cell 2 in Summer Versus Winter

	Summer (8/17/98)			Winter (2/19/98		
	Influent	Effluent	Percent removed	Influent	Effluent	Percent removed
T (°C)	19.0	19.1		2.0	2.7	
PH	6.80	6.84		6.42	6.73	
SC (μmho/cm)	872	1070		821	897	
Alkalinity	66.3	178		72	149	
DIC (mg °C/liter)	16.1	54.4		18.3	36.4	
Eh (mV)	476	−153		493	76	
As	8.6	11.9	(−38%)	7.3	5.2	28.8%
Cd	40.5	0.51	98.7%	32.7	0.52	98.4%
Cu	123	3.6	97.0%	216	1.0	99.5%
Fe	28.8	18.0	37.5%	171	528	(−209%)
Mn	6680	3640	45.5%	7070	8080	(−14%)
Zn (filtered)	9330	10.5	99.9%	9950	101	99.0%
Zn (unfiltered)	9850	42.9	99.6%	10,000	640	93.6%
Sulfate (mg SO_4/liter)	472	289	38.8%	336	313	6.8%
H_2S (mg H_2S/liter)	0.0	34.1		0.0	2.0	

Heavy metals = filtered (0.45 μm), μg/liter; alkalinity = mg $CaCo_3$/liter.
Source: From Seasonal Influences on Heavy Metal Metal Attenutation in an Anaerobic Treatment Wetland, Butte, Montana, 1999.

portion of the wetland and a decrease in the activity of SRBs during cold temperatures (15).

It should be kept in mind that Cell 2 has a volume of 20,000 ft^3 and a flow rate of approximately 5 gpm. It was thought that, upon scaling up, this volume-to-flow rate ratio would require excessive land surface for treating the required flow rate for the Superfund site. Early results of the WDP1 performance led ARCO to modify its wetlands design for the second and third demonstration projects. The projected large volume and surface area requirements based on WDP1 performance resulted in large costs of substrate for subsurface cells and geosynthetic lining for all cells. Therefore, the next generation of demonstration wetlands utilized a greater surface water volume to subsurface volume ratio, and subsequent ponds were unlined. Additionally, piping and enclosed flow meters were eliminated wherever possible, as these utilities were shown to be prone to freezing and clogging in WDP1. Ditches and flumes were utilized in subsequent installations.

7.3 BRW Performance

BRW wetland consisted of three open ponds separated by two porous treatment walls, with flow passing sequentially through the five cells. It was hoped that the cobble treatment walls would develop an anaerobic treatment zone where SRBs would generate H_2S and metal sulfides would precipitate. Additional treatment mechanisms were described previously.

The following trends were observed during the first two years of operation of the BRW wetland (13,16).

Sulfate did not show any removal trends.

Arsenic was not removed, but concentrations are below anticipated cleanup standards.

Cadmium concentrations were reduced by about half but remained above anticipated cleanup standards.

Filtered and unfiltered copper concentrations were reduced significantly. Unfiltered concentration reductions varied from 39% (2/27/97) to 99% (6/19/97). Filtered concentrations varied from 33% to 93%.

Zinc was removed significantly, but not enough to meet treatment objectives. Most of the zinc was contained in the dissolved phase.

7.4 CT Performance

The CT wetland achieved around 99% removal of copper, iron, manganese, and zinc (17). Of this removal, 96% to 99% occurred in the first surface water treatment cell. Recall that the CT system includes injection of lime at the inlet. It is likely that pH adjustment led to metal hydroxide, oxide, and carbonate precipitation in the first cell. Mulholland (16) concluded: "Even though the results obtained by the wetlands are primarily due to pH precipitation, there still may be some advantages to this treatment technology over a traditional lime treatment operation." These advantages include:

Compaction of precipitates over time, resulting in a lower volume of sludge than obtained in a settling tank.

The longer residence time allows formation of kinetically slower, more stable minerals such as manganese oxides.

Reducing environments may eventually transform precipitates into denser, more stable metal sulfides.

8 CONCLUSIONS

Acid mine drainage from mining of sulfur-bearing ores seriously impacts surface and groundwater quality. While waste minimization and pollution prevention can prevent generation of AMD in current operations, centuries of historic mining has resulted in acid mine drainage from thousands of mine sites across the United States. The control of these pollution sources is mostly limited to treatment. This chapter has explained the production of AMD, summarized common solutions to the problem, and presented engineered wetlands as a possible new control of this very serious pollutant. A demonstration-scale wetlands was presented to demonstrate treatability of AMD.

After two years of operation, the three demonstration wetlands in Butte, Montana, are successful to some degree in removing metals from mining-impacted surface and groundwater. While the subsurface cells performed well in reducing metals concentrations, they require a large surface area and are prone to flow rate reduction with time. The BRW wetland used surface water cells with cobble treatment walls between the surface cells. Significant reductions were achieved, but the treatment objective could not be reached. The CT wetlands were similar in design to the BRW setup, but lime addition at the inlet to the wetlands allowed this facility to meet treatment goals. Lime addition, however, was at a much smaller dosage than conventional lime treatment for metals removal.

The Butte climate is extreme, with a long and cold winter. Piping, valving, and meters were prone to freezing, so recommendations are to use open channels to deliver water and flumes or weirs to measure flow rate. Wetlands must be scaled to account for freezing in winter. Freezing of the top 12 in. of the treatment cells will obviously reduce the flow area and decrease hydraulic retention time. Also, bacterial and plant communities become much less active during the cold season, and storage of contaminated water may be required during the winter until natural degradation processes are once again active in the spring.

Precipitation of contaminants such as metal sulfides is desirable, and theoretically achievable using natural processes in engineered wetlands. At two years of operation, the ARCO wetlands were not mature enough to document precipitation of metal sulfides.

Certainly, the ARCO wetlands showed promise for removing metals from water. The extent of contaminant reduction and long-term effectiveness must wait to be proven until the wetlands mature. Cold water temperatures (in the winter season) obviously reduce the effectiveness of the wetlands. At a minimum, engineered wetlands can serve as a passive polishing step. With the addition of a small amount of lime to the influent, the wetlands may be able, with time, to meet treatment goals.

ACKNOWLEDGMENTS

The information concerning passive engineered wetlands presented in this chapter would not have been generated without the support and guidance of the Atlantic Richfield Company and their project manager, Dr. John Pantano.

REFERENCES

1. G. A. Mealey, Grasberg, Freeport-McMoRan Copper & Gold Inc., 1996.
2. Berkeley Pit Public Education Committee, *Pitwatch*, Vol. 2, no. 2, Butte-Silver Bow Planning Department, Butte, MT.
3. V. L. Snoeyink and D. Jenkins, *Water Chemistry*. New York: Wiley, 1980.
4. F. F. Munshower, *The Practical Handbook of Disturbed Land Revegetation*. Lewis Publishers, 1994.
5. Educational Communications, Inc., for U.S. Environmental Protection Agency, Surface Mining and the Natural Environment, 1985.
6. L. McCloskey, D. B. Kelly, and J. Gilbert, Source Control—Surface Waste Pile Demonstration Project, 2000 Billings Land Reclamation Symp., March, 2000.
7. ARCO, Environmental Action Plan for the Clark Fork River Basin, Spring 1998.
8. F. D. Searle, Evaluation of Factors Affecting Heavy Metals in Discharge from Warn Springs Ponds, Master's thesis, Montana College of Mineral Science and Technology, Butte, MT, May 1981.
9. B. Duff, Atlantic Ridhfield Company, personal correspondence, June 2000.
10. Bureau of Mines, U.S. Department of the Interior, In-Mine Treatment of Acidic Drainage Using Anaerobic Bioreactors. *Technol. News*, no. 444, December 1994.
11. ARCO, Wetlands Demonstration Project 2 Butte Reduction Works Final Design Report, December 1996.
12. Montana Department of Health and Environmental Sciences and the U.S. Environmental Protection Agency, Progress—Clark Fork Basin Superfund Sites, May 1990.
13. A. Frandsen, Review of ARCO Wetlands Demonstration Projects: Most Significant Findings through January 1999, unpublished, April 1999.
14. ARCO, Draft Silver Bow Creek/Butte Area NPL Site Butte Priority Soils Operable Unit, Wetlands Demonstration Project 2 Butte Reduction Works Preliminary Design Report, Anaconda, MT, August 1996.
15. C. H. Gammons, W. J. Drury, and Y. Li, Seasonal Influences on Heavy Metal Attenuation in an Anaerobic Treatment Wetland, Butte, Montana, unpublished, 1999.
16. Montana Tech, Wetlands Demonstration Project Progress, September 1997.
17. T. P. Mulholland, A Study of the Mass Deposited, Removal Efficiency, and Types of Minerals Formed in Colorado Tailings Constructed Wetlands, Master's thesis, Montana Tech of the University of Montana, May 1999.

22

Fuel Blends and Alkali Diagnostics: European Case Study

Ingo F. W. Romey
University of Essen, Essen, Germany

1 INTRODUCTION

The main goal of the project, "Advanced Combustion and Gasification of Fuel Blends and Diagnostics of Alkali and Heavy Metal Release," was the investigation and development of data and technologies for commercial use of waste in industrial applications. The R&D program has been set up in three main clusters featuring textile waste co-processing, plastic waste co-processing, and alkali and heavy metal diagnostics.

Ten main items have been investigated, covering the following topics and working packages: Determination and characterization of the supplied production wastes with regard to physical and chemical characteristics

Investigations on special preparation methods for ecologically and economically effective waste preparation (for instance, sorting, separation, shredding, grinding)

Upgrading and blending with coal or biomass and variation of coal/waste biomass/waste ratios

Supply of representative fuel blends from waste and coal to the partners performing the combustion/gasification tests

Determination of logistic and economic principles of the fuel blend preparation (transport, storage, location for the preparation facilities), environmental aspects

Combustion and gasification of fuel blends

Residue valorization concepts

Alkali sensor market study

Alkali diagnostics—measuring campaigns and evaluation

Techno-economic assessment studies

Environmental aspects of co-combustion

A very important part of the work deals with the preparation of fuel blends, transport of the materials, and development of feeding systems. In addition to the combustion and gasification behavior of the materials, special attention was paid to the composition of the fuel and flue gases and the utilization of the ashes. Especially for the later commercial use of the fuel blends in advanced combustion and gasification technologies, it was necessary to develop on-line measurements for alkali and heavy-metal diagnostics. Three different systems have been developed and investigated in the project to guarantee an immediate analysis of alkalis and heavy metals in the flue gas stream and to ensure 100% protection of gas turbines in combined-cycle technologies as well as protection of the environment in advanced combustion and incineration technologies. The R&D program covered basic research work at the laboratory scale as well as large-scale tests in commercial units to guarantee quick transmission of the results for later commercial use.

Due to the large amount of waste plastics in Europe, thermal utilization could be an attractive option in a number of industrial applications; however, standardization and equalization of national laws and reduction of preparation costs must be considered and solutions must be developed.

2 WORKING PACKAGES

In the project, 20 partners have worked together in a very cooperative way to reach the main goals of the planned 10 working packages.

2.1 Upgrading and Preparation of Waste (Textiles and Plastics) for Further Utilization with Coal in Combustion and Gasification

2.1.1 Sächsisches Textilforschungsinstitut (STFI)

The aim of the research work of STFI in the overall project was the transfer of production waste (not suitable for material recycling) by means of mechanical preparation methods into such a form that it can be combusted after blending with coal in combustion plants.

The following textile waste with different material composition has been investigated:

Cotton, viscose, viscose-acetate, polyester, polyamide 6

and regarding their textile materials structure,

Dust, fluff, fabric selvages, heavy screen fabrics

The following methods of mechanical preparation and thermal treatment for selected textile waste have been investigated:

Grinding process, with knife mills
Cutting process, with cutting machines
Breaking process, with tearing machines
Agglomeration process, with a plast-agglomerator-system

Compaction/briquetting tests of textile waste from used drying screens from the paper industry (PA) and a mixture of fabric selvages consisting of 50% polyester and 50% acetate were carried out, followed by combustion tests of the briquettes and emission measurements of combustion gases as well as chemical analysis of ashes. The permitted emission values can be kept for further optimization of the combustion conditions. The obtained ashes cannot be used further and must be landfilled.

Furthermore, agglomeration of synthetic textile waste has been investigated. The waste samples contained used wipes for machines, fabric selvages, and reclaimed fibers from used carpets. The materials were treated by a cutting process and than reprocessed by a plast-agglomerator system. The wipes, consisting of 100% cotton, were blended with cut film waste material of polyethylene. The results in bulk density obtained are between 210 and 350 g/liter.

Knife mills have to be tested where very small (dustlike) particles can be produced. Blending and dosage of small particles requires special constructive changes of transport and dosage systems at combustion plants suitable for textile matters.

The different combustion or gasification technologies are connected to special feeding systems. For these feeding systems a defined degree of treatment of textile waste is necessary

In principle, a higher degree of treatment leads to higher costs. However, the heat value, depending on the materials, the composition of the waste and the theoretical costs for landfilling have to be considered.

Blends of textile waste with coal are very difficult to obtain, because of the great differences in the materials density and the bulk density.

Orientating briquetting trials have shown that a very large surface of the textile waste is necessary. The reduction of particle size from 30 mm to 10 mm is connected with a decrease of the throughput of the cutting machine of about 60%.

The obtained densities for textile briquettes without addition of coal is about 0.7 g/cm^3. A reduction of storage and transport volume of 1/7 can be achieved if there are no high requirements regarding the strength of briquettes. Briquetting with lignite requires a share of coal of about 60%. The densities are > 0.9 g/cm^3. Problems of blendability occurred in the trials. Segregation during the feeding of the compression mold was observed. This led to reduced briquette quality.

The investigated industrial waste textiles from the paper industry (H1-10 and H1-30) are applicable in combustion plants under practical conditions. The combustion should be a combined process because of the problems arising during monocombustion (melting, coke-like sediments). The waste textiles can be used as textile cuttings blended with a second fuel as well as like briquettes with coal. During a combustion trial with blended briquettes (70% coal/30% waste textiles) in a slow-burning stove (nominal capacity 5.6 kW), it was shown that blended briquettes have a stable roasting residue, i.e., the briquettes do not break down and form a coarse ember bed with uniform roasting residues. The problems that occurred in monocombustion of these textiles could not be observed (1).

The measured emissions are under control with regard to legal conditions, after-treatment of the gases, and optimization of the combustion conditions.

2.1.2 Thyssen Umwelttechnik Beratung GmbH (TUB)

TUB has compared different methods of separation of waste plastics from used cars (shredder light fraction, or SLF). Based on analysis of the results, options have been determined for the adjustment of a plant configuration for the given application.

The licensing procedure with the authority, Bezirksregierung Düsseldorf, led to a license according to Bundes-Immissionsschutzgesetz (Federal Immission Protection Act) being granted for the erection and operation of such a trial plant. In several discussions with the regulatory authority it was possible to lower the requirements thus making the process viable.

As the shredder light-weight fraction contains a large amount of plastics and as plastics producers have a joint responsibility for this share, the Verband Kunststofferzeugende Industrie e.V. (VKE, Association of plastic producing industry) carried out a study on thermal and material recovery. This report states the procedure TUB presented to the VKE to be the most appropriate system. It is the only system offering the most suitable option for utilization of the output material.

Under licensing aspects, there is no problem in using the organic fraction via Thyssen Schachtbau by adding coal and by palletizing, since the Thyssen blast furnace operation has a trial licence from the authorities; however, during the running time of the project it was not possible to reach final agreement with Thyssen Steel.

TUB also got in touch with the Belgian and French cement industry to discuss an alternative use for the plastic fraction, but, the approach has not led to a final result to date.

2.1.3 Thyssen Schachtbau Kohletechnik GmbH

TSK has mainly investigated ways for thermal utilization of waste plastics from cars (SLF). Due to its high content of inerts, ranging from 20% to 50% (dust, metals, glass, etc.), preparation has been carried out to recover the organic components.

The following aspects have been investigated:

Analysis of SLF and suitable coal components. Besides the immediate analyses of the accompanying components chlorine and metals such as Cu, Zn, and Pb, investigation with regard to identifying the fractions of the SLF where these components are accumulated was carried out.

Methods of blending and conditioning with hard coals, including evaluation of possible industrial methods in terms of technical and commercial aspects in order to blend and condition such diverse materials to get a homogenous, pneumatic-conveyable fuel. Industrial tests and investigations showed that one possible way to get a homogenous solid fuel to blend the SLF with dried coal in a conveyor screw mixer, followed by pelletizing in a pellet press.

Transport/pneumatic conveyance of pellets obtained from the tests has been done successfully under industrial conditions.

Combustion behavior of the coal-enriched shredder fraction (shredder carbon pellets, SCP) has been tested in a small combustion test rig at DMT (fluidized bed furnace).

In general, due to its composition, the industrial kilns favored for the thermal utilization of SCP are

Blast furnaces
Cement rotary kilns
Fluidized bed combustors

The work programs by Thyssen Sonnenberg Umweltberatung and Thyssen Schachtbau Kohletechnik have provided the basis for the design of a SLF preparation plant. A material balance with the average composition of SLF differentiated by material groups is shown in Table 1.

It is obvious that this inhomogeneous shredder composition "run of plant" results in an undefined calorific value which can range from 10–22 GJ/t, accompanied by undesired components for industrial kilns such as metals, glass, etc. Therefore, a serial investigation was carried out regarding the immediate analysis of SLF at different sizes (<10 mm, <5 mm, >10 mm).

Due to the high inert content in the fraction <10 mm, a future preparation plant will operate with a prescreening to <10 mm followed by a further preparation of the materials >10 mm. The second step to recover an acceptable material for thermal conditioning is to separate metals and other inorganic compounds. The program work done on this matter was carried out successfully in more detail by project partner Thyssen Umwelttechnik Beratung GmbH. The materials obtained from the preparation tests showed calorific values ranging from 16 to 22,000 kJ/kg, but with a typical consistency unusable for pneumatic transport.

In order to stabilize the calorific value and to decrease the metal and chlorine contents, coal has to be added to the SLF; therefore, technical operations such as combined grinding, briquetting, etc., have been evaluated in technical and commercial terms.

Tests to pelletize SLF showed that a significant improvement in pellet strength can be achieved by adding 20% pulverized coal. Photographs taken at millimeter scale proved the effect that the plastic compounds starts softening and thus binding the pulverized coal.

A semiindustrial test run on a 300-kg/h pellet press confirmed these results, and approximately 1 t of pellets were obtained for Thyssen Stahl AG to investigate the mechanical properties as well as pneumatic conveyance.

TABLE 1 Composition of SLF by Material Groups (Sommer, D., Grünenberg, H., Gotthelf, H.)

Material	Weight percent
Plastics	30–35
Elastomers, tyres	20–30
Glass	10–16
Textiles	3–5
Woods/grains	3–5
Lacs	3–5
Metals	0.5–4
Inerts (sand/dust)	10–20

The pellets showed good performance on both criteria so that plant design could be completed to a 5-t/h pellet output capacity.

Table 2 presents the expected fuel qualities. The combustion tests showed no particular differences when firing pure SLF pellets or shredder carbon pellets (SCP). In both cases emissions were kept to the same level, which at least proved a homogenous fuel suitable for fluidized bed combustion.

A suitable way to recover fuel from raw shredder light fraction for industrial kilns such as rotary kilns or blast furnaces is to separate the inerts, e.g., metals, and to pelletize the residual organic material in a pellet press after blending with pulverized coal.

The SCP are pneumatic conveyable and have a calorific value of at least 22,000 kJ/kg with ash contents below 20%. The chlorine content ranks below 1%, which allows combustion in cement kilns. The metal content, in terms of the critical parameters for steel works, Pb, Cu, and Zn, ranks below 1%, which allows blast furnace operations.

Firing in industrial fluidized bed furnaces seems to work without problems regarding emissions as well as feeding and dosing behavior.

2.1.4   Fechner GmbH & Co. KG

Fechner, in cooperation with Krupp Hoesch Stahl (now Thyssen Krupp Stahl), jointly investigated a concept for blast furnace co-injection of pulverized mixed plastics from postconsumer packaging materials and pulverized coal as a ready-to-inject fuel blend. The idea behind this fuel blends project was to assess the technical and commercial viability of a low-investment and low-development-

TABLE 2   Expected Fuel Qualities

	Component/blend			
Specification	Prepared SLF	Pulverized coal	Pellet 70/30	Pellet 80/20
Total moisture (wt%)	2	1	1	1
Ash (d.b. wt%)	23	7	18	20
Volatile matter (d.b. wt%)	60	25	50	53
Sulfur (d.b. wt%)	1,0	0,7	0.9	0.9
NCV (a.r. cal/g)	4,600	7,500	5,500	5,150
NCV (J/g)	19,200	31,400	23,000	21,700
Chlorine (d.b. wt%)	1	0	0.7	0.8
Pb (ppm)	3,000	50	2,100	2,400
Cu (ppm)	5,000	35	3,500	4,000
Zn (ppm)	9,000	6,100	6,000	7,000

risk concept for mixed plastics injection by utilizing existing PCI installations and operational experience at Krupp Hoesch steelworks.

Final evaluation of the operational and commercial viability of this fuel blends concept in comparison to the alternate concept of separate injection of granular plastics or mixed plastics pellets, which is also under investigation by Thyssen Krupp Stahl, was one of the major outcomes of the project part.

The technical program investigated in the project dealt basically with the pulverization of thermally agglomerated mixed plastics with a sizing between 0 and 10 mm to be milled down to a particle size of less than 2 mm.

Following a critical assessment of candidate impact pulverizer systems, an air flow rotor impact mill (turbo-rotor mill) was finally selected for pulverizing mixed plastics agglomerates. The mill design throughput was 5 t/h of thermally agglomerated plastics from packaging waste, although the output capacity varied within the limits of 3–4 t/h of pulverized product.

The subsequent manufacturing of fuel blends with variable mixture ratios of pulverized plastics and coal dust in the range between 10 and 30 wt% of plastics was performed in the production, handling, and storage facilities at Fechner's Lünen works; and the ready-to-inject fuel blends were finally shipped by silo trucks to the blast furnace site at Dortmund.

The major technical objectives achieved were as follows.

> Pulverization of thermally agglomerated mixed plastics, including evaluation of different impact pulveriser systems, semitechnical-scale grinding trials with potential equipment suppliers, and selection and installation of a specific impact pulverizing mill for large-scale production of pulverized plastics. Laboratory scale investigations of the pneumatic conveying properties of coal dust and pulverized plastics mixtures with respect to the maximum/optimum plastics ratio in the fuel blends were carried out.
> Pulverized fuel blends manufacture, including evaluation of conceivable problems regarding the large-scale manufacturing of almost homogenous fuel blends; assessment of the dosing, mixing, and storage equipment requirements. Development of a low-cost approach using Fechner's available handling and mixing installations.

Sufficient fuel blending production trials with sieve analyzing and shearing resistance measurements in order to optimize the mixture ratios and the highest possible uniformity of the fuel blends were performed.

Blast furnace injection trials were carried out jointly with Krupp Hoesch; production of fuel blends for long-term blast furnace injection runs (up to the end of 1996 a total of 2500 tons of mixed plastics was injected) were carried out.

Based on the above technical assessments and operational experience gained from the mixed plastics grinding, fuel blends manufacturing, and blast furnace injection runs, the technical viability of the fuel blends concept has been demonstrated. However, a straightforward evaluation of the commercial viability of this concept for a normal routine blast furnace operation in comparison to the separate and direct injection of plastics agglomerates or pellets (mono-injection concept) needs still more large-scale investigation with the two different mixed plastics injection modes which have been investigated.

2.2 Co-combustion Behavior of Wastes (Textiles and Plastics) and Coal in Entrained Flow and Fluidized Bed Combustion

2.2.1 Instituto Superior Técnico (IST)

The work undertaken at IST dealt with the investigation of co-combustion of natural gas and textile waste entrained in a combustion air stream. The IST contribution can be divided into three main fields of research and action:

Design and construction of a textile waste feeding system and gas/textile waste burner
Simple gas combustion experiments
Mixed gas and textile wastes combustion experiments

Design and construction of a feeding system to continuously feed textile residues in a gas stream, for co-combustion with gas, was developed and tested.

The textile waste feeding system consists of a feed hopper that delivers pulverized textile waste into a vibrating chute and then to a balanced injector system in the primary air line, from where it is pneumatically delivered to the burner through a high-quality, metal-shielded weighbridge, so that the solids in the feed hopper can be weighed. The loss in weight is measured every few microseconds, this information being averaged and stored as a weight loss rate in a computer.

The design and construction of a burner for co-combustion of textile wastes with gas was successful.

IST has undertaken several flue gas measurements for co-firing two kinds of agriculture waste (pine shell and peach stone) and two types of textile waste with propane. The influence of thermal ratio (waste/propane) in the flue gas composition was sought. Detailed in-flame measurements for major local mean gas species (O_2, CO, CO_2, unburned hydrocarbons, and NO_x) and local mean gas temperatures for combined flames of gas + textile waste and gas + biomass were collected. Detailed flue gas measurements for different air staging configurations were also conducted.

Overall, the complete set of results allowed drawing the following conclusions:

1. NO_x emissions increase with the waste/propane thermal ratio regardless of the type of waste. The increase is remarkable in the case of one type

of textile waste owing to its high nitrogen content (fuel NO) and in the case of peach stone, probably because of its smaller particle size distribution (thermal NO).

2. In general, as the waste/propane thermal ratio increases, the CO and UHC emissions increase, particularly for the propane + biomass flames.

3. Attempts to co-fire pulverized coal with textiles were impossible due to their different physical proprieties.

4. NO_x emissions from propane + textiles and propane + biomass flames can be effectively controlled using air staging.

2.2.2   University of Bochum (LEAT)

LEAT investigated the combustion behavior of coal/textile blends in a CFBC.

The first step was the construction of a new heat exchanger for the CFBC test facility as a replacement of the old one, which had a low efficiency. For the collection of fly-ash samples needed for the development of valorization concepts at DMT-SysTec, an additional filter was built. The new devices were tested successfully.

The first combustion trials using blends from coal and shear dust failed due to the fact that the existing metering system was not able to feed the textile fibers continuously. The mixtures separated in the hopper, and also bridging occurred. So the most important task became the development of a reliable feeding system for the different kinds of textile waste, all of them having very specific physical properties. LEAT planned to integrate a separate metering system for textile waste into the test facility. In cooperation with the company Emde, a metering system employing a unique stirring device was developed. This system can feed most of the textile waste considered. The greatest advantage of two separate metering systems is that fluctuations of the textile mass flow have only a small effect on the operation of the facility. The reason is that the ratio of energy input of the textile waste is low in comparison to that of the coal. The dosing of coal using an independent metering system is very precise. In addition, fuel preparation is not necessary. Problems with the metering of textile waste do not lead to shutdown of the facility, since the energy input of the textiles can be easily compensated by additional coal. In view of the later planned industrial implementation of co-combustion of wastes, this is relevant.

Different kinds of textile wastes have been investigated, and the operating parameters of the CFBC in order to optimize the co-combustion of textile wastes have been improved.

The optimization of the burning process inside the CFBC is influenced by primary measures. Possible parameters such as air staging through different injection ports and variation of average combustion temperature were examined. In these series of investigations, shear dust I, consisting of 50% polyacrylics next to 50% cotton, shear dust II, consisting of 100% polyacrylic agglomerated textile waste TA, which is production waste of the car industry, and TD consisting of 100% polypropylene were applied. The major difference between those textile wastes was the nitrogen content (shear dust I, N = 14%; shear dust II, N = 24.34%; TA, N = 1%; and TD, N = 0%). This content establishes the formation of N_2O and NO_x.

The pollutant formation of fuel blends from coal and textile waste in a CFBC is influenced mainly by two factors. On one hand, the physical properties of the textiles lead to difficulties running the combustion process itself. The high CO and C_xH_y emissions are caused by the physical properties, especially the low bulk density of the textiles, because they lead to fast transportation through the combustion chamber without giving enough residence time for burning. On the other hand, the high nitrogen content of the synthetic fibers results in increased nitric pollution. The reduction of NO_x and N_2O is possible due to primary measures such as air staging, and variation of the air ratio and the combustion temperature. The dosability of textile blends and scraps are disadvantageous but excellent for agglomerates. The major disadvantage of burning the agglomerates is that the emissions lead to high CO and C_xH_y-emissions due to their physical properties.

Summarizing, the co-combustion of textile waste inside a CFBC is a possible but not favorable solution. If it is intended to employ this type of co-combustion, the textile fraction of the fuel ought to be minimal. For the future, more experiments applying cotton are needed. As soon as the CO problems has been overcome, the combustion of fuel blends from coal and cotton in a CFBC will be a good solution.

In a second field, LEAT investigated the use of textiles with a high nitrogen content as a substitute for NH_3 in the SNCR process. This textile waste consists of a milled agglomerated shear dust with a particle size smaller than 100 μm. These investigations were performed in a drop tube furnace with an electrical capacity of 50 kW. For simulation of the flue gas in a coal-fired power plant, dried flue gas of a natural gas burner with a fixed fraction of NO was used. The addition of the textiles was intended to reduce the NO_x emission.

Prepared textile dust with a high nitrogen content is an excellent substitute for NH_3 for the SNCR process. With the addition of NH_3, the NO_x emissions may be reduced to about 33%. The experimental results using textile waste as a substitute for NH_3 showed a reduction of NO_x up to 85%. The main conclusions are that the molar fraction of the fuel bound nitrogen from the additive, the gas temperature, and the residence within the furnace are the major parameters for NO_x reduction in the SNCR process. Further investigations are necessary in the drop tube furnace for exact predictions of the SNCR process. Variation of operating parameters such as gas temperature, molar ratio of the additive, and residence time within the furnace are required, accompanied by further experiments in technical-scale plants.

2.2.3 KEMA

KEMA has worked on reactivity measurements as selection criteria for combustion of mixed plastic waste. The activity consisted of several research steps, such as sampling, analysis, prediction of quality of waste products, prediction of emissions, and drop tube furnace experiments.

The topics on sampling/analysis were evaluated successfully, and tests dealing with quality of waste products, emission measurements, and drop tube furnace experiments have been carried out.

2.2.4 International Flame Research Foundation (IFRF)

The primary objective of the IFRF work has been the investigation at a semi-technical scale of firing solid fuels through the tuyeres of a blast furnace, as a means of reducing the coke content while maintaining and/or improving the reduction of iron oxide.

The program examined the possibilities of firing a blend of European coal with plastics in simulated blast air. The following parameters have been investigated:

Blending ratio
Fuel heat input
Fuel velocity
Flame length and flame boundary conditions
Flame penetration based on the kinetic energy of the blast
Burnout

The emphasis has been toward the testing of systems for the feeding of granulated and powder plastics. The results from these tests have defined the particle size that can be fed with the coal to make specific blends.

It is desirable for the blast furnace simulation that the particles are as small as possible, in order to expose a large surface area which should ensure complete burnout.

Plastics have been identified as possible fuels that can be used with coal of various rank to form a blended fuel that is suitable for injection into the blast furnace. Typical composition of plastics relative to other common fuels is given in Table 3.

A detailed evaluation of the combustion behavior of plastic blended with coal with different injection methods has been carried out. Furthermore, ash samples have been taken for ash valorization with other project partners.

Finally, the potential for plastic fuel as a blend fuel with coal for replacement of coke in the blast furnace has been ascertained.

TABLE 3 Typical Composition of Plastics Relative to Other Common Fuels

Element (% w/w)	Plastic scrap	Delayed coke	HVB/MVB coal	Municipal sewage sludge
Ash	1–2	<2	5–15	40–50
Water	0–15	<2	<3	<10
Volatiles	>95	10–13	25–40	50
C	60–80	92	70–78	30
H	6–14	3	4–5	4
S	0	1–3	1–2	1–2
N	0	0.5–1.5	1–2	3–5
Cl	0.1–1.5	<0.1	<0.2	<0.1
F	0–0.5	<0.1	—	<0.1
LCV MJ/kg (dry)	27–40	31–33	28–31	11–14

2.2.5 ICI Films

ICI, as one of the industrial project partners, took over the large-scale combustion of fuel blends consisting of coal and plastics film wastes in a 40-MW circulating fluidized bed boiler (CFB). Due to serious damage in the boiler, however, ICI decided to leave the project.

However, first trials in 1996 showed that the system after further optimization will work sufficiently. ICI has plastic waste samples available in a 100-t scale and was able to deliver the material for further combustion tests to the other project partners.

IST together with industrial partners in Portugal has taken over a project part and has carried out combustion tests in large-scale boilers.

In addition, IST has conducted detailed flue gas measurements of O_2, CO, CO_2, unburned hydrocarbons, and NO_x at the Portucel full-scale biomass boiler. During the tests around 40 t of plastics of four different types were burned. The plastics were burned in combination with biomass at different thermal ratios. From the data collected, it was possible to draw the following conclusions.

1. There is a value of plastics/biomass ratio that corresponds to a maximum value of NO_x emissions; above and below this value, NO_x emissions decrease. Consistently, this is the value to which corresponds a higher value of the mean boiler temperature.
2. SO_x emissions from the combined combustion of biomass + plastics are always insignificant. However, they increase marginally as the plastics thermal input increases because of the sulfur present in one type of plastic.

3. CO and UHC emissions were not detected in the flue gases for any of the tested conditions.
4. During extended and continuous boiler operation, firing small quantities of biomass plus high amounts of plastics, there was a propensity for the latter to melt on the grate, with the combustion efficiency suffering accordingly.

2.3 Co-gasification Behavior of Wastes (Textiles and Plastics) and Coal in Fluidized Bed Gasification

2.3.1 University of Lund

TU Lund has carried out work on gasification experiments of fuel blends of coal and textile and plastic wastes. The experience showed a need for development in mixing procedures for fuel blends and also a necessity for modifications to the existing feeding system.

Several successful and stable gasification experiments were done with coal and textile waste blends. The main goals in these experiments were

To study the effect of the various mixing methods on feeding process
To see whether different mixtures behave differently in the gasifier
To study the reliability of the gasifier and reproducibility of the gasification results

The experiments conducted showed a great improvement both in the gasifier operation and in the gasification quality; however, the intensive gasification of the coal blends in the gasifier resulted in high-temperature-induced material exhaustion of the reactor tube followed by melting of it and breaking down of two reactor furnaces. Therefore, further tests (after repairs) were carried out exclusively with biomass/textile waste blends.

Gas analyses showed a fluctuation in the product gas composition, which is a new phenomenon in the gas analyses for the Lund gasifier. It is believed that sometimes a separation of fuel blends results in an uneven oxygen distribution in the bed. The input oxygen reacts first with the textiles, which are more reactive than coal or biomass and results in total combustion of the textile fraction. In this case the amount of CO_2 in the gas increases at the expense of the combustible gases.

According to preliminary calculations, the volumetric heating value of the product gas varies between 3.5 and 4.5 MJ/Nm^3, depending on the operational conditions.

Study of the concentrations of trace gases shows that during coal/textile experiments more sulfur compounds are produced. The concentration of H_2S in these experiments exceeds 700 ppmv in the gas. The amounts of SO_2 and COS in the gas are also higher than those in the biomass/textile gasification.

The content of the sulfur compounds is not considerably affected by the textile mixing ratio.

Adding textile to biomass does not affect the composition of the product gas. However, at high mixing ratios the amount of unburned carbon in the fly ash increases.

There is a clear relationship between the biomass/textile mixing ratio and the concentrations of trace gases such as NH_3, HCl, and H_2S. While the S, Cl, and N contents of the textile are higher than those of the biomass, increased textile addition increases the amounts of these traces in the gas.

Ammonia formation in the gasifier is regulated by three important factors: the fuel nitrogen, the gasification temperature, and the ER value. The relationship between the fuel nitrogen and the ammonia concentration in the gas is linear. Higher temperature results in decrease of the ammonia. The effect of higher ER in ammonia decreasing is due to both gas dilution and also the improved oxygen availability in the gasifier.

The amount of the lighter PAHs in the tar is larger in the case of mixture gasification. Compounds with molecular weight larger than 168 (dibenzofurane) make up about 84–95% of the total PAHs for the pure biomass, while the corresponding value for the mixture is below 55% at its highest value.

Textile addition contributes to increased formation of light hydrocarbons such as benzene, toluene, ethane, and butane.

The heating value of the gas is affected mostly by the ER value. Including the combustion heats of the light hydrocarbons improves the gas heating value by 15–28%.

2.4 Analytics of Gaseous Emissions, Fly Ash, and Residues; Utilization of Residues such as Fly Ash and Ash in Subsequent Processes; Difference from Utilization and Treatment of Pure Coal Ashes

One partner has carried out valorization concepts for residues. Residues of combustion and gasification of fuel blends were delivered by other project partners working on co-combustion and co-gasification.

2.4.1 DMT-SysTec

In existing literature only a few reports are found in this context. One of these reports deals with thermal utilization of textile floor matting materials in a stationary fluidized bed (single-fuel) firing system. The composition of the resulting ashes with high calcium oxide concentrations should allow manifold applications in cement industries. Utilization possibilities quoted in the project are proportioning to concrete mixes as well as an additive for cement production feedstock. Supposedly, there are further utilization possibilities, such as production of sand-lime bricks and lightweight concrete. The ashes should contain only small concentrations of heavy metal and the elution rates of these heavy metals should be low as well, thus allowing environmentally benign and cost-effective disposal on mineral-matter dumps.

Further examples are the use of textile carpet waste as fuel or feedstock in blast furnaces or in a modified refuse incineration plant. Initial field trials have shown that the resulting blast furnace slags are particularly suited for producing cement qualities intended for underwater use, and the refuse incineration slags are suited for use as sand for various purposes in civil engineering and road construction. Dusts from dust separation systems as well as residual ashes need to be dumped.

Bochum University has carried out combustion tests in a fluidized bed combustion furnace with fuel blends of coal and textiles (blends of propylene and flax, TA, and TD). The investigations into the utilization possibilities obtained with these residues have shown that the levels of, for example, heavy metals are low enough and that they remain under the limit values for construction materials for restricted installation. Furthermore, it would be possible to use the residues as underground mortar, e.g., for backfilling or roadside packs. At the same time, the criteria to be fulfilled by the fluidized bed ash include the following: homogenous chemical composition and constituents, constant particle size distribution, sufficient strength, controlled carbon and CaO, and availability of adequate quantities

Special construction material tests to establish the suitability of the residues in certain construction materials, e.g., in mortars or mining mortars, or preparation tests, e.g., to separate interfering components, must be investigated with larger quantities of ashes.

Lund University has also conducted gasification tests (with different ER values and bed material) with biomass and textiles. The loss on ignition, some heavy metal levels, and the PAH values exceed the limiting values for utilization as construction material and for dumping on a mineral dump. Reducing the textile portion down to, for example, 5% does not produce any basic advantage. In addition, the chemical composition, with more than 75% MgO, also renders them unsuitable for such a use.

If the high MgO level is maintained with an optimization of the gasification process, however, it might be possible to recover the MgO at very high cost. Since the MgO exhibits a higher density than the other ash portions, consideration can be given to the procedures of wash table grading or grading in autogenous heavy media (upstream classification). It has not been possible to conduct examinations in this respect because the quantities of the samples were too small. The drawback of this procedure is the necessary drying of the separated material. The remaining residue with high carbon levels could then be returned to the reactor. In view of the small ash portions both in the biomass and in the textiles, only a small residue would have to be dumped in total. The advantages of this procedure would be a saving in fuel and bed material.

Further investigations were carried out with residues from the combustion of biomass and the co-combustion of biomass and high-ash-containing plastics (IST) and from the combustion of fuel blends from coal and plastics in a 1-MW plant (a pulverized dry firing system, KEMA). Without additional treatment the residues from the combustion of biomass/plastic and coal/plastic cannot be used in the construction materials domain due to, e.g., high Cl, sulfate, and heavy metal content, and partly due to very high electrical conductivity.

It is very laborious and costly to wash out the interfering constituents, e.g., to reduce the chlorine content, because the material washed out would then have to be dried and the contaminated solvent would have to undergo further treatment. In addition, it is not possible to say anything about the requisite plant concept and possible investment costs without more detailed knowledge of the quantities arising and the actual residue qualities.

One possible form of recycling would be use in a cement works as a supplier of fuel and minerals. For this purpose, however, the cement works would have to be equipped with the requisite gas cleaning facilities and the mineral composition of the residue would have to fit the raw materials concept.

A further possibility for recycling is integration of the ashes or polluting constituents in cement. In this case the total material would be evaluated in terms of its properties. On top of the installation costs, the integration would involve a high additional expenditure. Beyond this there is only the possibility of dumping on a household waste dump.

2.5 Market Study on the Worldwide Availability of Gas Analyzing Sensors, Especially Alkali Sensors, and Future Marketing Chances

All partners of the alkali measuring project group have contributed in the preparation of a market study for on-line measurement of alkali and heavy metal species. The study has been published under the title, "Diagnostics of Alkali and Heavy Metal Release" (EUR 18291 EN; ISBN 3-00-002948-6).

2.5.1 BTU Cottbus

In order to gain an overview about the current state of the art of the techniques and about the existing opportunities for the use of on-line alkali and heavy metal sensors, information was requested from more than 50 selected companies and institutions. The majority of written questionnaires was answered. The results of this survey, with phone calls and literature searches, contributed to the market study. It could be shown that on-line detectors are not yet available worldwide, although both in Japan and in the United States, there is ongoing research to develop corresponding measuring devices for commercial availability.

According to the survey, a potential market for on-line alkali measuring devices exists in the following areas: combined-cycle plants using pressurized fluidized bed combustion (PFBC, CPFBC); combined-cycle plants using integrated gasification cycles and hot gas cleaning (IGCC); combined-cycle plants using pressurized pulverized coal combustion (PPCC); plants for the thermal exploitation of waste, especially plastic waste (combustion, pyrolysis, gasification) and municipal solid waste (conversion); and plants for the thermal conversion of biomass. Correspondingly, potential buyers and users of on-line systems would be gas turbine producers, boiler producers, engineering companies, energy providers, and municipal waste combustion companies.

Together with DMT and the University of Heidelberg, a comparative test program with off-line measurements carried out by BTU Cottbus and the ELIF method of the University of Heidelberg was outlined. The measurements were carried out at DMT's atmospheric fluidizing reactor, ALFRED, at a temperature of about 850°C. Sampling was made with three gas washing bottles, and for the analysis of alkali species ion-selective measurements were chosen. The tests were carried out with two different coals (Westerholt bituminous coal and Rheinbraun lignite), both with and without additives (CH_3COONa, CH_3COOK), and measurements were made between the cyclone and hot gas filter and behind the filter.

Potassium off-line measurements are 2 to 200 times higher than on-line measurements, and sodium off-line measurements are 1.25 to 40 times higher than corresponding on-line measurements. A reason for the bad results was not identified.

Alkali concentrations measured behind the cyclone were much higher than those measured behind the filter. An explanation was supposed in the high amount of ash particles in the flue gas before filtration. There could have been alkali deposition on the surface of ash particles or there could have remained a water-soluble alkali fraction in the ash. In order to see if this assumption was right, BTU Cottbus tested several samples of ash for their content of water-soluble alkali compounds. At the cyclone there was only a very little amount of alkali compounds in the ash. It could be shown that the ash particles do not have a significant influence on alkali concentration before the filter. The alkali content at the filter was much higher than that at the cyclone. One explanation could be that alkali deposition took place at the filter in previous experiments and these deposits have been removed now.

Thermochemical Equilibrium Calculations.   These calculations were carried out with the program ChemSage, produced by GTT Technologies. GTT Technologies also delivered a data file which was originally created for the investigation of gas species and condensed phases formed during coal combustion processes and which was extended for the calculation of alkali release processes. This data file uses for the slag phase the quasi-chemical excess model for the slag phase. For the other phases it uses the ideal mixing model.

BTU Cottbus carried out calculations comparable to the on-line measuring results of the measurement campaigns at Foster Wheeler Energia Oy in Karhula/Finland carried out within this program. Furthermore, calculations have been carried out on the influence of temperature, limestone addition, and addition of various alkali getter materials.

Results of the calculations are 4 to 10 times higher than on-line measurements. At 800–1500 K, which is the temperature region of the Karhula tests, mainly NaCl and KCl are present, whereas at more than 1500 K, NaOH and KOH also show high concentrations. Na_2SO_4 becomes unstable at $T > 800$ K, and K_2SO_4 reaches its maximum at 1500 K.

To summarize the results, it can be said that SiO_2 is the only getter material of importance. Other materials probably act only via their content of SiO_2 (e.g., bauxite, $CaO \cdot MgO \cdot 2SiO_2$), and their use is only convenient if those materials are less expensive. But attention should be given to the fact that for the materials tested, more than the amount corresponding to pure SiO_2 is necessary.

2.6 Alkali Measurements Applying Three Different Analyzers in Flue Gas Streams of Combustion Facilities; Refinement and Extension of the Alkali Measurement Methods

2.6.1 University of Heidelberg (PCI)

The ELIF measurement method for PCI covered the following activities:

Design, construction, and testing of in-situ optical access for use in industrial-scale systems under realistic PFBC conditions. The new design allows coupling of the laser beam and collection of fluorescence at just one window. The durability of this optical access was demonstrated for conditions of 10 bar total pressure and ca. 800°C during continuous plant operation over periods of about two weeks.

Integration of a new detection setup in the ELIF measuring system using fiber optics to connect to the optical access. Simultaneous two-channel detection is also still feasible. The fiber optic cable allows the detection system to be placed together with the remaining electronics in a compact way and also much reduces adjustment effort. The components of the detection unit are in a fixed arrangement, requiring no adjustment except the simple rotation of a wheel to change neutral-density filters for different fluorescence intensities.

Redesign of evaluation software, allowing more efficient, user-friendly operation of the ELIF system and a fast overview of results.

In-situ acquisition of on-line data on alkali release from the 10-MW PFBC plant of Foster Wheeler Energia/Karhula, Finland, including observation of short-term (e.g., pulsing of hot-gas filter during cleaning, addition of limestone) and longer-term effects (e.g., variations in load and oxygen excess). Measurements were made during two periods of plant operation of 10–14 days each and were interrupted only briefly for refilling the laser (ca. 15 min) and securing the data. The ELIF system could be operated nonstop for at least 24 h. In addition, simultaneous measurements were made with the Tampere and Göteborg groups, using the PEARLS and SI techniques respectively, so that direct comparisons could be made. Measured concentrations are of the same order of magnitude for all three measuring techniques.

Calibration of the ELIF signals was performed on a facility designed and constructed by the group from Tampere.

Survey of possibilities for detection of heavy metals using the ELIF method, including assessment of collisional effects. Some detection schemes for Cd, Ni, and Zn compounds were selected, whereby single-photon excitation is to be preferred in the first instance, since lower laser energies will be required and interpretation of signals should be more straightforward.

2.6.2   University of Tampere (TUT)

The PEARLS alkali measuring instrument of the TUT group has been tested in the 10-MW PCFB of Foster Wheeler Energia Oy. The instrument has been used for simultaneous alkali measurements together with the ELIF and surface ionization instruments during different operations of the PCFB.

The PEARLS alkali measuring instrument has been developed by utilizing a new photodetector which substantially improves performance.

Improved detection limit below 1 ppb (previously 5 ppb)
Improved data acquisition rate, 2 values/s (previously 10 values/minute)

The improved data acquisition rate makes it possible to analyze fast phenomena, such as alkali behavior during the hot-gas filter cleaning pulse.

A pressurized and transportable test/calibration unit was designed and manufactured. The operation of the unit is based on the known vapor pressure of alkali salts at a determined temperature. The unit will help to improve the accuracy and develop the components of the measuring instrument.

TUT provided background information for other partners by establishing the format for preparative actions and sending a full copy of the technical drawings of the PEARLS instrument to the partners. The following features were utilized in the design of the ELIF and surface ionization instruments:

Identical DN150/DN50 reducer flange design for all instruments
Architecture of the instrument body (surface ionization)
Division of the sampling line to heating zones (surface ionization)
Protective arrangements of the sampling line (surface ionization)
Ball valve for window protection (ELIF)

The PEARLS instrument has been improved significantly in the test runs; comparison with the two other on-line instruments showed comparable alkali levels.

2.6.3   University of Göteborg (GU)

The main objective of the project has been to develop an on-line alkali measurement instrument for power plant applications based on surface ionization (SI). Alkali metal atoms are well known to ionize easily in contact with hot metal surfaces, and the instrument developed in this project is based on this principle. This is an almost unique property of alkali compounds, which implies that a sensitive and selective detector can be developed. The primary parts of the detector are a filament, generally made of platinum, and an ion collector. The platinum filament is supported between two electrodes and is resistively heated for alkali vaporization in ionic form. The filament is biased at a positive voltage of 500 V to repel the positive alkali ions formed. The ion collector is situated close to the filament and is grounded through a sensitive electrometer. The measured current is proportional to the arrival rate of alkali atoms onto the filament.

In this project an instrument consisting of a SI alkali detector, a hot-gas sampling line, and a control system has been constructed according to prevailing high-pressure and high-temperature standards. The instrument has been tested during two extended measurement campaigns in a PCFB combustion pilot plant at Foster Wheeler's R&D Center in Karhula, Finland. During the campaigns, the instrument was under operation for more than 500 h during pressurized combustion of coal. Simultaneous alkali measurements with the groups from PCI and TUT were performed on several occasions during the campaigns. The measurement campaigns have been evaluated in cooperation with PCI, TUT, and FW, and this work also included the calibration of the SI instrument. It can be concluded that the developed instrument has performed well and the experimental runs confirmed that the instrument can operate during extended times in the hostile environment present at a power plant.

After the campaigns the SI instrument was further developed. The SI instrument operation has been simplified by increased computer control, and the reliability of the measurement procedure has been further improved. The SI detector has also been tested in a laboratory-scale pressurized gasification unit, in order to analyze the requirements for instrument use in gasification. A description of the SI instrument, together with descriptions of the other detectors developed in the program, has been published in a market study on "Diagnostics of Alkali and Heavy Metal Release."

The developed instrument monitors the total alkali concentration (Na + K) in the hot flue gas of a pressurized system. In an extension of the technique, the instrument is also intended to differentiate between Na and K as well as discriminate between alkali species in molecular and particulate form. It was found that platinum might be a suitable material for the ionizing hot filament, considering both ionization efficiency and filament lifetime. A large number of other filament materials were also tested with respect to thermal stability and their surface ionization efficiencies for Na and K. A few of the tested materials can be considered as alternatives to Pt in future SI applications, including platinum alloys, high-temperature steels, and Kanthal. Based on the investigations, it can be concluded that all tested materials (except Pt) are oxidized at the high temperatures used in the instrument and are therefore not suitable for differential measurements of Na and K. It is suggested that the combination of SI detection and ion mobility spectroscopy or mass spectroscopy be investigated in future work on Na and K separation.

Laboratory investigations have been performed with alkali-containing particles in the size range 0.05–1 µm. The SI technique can be used to detect single particles in this important size range, and a selection between alkali in molecular form and alkali-containing particles can be based on this principle. The results also suggest that a new type of particle analyzer can be constructed based on SI technique. This simple and very sensitive device may prove ideal as an "early warning" system for measurements of increased particle concentrations in a gas turbine inlet.

2.6.4 DMT-FuelTec

DMT pursued various tasks in the project, including the adaptation and modification of the analyzers to applications in industrial plant. In particular, a rugged but versatile hot-gas extraction probe system permitting reliable sampling for accurate on-line measurements was designed and constructed. The retractable probe system included a pressure lock and incorporated a measuring chamber for the ELIF analyzer, providing a number of windows in order to let the exciting laser irradiation in and to detect the fluorescence signal, and providing an opportunity in its pressurized fluidized bed combustion facility, FRED, to test developments in analyzer improvements and to carry out combustion experiments at conditions most suited to instrument development needs. These runs complemented the field measurement campaigns of the instrument developing teams.

Based on observations and measurements with the ELIF method, drawing on experience with an extractive sampling line forming part of one of the three analyzers and including DMT expertise from other instrument developments, a new retractable hot-gas extraction probe system was conceived, designed, and constructed that incorporated the following features.

The system is strictly modular in design. The various functions of the system are allotted to individual components.

Components needing servicing or exchange/replacement are readily accessible.

The extracting gas tube is heated indirectly and can be replaced with tubes made from other materials if required.

The probe system is fitted to the plant via pressure lock. It can be withdrawn under pressure, isolated, and independently depressurized.

The measuring cell provides four rectangularly arranged access ports for observation. The measuring cell has its own individual heating system.

The sampling gas is passed, downstream of the measuring chamber, to a cooler, followed by a strainer, to the pressure letdown valve that also controls the total gas flow through the system. The gas can then be vented or used for other purposes, e.g., conventional gas analysis.

2.6.5   Foster Wheeler Energia Oy (FW)

The Foster Wheeler Pressurized Circulating Fluidized Bed Test Facility in Karhula, Finland, provided over 1000 h of operation on coal to be utilized for alkali measurements under industrial conditions during the two test segments in 1997. The measuring teas, from the University of Göteborg, Heidelberg University, and Tampere University of Technology, were able to take advantage of a large part of the operation. Technically, instruments of all three measuring teams showed remarkable potential for industrial use.

During summer 1998, the final results of the measurements of all teams were submitted to Foster Wheeler. The alkali emission results were reviewed together with steady-state process data, and possible correlations with process parameters such as temperature and pressure were investigated. Clear connections between the alkali emission and the amount of particulate in the flue gas as well as the flue gas temperature could be observed.

In general, with dust-free flue gas, the sodium and potassium emissions were in the range of 25 ppb (6% O_2) at temperatures of 800–850°C, and the range of 10 ppb is reachable at temperatures of 700–750°C already without special getters or process modifications with the coal used during the test campaign.

The main results of the project were published during the presentation of the alkali group and in the poster session of the final project meeting in Brussels on November 11–12, 1998.

A joint publication on alkali measurements was submitted together with the measurement teams to the organizers of the 6th International CFB Conference, held in Würzburg in 1999.

2.6.6 RWE Energie AG

The objective of RWE's participation in the project was to get actual and detailed information on possibilities of on-line measuring techniques of alkali in pressurized combustion or gasification processes.

To get results with the alkali on-line measuring systems at the pressurized fluidized bed combustion test facilities of Foster Wheeler, it was necessary to work out a main test program, in which the basic requirements (types of coal and additives to be used, constant and variable process parameters during the test runs) are described. This main test program has been discussed with all partners involved. The program has taken into account all important parameters and operating conditions of the plant and the needs of the three different measuring systems.

Concerning this main test program, RWE intended to participate in the following:

1. Participation in planning the experimental program for runs in different PFBC test units in cooperation with Technische Universität Cottbus and plant manufacturers
2. Adaptation and study of the requirements of the analyzing system in gasification plants

The requirements of the analyzing system for gasification plants have been discussed with the University of Heidelberg. The considered aspects for feasibility of the ELIF measurements are as follows:

Fluorescence quenching
Absorption of laser beam
Background emission
Dimensions and geometry of the system

The result was that measurements under the stated conditions (HTW gasification) will be feasible.

The potential applications relating to gasification of the measurement system are future power plant concepts with combined cycle such as "IGCC" or "PFBC second generation."

The PFBC process with extremely staged combustion could be one possible option for a high-efficiency power plant concept. The process will be investigated at the laboratory scale at different places and is in the very beginning of the development.

To get an impression what alkaline values are to be expected in a future combined-cycle process such as staged combustion, the University of Cottbus did some calculations with their program ChemSage for the use of lignite.

After finishing the investigation of the potential applications of the measurement system in gasification plants, it was discussed to use the systems in atmospheric combustion in conventional boilers. This had the following background.

The alkali content of the Rhenish lignite of some mining areas will increase during the next years. Important for RWE Energie is to obtain knowledge of the influence of higher ranges of alkali content. Suitable measuring systems need to be tested. Therefore theoretical investigations have been carried out concerning the possible application of the Elif system.

It is especially important to obtain a better understanding of alkali deposition and the incipience of fouling and corrosion.

The test employment of one or more measurement systems in conventional firing systems at RWE Energie is under consideration.

2.7 Database on the Occurrence of Textile Wastes in Industry and Their Characterization (Amount, Composition, Contamination, etc.)

Two project partners were involved in this topic.

2.7.1 University of Münster (FATM)

FATM carried out a database study on the occurrence of textile wastes from different textile branches in Germany. The main aim of this study was the calculation of the textile production waste that arises in the Federal Republic of Germany, according to production phases, in order to gain a basic idea about the amount of production waste that would be available for the combustion and gasification as an addition to coal. Typical waste quotas are determined for the individual production processes of the textile industry. In connection with the amounts and types of raw materials used, these quotas will make it possible to give a differentiated estimate of the amount of waste arising, according to its type and the raw material structure. These are basics for a following market study on the real and potential markets.

The calculation of the amount of textile production waste arising is based mainly on two sources. The first is the official statistics on production in German industry in 1995, which provides production data for the different branches of the textile industry. The other is a questionnaire sent to companies by the FATM in 1996, which was used to get and calculate waste quotas for different production processes.

In addition to waste quotas and waste volumes, a breakdown by types of fiber for the raw materials is given.

There is a significant concentration of textile mills in three German states. Taking the number of employees as a measure for the regional distribution of production, about 76% of the textile production waste in 1995, i.e., 83,000 t (83 kt), is concentrated in Nordrhein-Westfalen (NRW), 32 kt; Baden-Württemberg (BW), 16 kt; and Bayern (BY), 14 kt. At the county level, about 60% of the occurence of all textile waste is concentrated in only six regions: Münster (NRW), 12 kt; Düsseldorf (NRW), 12 kt; Oberfranken (BY), 8 kt; Freiburg (BW), 7 kt; Stuttgart (BW), 5 kt; and Kassel (Hessen), 5 kt. We find a similar result for the regional occurrence of textile dusts, which are of special interest due to problems in recycling and current prices of disposal on one hand, and opportunities in preparing the dusts for co-combustion with coal on the other hand. About 12 kt, i.e., 65%, of all textile dusts from production arise in only seven counties: Münster (NRW), Düsseldorf (NRW), and Köln (NRW), with a combined amount of dusts of 5.2 kt; Freiburg (BW), and Stuttgart (BW), with a combined amount of 4.4 kt; Oberfranken (BY), 1.4 kt; and Kassel (Hessen), 900 t.

Disposal Costs of and Profits from Textile Production Waste. Textile waste is handled as a commercial good. A positive price for a type of waste indicates a path for recycling the waste inside or outside the textile chain. If there is a negative price for textile waste, the textile companies have to pay for disposal and only in this case will textile waste be offered for co-combustion instead of landfilling. Table 4 offers information on the availability of textile waste for co-combustion with coal in regard to the market situation.

The operational result is that mainly textile dusts will be offered for co-combustion with coal, because fiber waste from the nonwovens production can be classified as textile dusts due to the very short length of residual fibers.

Collecting Systems and Remarks on the Logistics for Textile Waste, Especially Dusts. Concerning the regional occurrence of textile waste, a few com-

TABLE 4 Average Disposal Costs/Earnings of Textile Waste in the GE Textile Industry, 1995

	Textile dusts	Fiber waste	Yarn waste	Fabric waste
Spinning mills	−230 DM/t	+640 DM/t	+270 DM/t	—
Nonwoven mills	−320 DM/t	−320 DM/t	—	−320 DM/t
Weaving mills	−170 DM/t	—	+110 DM/t	+190 DM/t

Source: Personal calculation, based on FATM Questionnaire, 1996.

panies in these regions have specialized in buying and selling textile waste. It is therefore possible to refer to an existing logistics system for collecting textile dust from a great number of mills.

In spite of the very low density of textile dust we have a cost-intensive ratio of volume and weight in the transporting system. Since the three major waste-producing regions are located in three corners of Germany, a central point for preparing textile dust for co-combustion in a power plant nearby may not necessarily be the optimal solution with respect to transportation cost. Thus co-combustion plants in the main three regions might be the optimum solution with respect to logistics.

Conclusion. From a total of currently about 83,000 t per year of all types of textile waste from the textile production chain, only 18,000 t per year of textile dusts, including short fibers, are offered for co-combustion with coal. Although this type of textile waste has a high content of energy, it has to be collected from a great number of plants and prepared for co-combustion in a power plant with an extra feeding system. Technical and environmental problems with co-combustion, as well as the low rise of textile dusts over time in hundreds of mills located in several regions of Germany, seem to call for small technology concepts for using the energy content of textile dusts at the place of origin.

2.7.2 CITEVE

CITEVE worked on the calculation of the main textile production waste arising in Portugal, according to production phases of four representative industrial sectors in the country and, on other hand, the determination and characterization of the supplied production waste with regard to its fundamental characteristics. CITEVE also supplied representative wastes to the Portuguese partner (IST) performing the combustion tests (package 1).

The results developed, gathered from different companies, allowed the calculation of median percent figures of waste on production phases. It was possible to calculate figures for secondary raw material, primary waste, and secondary waste.

In terms of waste samples and quantities, samples were supplied to the University of Lund and to the partner IST consisting of about 1 t of seeds, dust, and very short fibers of cotton, 500 kg from a blowing at a spinning mill and the other 500 kg from the gig operation at finishing. This last kind consisted of short fibers (short woolens) with color and finishing/dyestuff.

Those two kinds of waste supplied to IST and another four, for a total of six main kinds of waste, have been identified as the most representative secondary waste in Portugal.

The kind and quantity of waste produced depends mainly on the production process. CITEVE collected information about different production processes of the most representative industrial sectors in the country: wool, cotton, knitting, and clothing. This waste was divided into three categories:

A: secondary raw material
B: primary waste
C: secondary waste

Since A and B waste can be recycled, only C waste were submitted to combustion. Also, the clothing and knitting industries do not produce A and B waste, so only the wool and cotton industries were analyzed in terms of category C waste.

Quantity of Waste Produced. The calculation of the amount of textile production waste arising was based mainly on the following sources:

Gathering of industrial quotes of waste per process by means of diagnostics on site
Statistics of Interlaine and *L'Atlas du Textile Mondial de Françoise Depin,* the first given for 1995 and the second for 1994, concerning the Portuguese textile production of the main four industrial sectors in the country.

Combining both sources, it was possible to approximate calculations of national textile production waste, as shown in Table 5.

It was established that in Portugal, in 1995, about 65,100 t of textile waste were produced. If we take into account just the most representative four textile sectors, considering nevertheless all kind of waste, either those which can be

TABLE 5 Amount of Waste/Year

Products	Gross production (t)	Total amount of waste (t)	Share of kind A (t)	Share of kind B (t)	Share of kind C (t)
Woolen yarns (worsted)	19,060	2,995	1,256	1,545	194
Woolen yarns (carded)	9,300	628		438	190
Woolen fabrics (worsted)	7,283	303		303	
Woolen fabrics (gig)	3,921	941		920	21
Cotton yarns	129,400	38,856	20,704	5,375	12,777
Cotton fabrics	49,000	1,945		1,515	430
Knitting fabrics	143,750	4,446		4,446	
Clothing	85,000	15,000		15,000	
Total		65,114	21,960	29,542	13,612

recycled or not, regarding the waste which cannot be raw material of another process (not recyclable), the amount is just about 13,600 t.

The regional occurrence of textile waste for combustion or gasification has to be determined under aspects of logistics optimization between regions. Regarding this aspect, the distribution of the Portuguese companies in land was found with the support of a study from CENESTAP, the Centro de Estudos Têxteis Aplicados, of 06/08/97, with about 1800 companies. With this study, it was possible to establish the distribution of the six more representative Portuguese districts, considering the number of companies per district (Braga, Porto, Guarda, Castelo Branco, Lisboa, and Setúbal). Lisboa and Setúbal are not considered as important as the others for the Portuguese economy, but we must refer to them if we look to the amount of waste produced by its companies, mainly clothing.

According to Table 4, the four main industrial sectors represent the production amounts in Table 6.

2.8 Techno-economic Study on Industrial Utilization of Fuel Blends Produced from Industrial Wastes and Coal

One partner has worked on this topic using data and test results from all other participants in the project.

2.8.1 University of Ulster (UUE)

The work of the Energy Research Centre of the University of Ulster was to perform technical, environmental, and economic assessment studies of a range of power generation technologies using coal/textile waste/plastic waste blends. Models have been developed for many different technologies, including combustion in fixed beds, bubbling beds, circulating fluidized beds, and entrained flow reactors, and gasification in a fixed bed gasifier with a gas engine, and a fluidized bed gasifier using a gas turbine in simple cycle, combined cycle, STIG, and ISTIG modes. These models have been used to integrate the results from the other partners. The specific work program called for:

> Collection of data on the wastes which are available and assessment of the cost and energy required to convert them to a feedstock suitable for co-processing with coal

TABLE 6 Production Amounts of Four Main Industrial Sectors

Clothing	85,000 t
Spinning/weaving (wool and blends)	39,564 t
Spinning/weaving/home textiles (cotton and blends)	178,400 t
Knitting	143,750 t

Validation of the pilot plant results and scale-up to a suitable size of operation

Performance of techno-economic assessment studies for the range of proposed feedstocks and the alternative co-processing technologies at a suitable size of operation

Performance of sensitivity analyses to determine the conditions required to make these alternatives competitive with existing disposal routes

Information was obtained from the other partners on the situation with regard to waste textiles arising in Germany and Portugal and plastic waste arising in Germany. Further information with regard to the geographic location was required in order to determine the relationship between cost of transportation and size of waste processing plant.

The technologies modeled included:

Small-scale combustion in a fixed bed furnace

Medium-scale combustion in a bubbling or circulating fluidized bed

Large-scale entrained flow combustion

Small-scale gasification in a fixed bed gasifier with a gas engine

Medium-scale fluidized bed gasification using either a gas engine or a gas turbine

Medium- to large-scale fluidized bed gasification systems using gas turbine topping cycle and steam turbine bottoming cycle

Medium-sized fluidized bed gasifier with STIG and ISTIG configuration

The overall results from the work using the ECLIPSE process simulator can be summarized as follows.

The addition of 5% plastic waste to a coal-fired PF power station reduces its efficiency by 0.3%, due to the high moisture content of the plastics. There is a small increase in capital cost associated with feeding the waste plastic to the boiler. For the overall economics to be the same as the coal-only case, the power station could afford to pay up to 20 ECU/dry t for the plastic waste, compared with 37 ECU/t for the coal. The disposal of plastic waste normally costs approximately 94 ECU/dry t. Therefore, provided the cost of transporting, handling, and preparing the plastic waste for use in the power station does not exceed 114 ECU/dry t, then it is economic for the power station to use plastic waste.

The addition of 20% plastic waste to a Texaco IGCC power station increases the net efficiency by 0.2%, due to the improved gasification qualities of plastic over coal. There is an increased capital cost for handling the plastic waste, which means that for the overall economics to be the same as the coal-only case, the power station could afford to pay up to 38 ECU/dry t for the plastic waste, similar to the coal price. Again, the disposal of plastic waste normally costs approximately 94 ECU/dry t and therefore, provided the cost of transporting, handling, and preparing the plastic waste for use in the power station does not exceed 132 ECU/dry t, then it is economic for the power station to use plastic waste.

While there are sufficiently large amounts of plastic waste available, the amount of textile waste suitable for co-combustion in Portugal and Germany is limited. If all the available textile waste in Germany is used in a 450-MWe British coal air-blown gasifier (BCABG), less than 2% of the coal thermal input can be replaced by textiles. This generally has a small positive benefit to the BCABG power station, which means that the textiles have a value of 32 ECU/t to this power station.

With textile waste suitable for co-processing of only 18,400 t/year in Germany, this imposes limits on the maximum size of a power station. When using a 5% textile share in a CFBC power station, a 122-MWe plant can be sustained with the available textile dust, while with a 20% textile blend this is reduced to 24 MWe. For textile waste arising in Germany, the ideal location for a power plant is Giessen for larger-scale plants, while Freiburg or Münster are best suited for smaller plants. Using textiles in a CFBC power station reduces the efficiency and increases the capital cost. However, the textiles have a positive value to the power station if it can be assumed that the textile disposal cost of about 135 ECU/t is avoided. Hence, although the textiles have a higher transport cost than coal, the overall fuel cost is improved when textiles are employed. As a result, a reduction in BESP can be achieved. The best reduction in electricity price for co-combustion of coal with textiles can be accomplished for a 5-MWe power station with a 20% textile share. For a 5% textile share the most advantageous size is a 20-MWe CFBC plant.

The co-combustion of a wide range of blends of waste plastic film with biomass was successfully modeled. The value of waste plastic as a feedstock was shown to be between 13 and 21 ECU/t greater than the equivalent ECU/GJ value of the biomass feedstock, due to its improved combustion performance. In the specific case studied, there are problems with the availability of suitable plastic waste and the high transport cost. However, it has been shown that this particular technology can successfully recover the energy from blends of waste plastic with biomass. Under more favorable waste plastic supply conditions and waste plastic transport distances, it has the potential to be economically viable.

3 CONCLUSION

Thermal utilization of residues and industrial waste in combination with fossil fuels and/or biomass will become a very important factor in future energy production from the viewpoint of environmental aspects, the careful use of resources such as fossil fuels, as well as developing/increasing new market segments for decentralized energy production. But also in existing markets and technologies such as iron and steel or cement production, the utilization of residues or waste might play a more important role in the future. The main goal of the project, "Advanced Combustion and Gasification of Fuel Blends and Diagnostics of Alkali and Heavy Metal Release," was the investigation and development of data and technologies for commercial use of waste in industrial applications.

The first aim of the project part "Utilization of Residues" was development of basic knowledge for characterization and assessment of these novel residues. On this basis, environmentally compatible valorization concepts for the various residues were developed for making up marketable products to be recycled to the economy. Residue samples (fly ashes, bottom ashes, filter material, etc.) were supplied by the partners performing the combustion/gasification tests of the fuel blends in their facilities and were subsequently investigated according to the work program. After analysis and characterization of the samples, valorization concepts have been developed or modified in comparison to known concepts. For developing the above-mentioned aims and for categorization of the residues in view of different valorization lines, exact analyses of the matters fed to the process via secondary fuels in the coal have been carried out. Due to the different properties and environmental impacts of the residues, specific preparation and valorization concepts were developed for each case. Above all, the fly ashes and filter cakes, which may be contaminated by organic noxious matters, in particular dioxins, furanes, and halogenated hydrocarbons, as well as by inorganic noxious matters, such as salts, desulfurization products, or heavy metals, have been investigated and the ashes have been classified for potential market segments, mainly in construction industries.

The conventional method of measuring alkali levels in hot gas consists of extracting a gas sample by means of a condensing probe followed by a number of impingers for the absorption of condensed aerosols. The probe is rinsed after taking the sample, and the aqueous liquids are then analyzed by flame photometry on alkalis. Sampling takes a long time because of the rather low sensitivity of the analyzing method and is prone to very many errors of unknown magnitude. No continuous monitoring of the alkali levels was possible at the beginning of the project, and plants had to be run at steady-state conditions during the sampling period, lasting several hours.

A market study has been performed concerning the availability of gas analyzing sensors, especially alkali sensors, and the future marketing possibilities promised after successful development of sensors. The study has shown on the one hand the worldwide novelty and uniqueness of the alkali sensor development to be undertaken in this project and on the other hand has clarified future marketing chances. In a joint effort, instrument developers, plant manufacturers, and operators have used worldwide business contacts and screen literature to meet the above-mentioned objectives.

It was possible to develop three different on-line alkali measuring devices which have been demonstrated under comparable conditions in a 10-MW plant, and the instruments might be ready for commercial use in industry.

The waste which has been investigated arose from different processing and production processes and varied in chemical composition and physical properties. Depending on their origin or their pretreatment, the waste was contaminated by heavy metals and organic compounds arising from colors and special coatings, e.g., for flammability reduction, and they occurred in the form of dust or pieces. One way for efficient energy recovery from those waste was co-combustion with coal, which is preferred compared with pure waste incineration, since this was not optimized for energy production and caused some problems with low melting points of the plastic waste. However, based on the results of our R&D program, it should be considered in later projects.

Three techniques for the on-line measurement of alkali species have been developed based on physical detection of alkali atoms. Simultaneously, all three analyzers in several individual experimental campaigns of the hot-gas duct of two pressurized fluidized bed combustion pilot plants of different design have been demonstrated and optimized. The project aim was to obtain an overview of alkali concentration levels in hot gas from coal combustion for a variety of fuels and plant operating conditions.

The alkali concentration levels have been correlated to the parameters set or found in the experimental programs and an attempt has been made to find relations between hot-gas conditions and alkali levels and to draw conclusions with respect to generalization of the results obtained. It has been shown that the risk alkali vapors present in the technology of pressurized fluidized combustion of solid fuels can be assessed on a fairly well-defined basis of data at the end of this project. Furthermore, it has been demonstrated that the accuracy and reliability of the three different alkali analyzers could be significantly improved and their capability to be used in commercial applications might be possible in future.

It can be stated at the end of the project that all 20 partners have worked together in a very cooperative way, that all groups have reached the main goals of their planned working packages, and that a number of results have been obtained for later technical/industrial use as described above. On the other side, some of the work has shown that there is still demand for additional development before using the results commercially.

Due to the large amount of waste plastics, thermal utilization will be an attractive option in a number of industrial applications; however, as already described, standardization equalization of national laws and reduction of preparation costs must be considered and solutions must be developed.

The textile waste market seems to be very limited, and it can be concluded that due to the limited amount of textile wastes for thermal utilization, the material will only be of interest for local application in grate firing in some regions; however, the material can be used in combination with biomass and can improve the combustion behavior of boilers.

The developed on-line alkali and heavy metal devices might be ready for technical application in advanced power generation technologies but also in conventional boilers, incineration plants, and in boilers where biomass is burnt.

REFERENCES

1. I. Romey, *JOULE-THERMIE Clean Coal Technology R&D, Fuels Blends and Alkali Diagnostics,* European Commission, 1999.
2. I. Romey, J. Barnish, and J. M. Bemtgen, (eds.), *Diagnostics of Alkali and Heavy Metal Release,* European Commission, 1998.
3. J. M. Bemtgen and I. Romey, *JOULE-THERMIE Clean Technologies for Solid Fuels R&D (1996–1998),* European Commission, 1996.
4. I. Romey, Co-utilization of Coal with Biomass and Wastes, *Proc. Final Conference, Volumes I, II, and III: Executive Summary, Final Reports,* APAS Clean Coal Technology, 1993–1994.

23

Best Practices for the Oil and Gas Exploration, Production, and Pipeline Transportation Industry

Bart Sims

Railroad Commission of Texas, Austin, Texas

1 INTRODUCTION

Many oil and gas exploration, production, and pipeline companies (E&P compa-nies), both large and small, have applied the concepts of waste minimization in their operations. As a result, E&P companies have achieved progress in pollution prevention by eliminating oil and gas wastes at the source. As well as pollution prevention efforts, E&P companies have implemented effective reuse and recy-cling alternatives for various oil and gas wastes.

Several E&P companies have provided case studies which highlight suc-cessful and beneficial waste minimization. The following case studies illustrate how E&P companies have applied waste minimization techniques, or "best management practices," to either reduce the quantity of waste generated at the source or to recycle, reclaim, or reuse waste streams that could not be reduced at the source. E&P companies have implemented equipment modifications, proce-dural or process changes, product substitution, and reuse of spent materials to achieve beneficial waste minimization. Importantly, the case studies demonstrate that, in addition to preventing pollution, waste minimization in oil and gas

497

operations can provide valuable benefits such as cost savings, increased revenue, improved operating efficiency, reduced regulatory compliance concerns, and reduced future potential liability concerns.

The following case studies have been provided to the Railroad Commission of Texas (RRC) Waste Minimization Program. Most are included in the RRC publication, *Waste Minimization in the Oil Field* (1). The case studies are presented for the various E&P operations in the following sequence: drilling operations, production operations, natural gas treating and processing operations, and pipeline operations. Each case study is presented as a "problem" which is addressed by a waste minimization "solution." The benefits gained close each case study.

2 CASE STUDIES OF SUCCESSFUL WASTE MINIMIZATION IN THE OIL AND GAS EXPLORATION AND PRODUCTION INDUSTRY

2.1 Waste Drilling Fluid in Conventional Earthen Pits at a Drilling Operation

A small independent operator was concerned about the volume of drilling waste in conventional reserve pits at his drilling locations. Waste management costs were a concern, as well as the costs associated with the impact of drilling fluid on adjacent land, due to earthen reserve pit failures. The operator was concerned about the potential for surface water or ground water contamination and the associated potential liabilities. As a solution, the operator implemented a procedure change, using a closed-loop drilling fluid system rather than conventional earthen reserve pits. The wells were to be drilled in relatively shallow, normally pressured strata and, therefore, were amenable to use of a closed-loop system. The operator negotiated with drilling contractors to obtain a turnkey contract that required the drilling company to use a closed-loop system and take responsibility for recycling the waste drilling fluid generated by the operation. The turnkey contract was incrementally more expensive. However, the operator realized a savings of about $10,000 per well because of reduced drill site construction and closure costs, reduced waste management costs; and reduced surface damage payments. Importantly, the drilling mud was reclaimed and reused, rather than disposed of by burial or land application. Also, the operator reduced the potential for environmental impact and associated potential liability concerns.

2.2 Used Lubricating Oil and Filters from Power Units on a Drilling Rig

This case history is taken from the Society of Petroleum Engineers paper titled "Monitoring Engine Oil" (2). A drilling company was concerned with the quantity of waste lubricating oil and filters generated by diesel power plants on its rigs, and the costs of new oil, new filters, and maintenance. The drilling company recognized that the problem stemmed from performing oil and filter changes at 500-h operating intervals as recommended by the engine manufacturer. The solution the company selected was a procedural change which resulted in source reduction, even though their main goal was to reduce operating costs. The company extended the operating interval between lube oil changes for the diesel power plants by performing sampling and analysis of the lube oil to determine when lube oil changes were actually needed. The company established threshold concentrations for specific analytes, such as contaminants, additives, and metals. Whenever a threshold value was exceeded, a lube oil and filter change was made. In any event, the maximum operating interval was set at 1250 h. This procedural change resulted in a decrease in oil costs from $64/day to $41/day, which translates to a 36% reduction in waste generation. Additional cost savings were realized due to decreased maintenance requirements, improved operating efficiency, and reduced waste management requirements. Importantly, no harm or unusual wear was experienced in the diesel power plants.

2.3 Coalescor Panels from Horizontal Separators in an Oil Field

An oil and gas company was concerned about the use of flow stabilizing coalescor pads (14 in. deep with 1-in. by $\frac{1}{2}$-in. openings) in over 300 horizontal oil/water separators used in a large field. The coalescor pads would plug with solids and ice, resulting in need for replacement. Occasionally, differential pressure would cause coalescor pad segments to break free and damage the separator's interior components (e.g., level controls and cathodic protection anodes), resulting in costly repairs. A typical coalescor pad replacement cost about $4000, and, if the separator was damaged, cost about $1000 to $5000 for repair. Frequent replacement of the pads resulted in high maintenance and waste management costs (15–20 separator coalescor change-outs would generate about 25 cubic yards of waste, costing about $1000 for disposal). Also, when a separator had to be serviced due to coalescor pad problems, production had to be shut-in. As a solution, the operator employed a simple equipment modification by designing a $\frac{1}{2}$-in.-thick fiberglass baffle perforated with 1-in.-diameter holes to replace the coalescor pads. The new baffle cost about $200 to fabricate and install in each separator. The new baffles performed well in the separators, providing flow characteristics adequate for optimum performance. The operator enjoyed impressive benefits from the equipment modification. The elimination of the old coalescor pads in turn eliminated all of the previously experienced separator maintenance, waste generation and management, and the associated costs. A simple $200 modification eliminated the waste disposal, repair, and maintenance costs. Therefore, it is apparent that the small capital investment was quickly recovered. Additionally, the operator gained benefits such as increased production, reduced regulatory compliance concerns, and improved worker safety due to reduced confined entry.

2.4 Crude Oil and Saltwater Releases from Flowline Ruptures in an Oil Field

A very small independent oil producer operated a shallow, 40-well oil field in an area subject to extremely cold winters. The operator's poly flowlines would freeze and rupture in the cold weather. Increased paraffin deposition also contributed to this problem. The flowline ruptures resulted in crude oil- and saltwater-contaminated soil that had to be cleaned up under state environmental regulations. Also, to prevent rupture of the flowlines, the rod pumps had to be shut down when freezing weather was predicted, resulting in loss of production and revenue. The operator employed an innovative and low-cost equipment modification to resolve the problem. The operator designed and installed automatic pump shut-off systems at a cost of about $75 per well, which included parts and labor. The total investment for installing the devices on the 40 wells was about $3000. The automatic shut-off system was made using an automotive brake light switch (pressure switch), copper tubing, hydraulic fluid, and a simple relay switch. The circuit was designed such that the pump had to be manually restarted after the cause of shut-off was determined. The operator's equipment modification resulted in impressive source reduction. Since installation of the automatic shut-off systems, the operator has not experienced a single flowline rupture. As a result, the operator eliminated the generation of contaminated soil and the associated cleanups and loss of production and revenue. The operator did not provide the specific economics for this project; however, it was clearly cost effective. The savings from reduced soil cleanups and increased revenue from more efficient production easily recovered the nominal capital investment.

2.5 Pump-Jack Gear-Box Lube Oil at an Oil Field

A small independent operator felt that the cost of replacing lube oil in pump-jack gear boxes, including maintenance and waste lube oil management, was excessive. The operator investigated opportunities to reduce these costs and the management of the waste lube oil. The operator's simple solution, which resulted in source reduction, was to contract a service company to service the pump-jack. The service company filters and treats (i.e., purifies) the lube oil on-site, and then returns the reclaimed lube oil to the gear box for reuse. The cost of this service was about $35–$40 per pump-jack in 1995. The operator's use of this service eliminated his generation and management of waste lube oil and the associated maintenance requirements. The change in procedure was cost effective. New replacement lube oil costs about $175 per pump jack; therefore, a savings of about $135 per pump jack was realized. Additional savings were realized because of reduced waste management and maintenance costs.

2.6 Produced Water Filters at an Enhanced Oil Recovery Project

A small independent operator generated a large quantity of waste filters from a produced water injection system used in an enhanced oil recovery project. About 14,000 bbl of produced water were injected each day. The operator was replacing two produced water filter units at each of 36 injection wells twice per month, resulting in about 1700 waste filters per year. The operator spent $4148 per year for new replacement filters. Additional expense was incurred from waste filter management and maintenance. The operator made a simple equipment modification and procedure change to achieve significant source reduction. The operator began basing filter replacements on differential pressure rather than on a twice-monthly schedule. The operator installed valves on each filter inlet and outlet to accommodate a temporary pressure gauge hookup for differential pressure measurement. A capital investment of $1800 was required for installation of the valves. The operator's simple equipment modification and procedure change resulted in a significant reduction in the quantity of waste filters. In the year following the change, a total of 28 waste filters were generated—a 98% reduction in waste generation. The change was very cost effective due to reduced maintenance requirements, reduced waste management, and reduced filter replacements. The operator saved about $4000 per year due only to the reduced filter replacement costs. The capital investment was recovered in less than five months.

2.7 Heat-Medium Oil Filters at a Natural Gas Processing Plant

A natural gas processing plant operator was generating a large quantity of conventional filters from the operation of a heat-medium oil system. The system had three sets of filters: a 36-filter set changed weekly; a 3-filter set changed bi-weekly; and a 54-filter set changed bi-monthly. The costs associated with the filter change-outs, new filters, and lost oil were high. Also, the plant had changed to a new type of heat-medium oil that more effectively cut deposits in the system, resulting in the need for more efficient filtering. To address the waste filter generation problem, the operator made an equipment modification. The operator installed spinner (i.e., centrifugal) filter units in place of the conventional 36-filter set and 3-filter set. The heat-medium oil is now circulated out of the system, cooled, and run through the spinner filter units. The spinner filter units require clean-out three times per week, rather than removal and disposal of a filter casing and media. By eliminating the conventional filter sets, the operator eliminated the generation of 1950 waste filters per year. The cost savings were significant. Approximately $18,500 per year was saved due to reduced filter replacement costs, reduced waste management costs, reduced labor and maintenance costs, and reduced lost oil costs.

2.8 Screen-Type Lube Oil Filters on Drive Engines at a Natural Gas Processing Plant

A natural gas processing plant operator was concerned about the quantity of waste sock lube oil filters and lost lube oil created by filter changes on the plant's 16 large drive engines. The seven sock filters on each drive engine were changed four times per year or every 2000 operating hours. Each filter change required about 40 gal of new lube oil, as well as new filter installation and spent filter disposal. The total filter management cost per year for the 16 units was $10,315.00. The operator made a simple equipment modification. Two reusable screen-type filters were installed in parallel with each other and in series with the sock filters on each of the drive engine's lube oil systems. The screen filters could then be alternately bypassed for cleaning while the unit continued to operate. As a result, the sock filters required replacement only once per year. This simple equipment modification was cost effective. The reduced sock filter changes saved $7736.00 per year, not including the savings from reduced management and disposal costs for the eliminated waste sock filters. The operator also gained the benefit of reduced regulatory compliance.

2.9 Soda Ash Reagent in a Sulfur Dioxide Scrubber at a Natural Gas Processing Plant

This case history is taken from the Society of Petroleum Engineers paper titled "Reuse of Spent Natural Gas Liquid Sweetening Solutions" (3). A natural gas company was using sour gas fuel in their operations and necessarily operated exhaust gas scrubbers to control sulfur dioxide (SO_2) emissions. The company conducted a study to determine if partially spent caustic natural gas liquid sweetening solution could be used in place of soda ash solution as the reagent in the SO_2 scrubbers. To achieve acceptable performance using partially spent caustic solutions, they found that necessary changes to the scrubber operation were reagent feed rate, scrubber liquid pH and specific gravity, and blowdown rate. The company found that the SO_2 scrubber could be operated using the partially spent caustic sweetening solution as the reagent without negative effects on performance, regulatory compliance, or operating costs. The reuse of the sweetening agent eliminated its disposal as waste and eliminated the purchase and handling of soda ash. As a result, the company realized the benefits of cost savings and reduced regulatory compliance concerns.

2.10 Spent Sand-Blasting Media at Pipeline Compressor Stations

A natural gas pipeline company was using conventional sand-blasting to prepare compressors and other equipment surfaces for painting. This procedure had generated relatively large quantities of spent sand-blast media. In this particular instance, the sand-blasting of a compressor generated 27 drums of spent sand-blast media, which was characteristically hazardous waste due to lead toxicity. The result was significant waste management concerns and costs. First, the cost of disposal was $6426 for the 27 drums of hazardous waste ($238 per drum). Also, the operator was required to register as a large-quantity generator (LQG) of hazardous waste, resulting in compliance with the state's stringent hazardous waste management requirements and a waste generation fee of $2000. To address the high waste management costs and the regulatory compliance concerns, the operator implemented a procedural change. Rather than use conventional sand-blasting, the operator began using pneumatic needle scalers to remove old paint and prepare surfaces for painting. The process was somewhat slower, but it resulted in much less waste. And labor requirements were actually reduced, because less cleanup was required. As a result of this procedural change, the operator realized a cost savings of $8188. Much of the cost savings were obtained by reducing the quantity of hazardous waste from 27 drums to one drum. This source reduction also significantly reduced the operator's regulatory compliance concerns with respect to hazardous waste management.

3 CONCLUSIONS

These case studies are clear evidence that waste minimization efforts in the oil and gas exploration and production industry can provide impressive benefits, as well as environmental protection. The case studies provide excellent examples of source reduction through equipment modification, procedural change, preventive maintenance, and excellent examples of waste minimization through recycling and reuse. In each instance, the oil and gas company has saved money, improved operating efficiency, reduced regulatory compliance concerns, and reduced potential future liability. More and more oil and gas companies are recognizing that waste minimization can obtain the environmental benefit of pollution prevention and be a good business decision.

REFERENCES

1. Railroad Commission of Texas, *Waste Minimization in the Oil Field*. Austin, TX: RRC, revised April 1999.
2. Fullerton et. al., Monitoring Engine Oil, Society of Petroleum Engineers Paper 18663. Proc. SPE/IADC Drilling Conference, New Orleans, LA, February 28–March 3, 1989.
3. Hahn et al., Reuse of Spent Natural Gas Liquid Sweetening Solutions, Society of Petroleum Engineers Paper 29733. Proc. SPE/EPA Exploration & Production Environmental Conference, Houston, Texas, March 27–29, 1995.

Index